SCOPE 26 ... n and

SCOPE 27: Climate ... imate and Society, 1985, 650 pp

SCOPE 28: ... , 400 pp ... 85, 563 pp

SCOPE 29: ... systems, 1986,

SCOPE 30: ... of Chemicals,

SCOPE 31: ... ronment, 1987,

SCOPE 32: ...

SCOPE 33: ... , 1988, 478 pp

SCOPE 34: ... namic Change

SCOPE 35: ... Variability in ... pp

SCOPE 36: ...

SCOPE 37: ... 528 pp

SCOPE 38: ... ce to Hot and

SCOPE 39: ... Cycle, 1989,

SCOPE 40: ... rom Chemical

SCOPE 41: ... Effects, 1990,

SCOPE 42: ... 56 pp

SCOPE 43: ... ulphur in the

SCOPE 44: ... sms into the

SCOPE 45: ...

SCOPE 46: ... d Non-Human

SCOPE 47: Long-Term Ecological Research, 1991, 320 pp

Funds to meet SCOPE expenses are provided by contributions from SCOPE Committees, an annual subvention from ICSU (and through ICSU, from UNESCO), an annual subvention from the French Ministère de l'Environment, contracts with UN Bodies, particularly UNEP, and grants from Foundations and industrial enterprises.

SCOPE 47

Long-term Ecological Research

SCOPE 47

Long-term Ecological Research

An International Perspective

Edited by

PAUL G. RISSER

University of New Mexico, Albuquerque, New Mexico, USA

Published on behalf of the Scientific Committee on Problems of the Environment (SCOPE) of the International Council of Scientific Unions (ICSU)

by

JOHN WILEY & SONS

Chichester · New York · Brisbane · Toronto · Singapore

Published in 1991 by John Wiley & Sons Ltd,
Baffins Lane, Chichester,
West Sussex PO19 1UD, England

Other Wiley Editorial Offices

John Wiley & Sons, Inc., 605 Third Avenue,
New York, NY 10158-0012, USA

Jacaranda Wiley Ltd, G.P.O. Box 859, Brisbane,
Queensland 4001, Australia

John Wiley & Sons (Canada) Ltd, 22 Worcester Road,
Rexdale, Ontario M9W 1L1, Canada

John Wiley & Sons (SEA) Pte Ltd, 37 Jalan Pemimpin #05-04,
Block B, Union Industrial Building, Singapore 2057

Library of Congress Cataloging-in-Publication Data:

Long-term ecological research : an international perspective / edited by Paul G. Risser.
p. cm. — (SCOPE ; 47)
Includes bibliographical references and index.
ISBN 0 471 93005 9 (cloth)
1. Ecology—Research—United States—Congresses. 2. Ecology–Research—Congresses. I. Risser, Paul G. II. International Council of Scientific Unions. Scientific Committee on Problems of the Environment. III. Series : SCOPE report ; 47.

QH541.26.L66 1991
574.5′072—dc20 91–7557
CIP

British Library Cataloguing in Publication Data:

Long-term ecological research: An international perspective. — (Scientific Committee on Problems of the Evironment (SCOPE))
I. Risser, Paul G. II. Series
574.5

ISBN 0 471 93005 9

Printed and bound in Great Britain by Courier International, East Kilbride

International Council of Scientific Unions (ICSU)
Scientific Committee on Problems of the Environment (SCOPE)

SCOPE is one of a number of committees established by the nongovernmental group of scientific organizations, the International Council of Scientific Unions (ICSU). The membership of ICSU includes representatives from 75 National Academies of Science, 20 International Unions, and 29 other bodies called Associates. To cover multidisciplinary activities which include the interests of several unions, ICSU has established 13 Scientific Committees, of which SCOPE is one. Currently representatives of 35 member countries and 21 Unions, Scientific Committees and Associates participate in the work of SCOPE, which directs particular attention to the needs of developing countries. SCOPE was established in 1969 in response to the environmental concerns emerging at the time: ICSU recognized that many of these concerns required scientific inputs spanning several disciplines and ICSU Unions. SCOPE's first task was to prepare a report on Global Environmental Monitoring (SCOPE 1, 1971) for the UN Stockholm Conference on the Human Environment.

The mandate of SCOPE is to assemble, review and assess the information available on man-made environmental changes and the effects of these changes on man; to assess and evaluate the methodologies of measurement of environmental parameters; to provide an intelligence service on current research; and by the recruitment of the best available scientific information and constructive thinking to establish itself as a corpus of informed advice for the benefit of centres of fundamental research and of organizations and agencies operationally engaged in studies of the environment.

SCOPE is governed by a General Assembly, which meets every three years. Between such meetings its activities are directed by the Executive Committee.

R. E. Munn
Editor-in-Chief
SCOPE Publications

Executive Secretary: V. Plocq-Fichelet

Secretariat: 51 boulevard de Montmorency
75016 Paris

Contents

Scientific Advisory Committee

List of Contributors*

Darwin W. Anderson
Saskatchewan Institute of Pedology, University of Saskatchewan, Saskatoon S7N 0W0, Canada

J.M. Briggs
Division of Biology, Ackert Hall, Kansas State University, Manhattan, Kansas 66506, USA

James T. Callahan
Ecosystems Studies Program, National Science Foundation, 1800 G Street NW, Washington, DC 20550, USA

K. L. I. Campbell
Serengeti Ecological Monitoring Programme, PO Box 3134, Arusha, Tanzania

Henri Décamps
Centre d'Ecologie des Ressources Renouvelables, CNRS, 29 rue Jeanne Marvig, 31055 Toulouse Cedex, France

Madeleine Fortuné
Centre d'Ecologie des Ressources Renouvelables, CNRS, 29 rue Jeanne Marvig, 31055 Toulouse, Cedex, France

James R. Gosz
Biology Department, Castetter Hall, University of New Mexico, Albuquerque, New Mexico 87131, USA

W.D. Grossmann
Austrian Academy of Sciences, Kegelgasse 27, A-1030 Vienna, Austria

O. William Heal
Institute of Terrestrial Ecology, Edinburgh Research Station, Bush Estate, Penicuik, Midlothian EH26 0QB, Scotland

*For contributions by three authors or more, only the first author is listed.

A.E. Johnston
Division of Soils and Crop Production, AFRC Institute of Arable Crops Research, Rothamsted Experimental Station, Harpenden, Herts AL5 2JQ, England

John J. Magnuson
North Temperate Lake LTER Site, Center for Limnology, 680 N. Park Street, University of Wisconsin-Madison, Madison, Wisconsin 53706, USA

Sam J. McNaughton
Biological Research Laboratories, Syracuse University, 130 College Place, Syracuse, New York 13244-1220, USA

Steward T.A. Pickett
Institute of Ecosystem Studies, The New York Botanical Garden, Mary Flagler Cary Arboretum, Box AB, Milbrook, New York 12545, USA

Paul G. Risser
University of New Mexico, Albuquerque, New Mexico 87131, USA

Tim R. Seastedt
Division of Biology, Ackert Hall, Kansas State University, Manhattan, Kansas 66506, USA

H.H. Shugart
Department of Environmental Sciences, University of Virginia, Charlottesville, Virginia 22903, USA

E.H. Trotter
Department of Biology, Castetter Hall, University of New Mexico, Albuquerque, New Mexico 87131, USA

Mark Westoby
School of Biological Sciences, Macquarie University, NSW 2109, Australia

1 Introduction

PAUL G. RISSER
University of New Mexico, Albuquerque, New Mexico 87131, USA

1.1 INTRODUCTION

Long-term ecological studies have been focused around the relatively straightforward notion that certain processes, such as succession, the natural frequency distributions of climate regimes, or disturbances exemplified by treefalls and fires, are long-term processes and must be studied as such (Likens, 1983; Lawes Agricultural Trust, 1984; Wiens, 1984; Coull, 1985; Greenland, 1986; Munn, 1986; Strayer, 1986; Strayer *et al.*, 1986; Peterken and Jones, 1987). Indeed, there are numerous examples in the literature in which data trends from short-term interpretations yield significantly different information from longer-term analyses. The belief has been that the collection of data over long periods of time is necessary to permit generalizations and theory over sufficiently large spatial and temporal scales to evaluate the ecological consequences of such events as fire, grazing, deforestation, acid deposition, and changes in trace gas fluxes (Callahan, 1984).

Analyses of existing long-term ecological studies show several conditions under which these types of studies are particularly beneficial. That is, long-term studies are useful when the phenomena are themselves long term in their dynamics; when the phenomena are episodic, rare, complex or subtle, and long-term measurements are needed to isolate their dynamics and the control processes; when the phenomena are poorly understood and cannot be predicted from short time scales; and when long-term records are needed for making policy decisions. Despite the documented need for long-term studies, there are relatively few institutionalized comprehensive long-term investigations. Those that do exist include the agricultural plots at Rothamsted in the UK, forest productivity research in Sweden, and watershed studies at Coweeta and the Hubbard Brook experimental forests in the United States. In addition to these relatively large studies, there are numerous examples of individual scientists who have collected data on a phenomenon or site over long periods of time. In many instances, successful long-term studies have depended upon such individual scientists, who have been dedicated in their pursuit of measurements, over many years.

Long-term Ecological Research. Edited by Paul G. Risser

The relative paucity of long-term ecological studies is the result of several obstacles. Among the most important impediments are the following:

(1) Long-term measurements may not be considered at the forefront of innovative science, and thus obtaining continuous funding is difficult.
(2) The site where the measurements are made may be changed, making the long-term results meaningless or difficult to interpret.
(3) The experimental design may be too ambiguous for long-term consistent measurements or may not include adequate auxiliary studies for unravelling controlling processes.
(4) The resources or incentives for the dedicated individual scientists or research leaders to continue the measurements may be inadequate.
(5) Evolution in instrumentation may render the current methodology obsolete, and there may be insufficient attention to calibrating the old and new technologies.
(6) New scientific advances may make the original question or hypothesis uninteresting or may provide a definitive answer; in either case, the measurements need not be continued.

This book summarizes the results of a project that has analyzed long-term ecological research in an international context. Conducted under the auspices of the Scientific Committee on Problems of the Environment (SCOPE), the project consisted of two workshops. The first was held 18–22 September 1988, in Berchtesgaden, Federal Republic of Germany (see Table 1.1 for the list of participants). This workshop had three objectives:

(1) To provide a forum for describing and analyzing selected existing long-term ecological research sites and networks from various countries;
(2) To further refine the rationale for long-term ecological research and identify important existing and emerging scientific questions which could most appropriately be addressed at these sites, particularly those that relate to environmental changes at the global scale; and
(3) To begin to formulate a plan for international communication and co-ordination among long-term ecological research sites.

This workshop primarily considered national or regional programs rather than long-term ecological projects of individual investigators. Designated and protected research sites offer several potential advantages, e.g. economy in site dedication and protection, sharing of equipment and data sets, and the synergism brought about by interactions among scientists, especially those from different disciplines. Nevertheless, many of the conclusions of the workshop are equally applicable to long-term studies conducted by individual scientists.

The following twelve chapters describe and discuss long-term studies from several countries: Australia, Canada, England, Federal Republic of Germany, France, Kenya, Scotland, and the United States. There are also discussions of

Table 1.1 Participants in an International Workshop, 'Long-term Ecological Research: A Global Perspective', 18–22 September 1988, Berchtesgaden, Federal Republic of Germany

Dr Jerry Melillo (Chair)
Ecosystem Center
Marine Biological Laboratory
Woods Hole, MA 02543
USA

Dr Darwin Anderson
Department of Soil Science
University of Saskatchewan
Saskatoon S7N 0W0
Canada

Dr James Callahan
Ecosystem Studies Program
National Science Foundation
1800 G Street NW
Washington, DC 20550
USA

Dr Henry Décamps
Centre d'Ecologie des Ressources
Renouvelables
29 Rue Jeanne Marvig
31055 Toulouse Cedex
France

Dr Melvin I. Dyer
Route 2, Box 330A
Lenoir City, TN 37771
USA

Professor Dr O. Fränzle
Geographisches Institut der
Universität Kiel
Olshausenstr. 40-60
2300 Kiel
Federal Republic of Germany

Dr Francisco Garcia Novo
Departamento de Ecologìa
Facultad de Biologìa
Universidad de Sevilla
Apartado 1095
41080 Sevilla
Spain

Min R. W. Goerke
Deutsches Nationalkomittee fur
das UNESCO-Programm
'Der Mensch und die Biosphäre'
Bundesministerium für Umwelt,
Naturschutz und Reaktorsicherheit
Postfach 120 629
D-5300 Bonn 1
Federal Republic of Germany

Dr James Gosz
Biology Department
Castetter Hall
University of New Mexico
Albuquerque, NM 87131
USA

Dr W.D. Grossmann
Kegelgasse 27
A-1030 Vienna
Austria

Professor Dr W. Haber
Lehrstuhl für Landschaftsökologie
der Technischen Universität München
D-8050 Freising-Weihenstephan
Federal Republic of Germany

Dr O. William Heal
Institute of Terrestrial Ecology
Bush Estate
Penicuik
Midlothian EH26 OQB
Scotland

Dr Rafael Herrera
Center for Ecology and Environmental
Sciences IVIC
Apartado 21827
Caracas 1020A
Venezuela

continued overleaf

Table 1.1 *(continued)*

Mr A.E. Johnston
Head of Division of Soils and Crop Production
Rothamsted Experimental Station
Harpenden
Herts AL5 2JQ
England

Dr Kari Laine
Dept of Botany
University of Oulu
SF-90570 Oulu
Finland

Dr John Magnuson
Center for Limnology
680 North Park Street
University of Wisconsin-Madison
Madison, WI 53706
USA

Dr S.J. McNaughton
Biological Research Labs
Syracuse University
130 College Place
Syracuse, NY 13244-1220
USA

Dr Jean-Claude Menaut
Laboratoire d'Ecologie
Ecole Normale Supérieure
46 rue d'Ulm
75230 Paris Cedex 05
France

Dr Steward T. Pickett
Institute of Ecosystem Studies
The New York Botanical Garden
Cary Arboretum Box AB
Milbrook, NY 12545
USA

Dr F. Precht
Deutsche UNESCO-Kommission
Colmantstr. 15
D-5300 Bonn 1
Federal Republic of Germany

Professor Dr K. Reise
Litoralstation List
Postfach 60
D-2282 List/Sylt
Federal Republic of Germany

Professor Dr H. Remmert
Fachbereich Biologie/Zoologie
Universität Marburg
Karl-von-Frisch-Strabe
D-3550 Marburg
Federal Republic of Germany

Dr Jane Robertson
UNESCO
Division of Ecological Sciences
Place de Fontenoy
F-75700 Paris
France

Dr J. Schaller
Firma ESRI
Environmental System Research Institute
Ringstrabe 12
D-8051 Kranzberg
Federal Republic of Germany

Dr Tim Seastedt
Division of Biology
Ackert Hall
Kansas State University
Manhatten, KS 66506
USA

Dr H.H. Shugart
Department of Environmental Science
University of Virginia
Charlottesville, VA 22901
USA

Dr W.G. Sombroek
Director, International Soil Reference
and Information Centre (ISRIC)
9 Duivendaal
PO Box 353
6700 AJ Wageningen
The Netherlands

Table 1.1 *(continued)*

Professor Li Wenhua Secretary-General National Committee of MAB for China Commission for Integrated Survey of Natural Resources PO Box 767 Beijing China	Dr Mark Westoby School of Biological Sciences Macquarie University NSW 2109 Australia
Dr Robert G. Woodmansee Director Natural Resources Ecology Laboratory Colorado State University Fort Collins, CO 80523 USA	

general topics, such as administrative structure, role of sites, conceptual design, field measurement techniques, and modeling in the context of long-term ecological research. These chapters satisfy the first workshop objectives, namely analyzing existing studies on an international basis and refining the role and value of long-term ecological study. In addition, the workshop compared the international research effort with the Long-term Ecological Research Program supported by the National Science Foundation in the United States. This program involves 17 research sites and has been in existence for approximately ten years. The objectives of this program are (Callahan, 1984):

(1) Long-term analysis of site-specific ecological phenomena;
(2) Comparison of observations across diverse ecosystems and in terms of general ecological systems theory; and
(3) Provision of large, secure, ecologically diverse sites with well-developed support capabilities.

Many of these sites have supported long-term research for decades, and as such, represent a valuable experiential data base for further developing the concepts and processes for long-term studies.

The second workshop specifically addressed the third of the original objectives, that is, to begin to formulate a plan for international communication and co-ordination among long-term ecological research sites. Participants (see Table 1.2) were selected to represent active research programs in three biomes: arid to semi-arid, tundra-boreal, and temperate forest. During the workshop, specific plans were developed for international collaborative long-term ecological research, and working groups were identified for conducting experiments. The questions to be addressed and the proposed investigative approaches are summarized in Chapter 14.

International collaborative long-term ecological research requires not only the site-specific characteristics discussed above, but also extensive communication

Table 1.2 Participants in an International Workshop, 'Long-Term Ecological Research: International Workshop II', 2-4 October 1989, Albuquerque, New Mexico, USA

Dr James Gosz (Chair)
Biology Department
Castetter Hall
University of New Mexico
Albuquerque, NM 87131
USA

Dr Peter Beets
Forest Research Institute
Private Bag
Rotorua, New Zealand

Dr Mike Farrell
Oak Ridge National Laboratory
PO Box 2008
Mail Stop 6335
Oak Ridge, TN 37831
USA

Dr Jerry Franklin
College of Forest Resource, AR-10
University of Washington
Seattle, WA 98195
USA

Dr Francisco Garcia Novo
Departmento de Ecologìa
Facultad de Biologìa
Universidad de Sevilla
Apartado 1095
41080 Sevilla
Spain

Dr W. D. Grossmann
Austrian Academy of Sciences
Kegelgasse 27
A-1030 Vienna
Austria

Dr O. William Heal, Director
Terrestrial Ecology
Institute of Terrestrial Ecology
Bush Estate, Penicuik
Midlothian EH 26 OQB
Scotland

Dr Kari Laine
Department of Botany
University of Oulu
SF-90570 Oulu
Finland

Dr William K. Lauenroth
Natural Resource Ecology Laboratory and Range Science Department, CSU
Fort Collins, CO 80523
USA

Dr Jerry Melillo
Ecosystem Center
Marine Biological Laboratory
Woods Hole, MA 02543
USA

Dr C. Montana
Instituto de Ecologìa
Apartado Postal 263 'B'
35070 Gomez Palacio
Durango
Mexico

Dr William Parton
Natural Resource Ecology Laboratory
Colorado State University
Fort Collins, CO 80523
USA

Dr Paul G. Risser
Provost and Vice President for Academic Affairs
Scholes Hall 108
University of New Mexico
Albuquerque, NM 87131
USA

Dr Uriel Safriel
Jocob Blaustein Institute for Desert Research
The Mitrani Center
Sede Boquer
Israel

Table 1.2 (*continued*)

Dr J. Schaller Firma ESRI Environmental System Research Institute Ringstrasse 7 D-8051 Kranzberg Federal Republic of Germany	Dr R.W. Wein Boreal Institute of N Studies University of Alberta Edmonton, Alberta T6G 2E9 Canada
Dr Gaius Shaver Ecosystems Center Marine Biological Laboratory Woods Hole, MA 02543 USA	Dr Walter Whitford Department of Biology New Mexico State University Las Cruces, NM 88003 USA
Mr Steven Storch BBN Systems & Technologies Corp. 10 Moulton St. Cambridge, MA 01238 USA	Diane Wickland Land Processes Branch Earth Science & Applications Division NASA Headquarters (Code EEL) Washington, DC 20546 USA
Dr Don Strebel Versar Inc. 9200 Rumsey Rd. Columbia, MD 21045-1934 USA	Dr John Yarie USDA, Forest Service Institute of Northern Forestry Fairbanks, AK 99775-0082 USA

among the investigators. Therefore, at the second workshop, considerable attention was devoted to discussion of the newest electronic technologies for data management and communication. During the workshop, two sessions were held in a studio and the discussions were sent via satellite to more than 100 scientists at several sites across the United States. These scientists at remote sites sent questions and comments that were then discussed by the workshop in a second televised session.

These two SCOPE workshops brought together scientists from 15 countries to analyze the approaches most likely to facilitate long-term ecological research. The following chapters describe existing international long-term research efforts, evaluate facilitating mechanisms, propose important research questions that require international co-operation, and discuss technologies that will assist in this collaboration.

1.2 ACKNOWLEDGMENTS

Funding for this project was provided by the US National Science Foundation and is gratefully acknowledged. Also, the National Committee of the Man and the Biosphere Program, Federal Republic of Germany, provided partial support for the first workshop.

1.3 REFERENCES

Callahan, J.T. (1984). Long-term ecological research. *BioScience*, **34**, 363–367.

Coull, B.C. (1985). The use of long-term biological data to generate testable hypotheses. *Estuaries*, **8**, 84–92.

Greenland, D. (1986). Standardized meteorological measurements for long-term ecological research sites. *Bulletin of the Ecological Society of America*, **67**, 275–277.

Lawes Agricultural Trust (1984). Rothamsted: the classical experiments. Rothamsted Agricultural Experiment Station, Harpenden, England.

Likens, G.E. (1983). A priority for ecological research. *Bulletin of the Ecological Society of America*, **64**, 234–243.

Munn, R.E. (1986). Environmental prospects for the 21st century: implications for long-term policy and research strategies. Symposium on New Directions in International Research, Education and Practice. Ohio State University, 5–6 December 1986. Columbus, Ohio.

Peterken, G.F. and Jones, E.W. (1987). Forty years of change in Lady Park Wood: the old-growth stands. *Journal of Ecology*, **75**, 477–512.

Strayer, D. (1986). An essay on long-term ecological studies. *Bulletin of the Ecological Society of America*, **67**, 271–274.

Strayer, D., Glitzenstein, J.S., Jones, C.G., Kolasoi, J., Likens, G.E., McDonnell, M.J., Parker, G.G. and Pickett, S.T.A. (1986). Long-term ecological studies: an illustrated account of their design, operation, and importance to ecology. Occasional Publication of the Institute of Ecosystem Studies, No. 2. Millbrook, New York.

Wiens, J.A. (1984). The place of long-term studies in ornithology. *Auk*, **101**, 202–203.

2 Long-term Ecological Research in the United States: A Federal Perspective

JAMES T. CALLAHAN
National Science Foundation, Washington, DC 20550, USA

2.1 ECOSYSTEM RESEARCH: THE CORE SCIENCE OF US/LTER

The fundamental scientific concept underpinning the United States' Long-Term Ecological Research (US/LTER) is that of the ecosystem as a describable entity. This concept is strengthened by the hierarchical view of levels of resolution above and below the particular ecosystem. The concept includes the notion that ecosystems and components of ecosystems interact in both temporal and spatial dimensions. Subordinate and superordinate hierarchical levels need not be viewed as cryptic; they may be simple scalars like decimeters and decameters bracketing the meter or weeks and years around the season. In that context, an ecosystem property

Long-term Ecological Research. Edited by Paul G. Risser

like primary production might be assessed at the level of the tree, the stand, or the forest, or for minutes, days, or years. Arguments about possible cybernetic qualities of ecosystems have been set aside in favor of furthering our collective ability to analyze ecosystem component entities and processes, to understand their interactions, and to represent them to others.

This is a systems ecology paradigm which demands integral participants from related disciplines and subdisciplines. This paradigm seeks actively to promote the synthesis of knowledge both within and among levels of resolution. It gains strength from a diversity of scientific inputs, and it challenges scientists to re-evaluate earlier conclusions.

2.2 GENESIS OF US/LTER

The United States' effort in Long-term Ecological Research (LTER) began formally in October, 1980, with six research grants. These were followed by five grants in January, 1982. After an interim of nearly six years, five projects were started in 1987, and three began in 1988. Two of the original studies have been discontinued. Seventeen projects are currently active (Table 2.1), as the result of four competitions for new proposals and two reviews of renewal applications.

The analysis provided here of the US/LTER is a qualitative one and is presented from the point of view of a science administrator and planner. Certain characteristics will be noted as either contributing to or detracting from an interactive network of LTER projects. Prospects for future directions will be enumerated. An assumption of primary importance is that it is unnecessary to justify the necessity of long-term

Table 2.1 US Long-term Ecological Research Projects (LTER site name; state; ecosystem type)

H.J. Andrews Experimental Forest; Oregon; temperate coniferous forest.
Arctic Tundra; Alaska; tundra, lakes, streams.
Bonanza Creek Experimental Forest; Alaska; taiga.
Cedar Creek Natural History Area; Minnesota; temperate deciduous forest and savannah.
Central Plains Experimental Range; Colorado; shortgrass prairie.
Coweeta Hydrologic Laboratory; North Carolina; temperate deciduous forest.
Harvard Forest; Massachusetts; temperate deciduous forest.
Hubbard Brook Experimental Forest; New Hampshire; temperate deciduous forest.
Jornada; New Mexico; sub-tropical desert.
Kellogg Biological Station; Michigan; row-crop agriculture.
Konza Prairie; Kansas; tallgrass prairie.
Luquillo Experimental Forest; Commonwealth of Puerto Rico; tropical rainforest.
Niwot Ridge/Green Lakes Valley; Colorado; alpine tundra, forest, lakes.
North Inlet; South Carolina; coastal forests, marshes, estuary.
North Temperate Lakes; Wisconsin; lakes, temperate deciduous forest.
Sevilleta National Wildlife Refuge; New Mexico; conifer woodlands, shrub-steppe, shortgrass prairie, desert.
Virginia Coast Reserve; Virginia; coastal barrier islands, marshes, estuary.

ecological research, the validity of arguments in support of it, or the utility of results from it. Discussion of these parameters can be found in another publication (Callahan, 1984).

2.2.1 SETTING THE TONE

Planning for US/LTER began in 1976 and involved the research community along with National Science Foundation staff. The planning process was augmented in 1977 with the first of a series of three annual workshops. Nearly 100 members of the ecological sciences community participated in the workshops, including academic research scientists and scientists and science administrators from private and governmental sectors. Each of the workshops produced a report to the National Science Foundation (NSF) (NSF, 1977, 1979, 1980). These reports became an important documentary base in the development of fiscal and administrative resources for the support of LTER.

It is fair to say that LTER is now well known by a large portion of the community, and that it stands as an identifiable entity. Nonetheless, LTER has been the recipient of critical attitudes from the past. For example, it has been necessary to remind some observers that LTER is not a monitoring effort such as the National Weather Service, nor is it an inventory such as might be undertaken in a biological survey. Perhaps most difficult has been convincing critics that LTER is not the re-invention of the International Biological Program (IBP).

The societal context of, and demand for, long-running ecological studies is diverse. The federal mandate for such an approach derives from law (for example, the National Environmental Policy Act) and from numerous executive-level reports. Within the United States' federal research structure there are a substantial number of antecedent programs of relevant character, such as National Environmental Research Parks (Department of Energy), Research Natural Areas (multi-agency), and Biosphere Reserves (US/Man and the Biosphere). Scientific interest in and demand for LTER was reaffirmed by the attendees of the NSF-sponsored workshops, who argued that a large portion of (perhaps most) ecological behavior is either not observed or is not accurately observed in the context of traditional two- or three-year grant-supported research projects. Another main supporting argument stated that comparisons among the same kinds of variables across a diverse array of ecosystem types should lead to more robust conclusions about ecological behavior. It is incumbent upon all LTER participants to recognize that the societal context is no more static than the scientific context for LTER: both change, although not necessarily in harmony. An adaptive approach is required, which can identify emerging scientific opportunities and successfully tie them to societal interests; for example, the logic of linking together Congressional interest in climate change with the emerging International Geosphere/Biosphere Program to derive the NSF's budget initiative focused on Global Geosciences. Within the Global Geosciences initiative, LTER has been successfully put forward as the locus of studies concerned with ecosystem dynamics.

2.2.2 CREATING A STRUCTURE

A decision made early in the genesis of US/LTER has proved to be unusually foresighted. That was the decision to provide external means for co-ordination and collaboration among the projects. It was recognized that the activities would be better accomplished if the costs of the activities were borne outside the individual project research budgets. Soon after the first projects began, a proposal was made by the associated project leaders, the LTER Co-ordinating Committee, that the NSF provide separate support for co-ordinating activities. Within a year of the first individual LTER project award, the first LTER co-ordination grant was made. Research groups who had only recently been competitors for LTER funding became co-operators and collaborators.

Two other decisions concerned the unity of the effort. First, LTER would be a fiscal undertaking of the NSF alone, and would not be dependent upon the behavior or budget-generating power of other agencies. Furthermore, within the NSF, one official would speak for LTER, both to the principals of the projects and to the community at large. However, unity was not intended to connote exclusivity. From the beginning, within the administrative structure of the NSF and in the projects, LTER has encouraged input from, and collaboration with, scientists who are not directly supported LTER participants and from other interested agencies.

2.2.3 SETTING STANDARDS

LTER specifications have consistently required that proposals address five core research areas drawn from the earlier series of workshop reports. The five core areas are:

(1) Pattern and control of primary production;
(2) Spatial and temporal distribution of populations selected to represent trophic structure;
(3) Pattern and control of organic matter accumulation in surface layers and sediments;
(4) Pattern of inorganic input and movement through soils, groundwater, and surface waters; and
(5) Pattern and frequency of disturbance to the research site.

How these core areas are treated in terms of conceptual approaches, specific questions, and analytical methods have been the primary foci of the peer scientific review of LTER proposals.

Project organizational structure for management of the science has also been an important determinant in the evaluation process. Although no fixed recipe has ever been dictated, certain more successful templates have been identified. Budget levels, in terms of maximum annual rates and future year amounts of funding, have always been specified by the NSF and enforced upon the projects except in unusual circumstances.

Data management was specified as a topic for attention in all announcements calling for proposals. As LTER has developed, data management has risen to a higher level of importance in the evaluation of both new and renewal proposals. Realistically, data management now has equal rank to the five core research areas, and requires an equivalent proportional allocation of fiscal, physical, and personnel resources.

2.2.4 FOSTERING COLLABORATION

As stated earlier, collaboration among LTER scientists has been a part of the plan from the beginning. In fact, it is a fundamental element of the LTER philosophy with its roots in the original workshop reports and preceding discussions. However, prospective project groups have encountered many unknowns as they have tried to find ways of satisfactorily and productively implementing the philosophy. In the first place, everyone can correctly insist that intense collaboration is already taking place within each LTER group as the project is conducted through the various subprojects, and certain types of interproject collaboration come easily as specialists in individual projects get to know each other and discover mutual interests. Through the LTER Co-ordinating Committee and its activities, scientists have been able to share their work and ideas in workshops, symposia, and other meetings on a fairly regular schedule. Nevertheless, persons at all levels of the US/LTER activity sense that collaboration among projects and scientists has not yet been fully developed. The missing level of development, which has come to be called 'networking', has become a major objective on all planes of LTER operations. The word 'network' is used to define an intensity of interaction that should far exceed the occasional exchange of written correspondence or telephone calls, the citation of papers, or even the conduct of parallel experiments. The definition includes all those things, but the 'network' is intended to be both more extensive and more intensive.

2.3 EVOLUTION OF US/LTER

LTER in the United States has not been a static entity. It has undergone an evolutionary process on at least three levels: organizational, conceptual, and technical. The conceptual level has already been discussed at some length, and it needs only to be restated that the central LTER concept is the ecosystem.

The development of LTER has combined a carefully measured pace with continuity at the executive level. We have tried to be assertive in developing fiscal resources that would remain part of the agency's base funding and would not stand out as a special (and possibly ephemeral) program. However, in developing the LTER portion of the budget, great emphasis has been placed on predictability as well as on opportunity. Predictability refers to the annual core operating budgets for LTER projects, which are set in advance on an extended time scale of several years. These budgets have provided the ability to request and defend the essential

LTER fiscal resource in the most reliable way. Opportunity refers to the sum of efforts to generate more money for the support of LTER by the identification of initiatives that fit well with the overall concept and capabilities of LTER.

Essentially, all funding for LTER has come to the NSF as new money and not by reprogramming existing base funds. To take this approach, it is necessary to identify new initiatives or to participate in new opportunities. As an example, the LTER concept fits with the emerging planning, within NSF and the US federal science establishment, for US participation in the International Geosphere/Biosphere Program (IGBP). NSF also tries to respond to the signals being sent by project leadership and participants with regard to their technological and personnel needs for emerging scientific challenges. These signals can often be translated into fiscal initiatives within the agency. An appropriate example is NSF's recently begun initiative to provide technologies which are pertinent to Remote Sensing and Geographic Information Systems (GIS), tailored to meet the needs of individual projects and with an obvious tie to the predictable needs of IGBP.

In an organizational or management concept, the US/LTER operates in a mode of mutual induction. The NSF tries to induce the projects to generate salable output (salable in the sense of having the power to influence the budget process in a positive manner) while the projects attempt to induce the NSF to provide them with more money for research. As long as both parties subscribe to the mutuality of the relationship and the national economic/political environment remains favorable, this concept remains valid. However, changing times may require a change of tactics within the overall strategy, which is aimed at increasing the ecological sciences community's identification with, and subscription to, LTER as a productive means of supporting ecological research.

With interproject collaboration as a main goal in LTER, the NSF has wisely looked for ways to promote harmony, and avoid disharmony among projects. The NSF has provided for a separate co-ordination budget to take most of the financial burden for their activities off the individual project budgets. With regard to the project budgets themselves, NSF has gone even farther: project budget totals are specified in advance at identical levels for all projects. Within those constraints, the leadership of each project is free to allocate the resources as they deem most appropriate. When opportunities are developed for extra money, the approach has been to make the funds available to all projects as supplements to their base budgets. Perhaps the most useful way to promote harmony and to augment the total research support at an LTER site is through the development of LTER-related satellite projects. Whether funded by NSF or by some other agency, such projects add structural strength, diversity, and resources to the available pool.

National Science Foundation staff support for LTER has been minimal. Recent additions in staff have included one person whose full-time efforts are directed toward interproject co-ordination, including the identification and promotion of opportunities for collaborative research. Approximately half of the effort of another new staff position is directed at developing electronic networks among projects.

For the last year, an internal LTER management group, consisting of the five most pertinent scientific staff, has met regularly to address planning issues such as technology acquisition, networking opportunities, and budget development.

The decision was made early on that LTER proposals would be subject to the same type of review that is imposed upon all other competitively submitted research proposals by the Division of Biotic Systems and Resources (BSR). For the first group of new LTER proposals, we followed our standard pattern of requesting ad hoc reviews by mail and convening an advisory panel to review the proposals. All subsequent rounds of proposal review have utilized only an advisory panel. The experience with ad hoc reviews was disappointing in that opinions were uncharacteristically divergent, often missed the point, and were generally not useful. An assessment of the problem was that the proposals were of a new type and unfamiliar to the reviewers, and were therefore difficult to judge in any comparative sense. Proposal success rate in the LTER review process has been similar to that for general research proposals, about 20%.

2.3.1 BUILDING A CONSTITUENCY

The dominant role of non-LTER scientists (i.e. not directly participating in an LTER project) in the peer review process, coupled with their participation in the early formative period, constitutes a substantial base of broad community input to LTER. However, these kinds and amounts of participation are not viewed as sufficient to give major substance to the element of US/LTER philosophy which regards LTER projects (their sites, scientists, data bases, and general working environment) as resources available to the entire ecological research community. Several more direct mechanisms have been identified to induce higher rates of collaboration:

(1) The leaders of each project must clearly state that the project welcomes such collaboration, and they must show that they are prepared to assist collaborators;
(2) Project groups must have, or strive to create, physical facilities to accommodate collaborators from outside the parent institution;
(3) The LTER co-ordination effort includes a specific operational mandate to support the participation of other scientists in LTER-sponsored topical workshops;
(4) LTER-sponsored and LTER-related symposia are regularly organized in the context of major national professional meetings;
(5) Under the auspices of the LTER/Co-ordinating Committee, two publications are distributed to the broader community: the LTER Network News and a brochure containing descriptions of all the research sites.

These mechanisms are all more or less informational approaches. Special funding opportunity programs of the NSF are regularly used to encourage the conduct of research at LTER sites by scientists not directly supported by LTER grants. Women's programs, small college faculty programs, post-doctoral fellowships, and mid-career fellowships are all being utilized in part to expand the LTER-associated community. A new opportunity has recently been announced as available to all current grantees of the BSR Division. This opportunity is for funding supplements, as extra money available to projects already in place, and for research at LTER sites to expand the scope of the research already funded to include samples from or comparisons with LTER sites. The community response to this new opportunity was enthusiastic. Of 31 applications received, 12 supplementary awards of funds were made. The intent is to continue this special opportunity for at least two more years.

Numerous related and proactive mechanisms are being employed by the National Science Foundation together with LTER project leaders to expand the community of scientists who rely on LTER as a resource for the productive pursuit of their research interests. In the sense of scientific infrastructure, it is an explicit goal to make LTER a widely accepted resource for research, a resource that might be characterized as a new kind of ecological research environment. It is to be hoped that this new research environment will promote collaboration among scientists, challenge many of them with new concepts and approaches, allow them to ask questions that in other contexts could not be asked (or answered), and lead them to synthesize information in new ways. In pursuing the goal of developing LTER projects as a resource for research in many aspects of environmental biology, and in ecosystem research itself, a unique LTER identity is being created as a broad community of scientists is called upon to collaborate in ways that would not otherwise be likely.

2.3.2 ORGANIZATION OF THE PROJECTS

Among the US/LTER projects there are as many variants of internal project organization as there are projects. The NSF has not attempted to create formal specifications for organization; however, the most successful of the projects share certain characteristics in their organization and in their mode of operation. The foremost of these characteristics is that of leadership. Those projects that have multiple (i.e. three to five) intellectually strong leaders are the most effective ones.

The second important feature of the most successful projects concerns the creation of a project team which is both philosophically and operationally compatible. One group has even gone so far as to include an organizational psychologist in the project cadre who actively engages in team building and conflict-resolution exercises. Whatever the mechanism, it has become apparent that research team identity is of great importance.

Careful attention must be paid to intraproject communication. In spite of individual protestations to the contrary, it seems unlikely that too much effort can be spent in this way. The LTER groups that are most effective have regularly scheduled, all-participant meetings. A major addition to intraproject communication can come from encouraging (even requiring) participants to assist others in scientific areas that are far removed from their own specialties or central interests.

The pattern of intraproject allocation of fiscal resources is vital to the maintenance of team spirit. Effective project leaders will apportion their budgets in accordance with the diverse requirements of the entire integrated research effort. The allocation of salary, support personnel, and equipment for the singular benefit of the most senior team members can be counted on to generate dissatisfaction and dysfunction.

2.3.3 LTER NETWORK ORGANIZATION

The first manifestation of organizational structure at the LTER network level was the Co-ordinating Committee mentioned earlier. The first major act of the committee was to develop a proposal for separate co-ordination support. The committee also assumed the function of sponsorship of an array of technical committees, of which data management and meteorology were among the first, with modeling a more recent addition. These committees and more *ad hoc* topical groups have organized many technical workshops for the benefit of LTER and the community at large.

A more recent creation of the Co-ordinating Committee is the Executive Committee. This four-member subgroup recognizes the need for the Co-ordinating Committee to act more expeditiously now that the network has grown to include 17 projects. The Executive Committee provides a mechanism for timely interaction between the NSF and the LTER network. For several years, a modest 'office' structure has assisted the Chairman of the Co-ordinating Committee to carry out the operational functions of the Committee. That office is now completing a major expansion of its duties and staffing.

The NSF has begun to provide funds for physical networking among projects in order to promote and support scientific collaboration. When completely realized, this plan should provide an everyday, openly accessible means of communication, not only text communication, but also fully user-transparent support for joint activities, from experimental design through interactive graphics in data analysis and manuscript preparation. The system will be computer supported at all levels, perhaps utilizing the NSFnet supercomputer network, and should provide ready access to distributed data bases, support the analysis of remotely sensed data, and employ the best generally available geographic information system (GIS) technology.

2.3.4 US/LTER TECHNICAL METHODS

No directed attempt has been made to require all LTER projects to practice the same data acquisition or analytical methods. However, there has always been an element of the LTER philosophy which states that individual projects should strive to make their data usefully comparable with similar data from all other LTER projects. Comparability will be stronger if it derives from the excellent quality of documentation of the conditions under which, and the methods by which, data were collected, and if the data sets meet closely matched statistical indices of reliability. That is, comparability will be better if the methods are designed to fit the ecosystems being addressed and the people employing the methods.

It is the attitude of the NSF toward the means by which data are generated, and the protocols by which the data are maintained and accessed for synthesis and reporting purposes, that the projects should use the best generally available technology for gathering and handling data, and should be aggressive in the identification and testing of new technologies which may be adaptable to LTER needs. In our accumulated experience, the two main impediments to progress along those lines are the limited ability of the NSF to provide new money above project base budget levels, and the lack of enthusiasm by investigators to spend new money on switching methods when they would rather spend it in other ways. So far the push–pull interaction has worked fairly positively, and several new technologies are being pursued. However, while the NSF wishes the projects to utilize new methods effectively, it does not necessarily require that they invent them.

2.4 EVALUATION

2.4.1 DIFFICULTIES

It is unfortunate that the US/LTER operation did not recognize earlier the formal organizational structure that would be required for interproject collaborations in the network of projects, a network which grows increasingly complex. It is the lack of structure which has caused the slow rate at which interproject collaborations have been identified and actuated. Similarly, the network is now beginning to realize the need for the adoption of standards of performance with regard to some types of data, and especially with regard to how and in what formats the data are maintained.

Some difficulties have been attributable to NSF and its ability to do its part of the job. Chief among these has been our inability to bring to bear sufficiently regular and critical on-site reviews of projects. The long gap (nearly six years) between the second and subsequent rounds of competition for new LTER funding was not good. Too much catching up has had to be done by newer project groups, whose input would have been easier to assimilate earlier. Both of these difficulties relate to internal constraints on resources.

Establishing LTER as a recognizable entity with a character of its own in the scientific community has not been easy. Establishing an identifiable profile on the outside while avoiding a 'special' designation on the inside are not easily compatible tasks. However, continuity is essential in that it is the key to maintaining a functional identity and to creating a constituency that is willing to speak positively for the effort.

It has been difficult to guide the effort without appearing to direct it with a heavy hand. The freedom to do whatever research is appealing by whatever means seem appropriate is zealously defended by the academic research community. That is as it should be. It is also important that the agency managers be able to choose the best tactics for maintaining and increasing the fiscal resources available to support the effort. The effective meshing of these two considerations has required a willingness to accommodate and compromise on both sides. The relationship is now a productive one.

2.4.2 POSITIVE, TIMELY DECISIONS

Some of the best decisions that were made in the creation of the LTER project, which have already been presented in some detail, although in a diffuse pattern, in this chapter, are summarized here:

(1) Single agency core funding for LTER;
(2) All new money for LTER;
(3) Single program officer responsibility for LTER;
(4) Pre-defined core areas for organizing proposals;
(5) Separate support for most co-ordination efforts;
(6) Close interaction between the Foundation and projects; and
(7) Efforts to broaden the LTER-affiliated community.

Perhaps the soundest decisions were those that pertained to maintaining US/LTER as a genuinely competitive grants emphasis. That is, LTER has not been given an inside track, and all projects in the course of their conduct and all proposals in the quality of their content must meet the highest competitive standards of evaluation. Nothing has been formally institutionalized, and individual projects that fail to meet competitive standards can be, indeed have been, discontinued.

2.5 PROSPECTS FOR THE FUTURE

Constraints upon the prospects for growth in the number of LTER projects include the ability of the scientific community to perform, the availability of appropriate research sites, and the capacity of the sponsoring agency to administer the needs of large, complex research projects. The Foundation has accepted the fact that there must be some functional maximum for concurrently active LTER sites. This does not necessarily mean that there can be no new LTER projects once that ceiling is reached, if it is reached. We have already witnessed the occurrence of some attrition among projects. It should be possible to accommodate limited numbers of new applications in the future as long as the total number of operant projects remains below the effective ceiling.

Assuming that LTER continues to prove its worth in the highly competitive arena for budgetary attention, there are several augmentations that have appeal.

The acquisition and operation of technologies to promote 'on-line' collaboration among scientists has been mentioned previously. Much can be accomplished in that manner. Continuing the effort to identify and adapt new analytical technologies to LTER objectives deserves increased attention; the distribution and implementation of productive technologies will command major support.

Expansion of the web of LTER affiliations requires significant investments of both human and fiscal resources. Domestically, LTER could grow to become significant in the area of graduate education as well as in research. Fulfilment of the vision of LTER as a resource for diverse kinds of ecological and related research will probably command investments in physical plant that have not yet been made. At the international level, US/LTER must be somewhat more aggressive in identifying and supporting joint endeavors and exchanges among research groups. Application of LTER data and syntheses to the resolution of major global environmental questions will require special efforts and expansion in new directions.

It is useful to ask, 'What if we had it to do all over again, would we do it differently?' There is no single-dimension answer to these questions, but a few of the obvious partial answers are worth discussing. It would not be wise to constrain the applications to a suite of biogeographically predetermined ecosystem types. Such a constraint would probably act negatively on the pool of potential LTER research groups and on their ability to create competitive proposals. The 'core areas' have served as a useful organizing tool for LTER proposals and projects, and they should not be viewed as having limited the investigations. These areas have been the foci around which subprojects are organized, and from which independently competitive proposals have been derived for more detailed studies of, for example, organic decomposition or the role of animals in imposing more complex structures on ecosystems. Neither of these characteristics would be productively changed.

It is difficult to address the question of whether LTER has caused the discovery of anything of a scientifically fundamental nature at this time. To do so is to appraise the long term after only a short period. However, a positive partial appraisal is possible. LTER has brought considerable attention to bear on the myth of 'average' or 'normal' conditions of ecosystem behavior. I am tempted to say that almost any variable that falls within close proximity of a long-term average value should be regarded with suspicion. Of course that is not true, but it does now appear that many apparent annual averages for both biotic and abiotic variables are of severely limited utility in characterizing ecosystems. Inherent variability is simply too great. Similarly, the occurrence of significant disturbance in ecosystems appears to be frequent. Many scientists would assert that disturbance is the major determinant of ecosystem character in that it keeps the system 'stirred up' by imposing spatial variety which promotes biotic diversity. It can also be noted that many ecological conclusions have previously been founded upon transitory behavior that is often directly counter to longer-term ecosystem changes. These are not 'discoveries' *per se*; however, more attention to these questions has been brought to bear more quickly than would probably have occurred under pre-LTER conditions.

As judgments are made of the US/LTER undertaking, we should hope that they are made with the understanding that LTER is an experiment. It represents a departure from past ways of doing things. Evaluators should strive to understand that LTER is not a static entity but a dynamic one. The visage of LTER that exists today is not the same as that of eight years ago, nor is it the same as will exist in a few more years. With regard to paradigmatic, technological, and organizational changes, more scientists are likely to move forward more rapidly than they would without LTER.

Beyond the level of basic ecological science, LTER will play an increasing role in providing inputs to the management of natural resources. At the very least, LTER projects, by their very existence, create identifiable centers of expertise for particular types of ecosystems. In addition, LTER projects have from the beginning encouraged and benefited from the participation of scientists employed by resource management agencies.

2.6 ACKNOWLEDGMENTS

Expressing appreciation to individuals who have played significant roles in the creation of US/LTER would be an almost endless task. In lieu of that, my thanks go to the ecological sciences research community of the United States and to the staff of the National Science Foundation.

2.7 REFERENCES

Callahan, J.T. (1984). Long-term ecological research. *BioScience*, **34**, 363-367.

National Science Foundation (1977). *Long-Term Ecological Measurements*. Report of a Conference. Woods Hole, Massachusetts.

National Science Foundation (1979). *A New Emphasis in Long-Term Research*. Woods Hole, Massachusetts.

National Science Foundation (1980). *Long-Term Ecological Research (LTER)*. Indianapolis, Indiana.

3 The Role of Study Sites in Long-term Ecological Research: A UK Experience

O. WILLIAM HEAL
Institute of Terrestrial Ecology, Edinburgh Research Station, Bush Estate, Penicuik, Midlothian EH26 OQB, Scotland

3.1 INTRODUCTION

In the UK consideration is being given to establishing a network of long-term ecological references sites. Why? Questions are asked with increasing frequency about changes in the environment (levels of pollutants, climate, species introductions, habitat gains and losses, shifts in land use, etc.). Answers to the questions are usually cautious because we lack information on the past, we cannot distinguish trends induced by man from natural variations or attribute cause to any observed change, or because the available information is not appropriate to the particular conditions in question.

Instead of setting up new studies to answer each question as it arises, the approach being considered in the UK is to make use of well-established sites with existing data to establish a network of long-term ecological reference sites, with ongoing studies to help answer questions in the future. How suitable are the present sites? The sites have been maintained by different organizations which use them for their own purposes, and many of them are not known to others. However, when combined, will these sites constitute a network which adequately represents the range of environmental and management conditions in the UK?

Long-term Ecological Research. Edited by Paul G. Risser

Is the information available from these sites useful in answering questions about environmental change? These are the questions being asked and discussed by a working party established by the Natural Environment Research Council.

Although based on the working party's considerations, this chapter represents a personal view of the general issues. Selected examples are presented, in this chapter, and some other examples are given in Chapter 6 of this volume.

The main rationale for proposing a network of sites is as follows:

(1) Study sites are an essential component of ecological research. To answer questions on changes in the environment, we need sites which represent the main environmental, ecological, and management variations in the UK While studies of some individual topics will require other sites with particular characteristics, a 'core' network will provide the opportunity to use existing information on the related topics and facilities.
(2) Long-term studies are required to monitor changes external to the system which take place gradually or at infrequent intervals. Responses to those changes may occur through species or processes which have a slow turnover time or through a series of linked short-term events, the results of which are only apparent in a long-term study (Likens, 1989).
(3) In addition to delayed and serial responses, it is also necessary to distinguish between the different factors which cause, or interact to cause, change. For these reasons it is important to have sites with integrated or multi-media monitoring, and to carry out both observational and experimental research.
(4) The scientific case for a network of long-term study sites in the UK is strong. Information on environmental changes and on their consequences is a serious need in government. By concentrating on established sites, the cost of creating such a network can be kept to a minimum.

This chapter considers three aspects of the role of study sites in long-term research and monitoring: (1). the ecological questions to be addressed, (2). the separation of natural and man-induced phenomena (noise from signal), and (3). strategies in the selection of sites.

3.2 QUESTIONS TO BE ADDRESSED

Questions asked by those concerned with environmental policy or management test the scientists' ability to detect changes in the state of the physical, chemical, and biological environment, and to distinguish cause from effect. The basic questions ask what is happening, where it is happening, what the consequences are, whether there are thresholds, and whether the thresholds are reversible. Apart from responding that more research is needed or that it all depends on the conditions, most ecologists will turn to their favorite site and, using their favorite methods, will measure as intensively as labor allows. While many of the questions can be explored in the laboratory, we must use field observation and experiment to evaluate extrapolation from particular laboratory conditions to the complexity of the field, and also to study phenomena which cannot be examined in the laboratory, test predictions of responses to control measures, and provide early warning of new changes. It follows that the sites and conditions in which measurements are made must be selected in relation to the questions asked. Here lies a major problem. We know what today's questions are, but what are the questions of the future towards which long-term research and monitoring should be directed? Although we cannot specifically predict future environmental concerns, we do recognize the four generic subjects of climate change, chemical pollutants, management effects, and invasions or extinctions. Each of these makes particular demands upon a site, and in a study will require particular types of site, measurements, and experiments.

3.2.1 CLIMATIC CHANGE

Monitoring the physical climate is reasonably straightforward, and is systematically undertaken by the Meteorological Office. Prediction of climatic change is much more difficult, especially at the regional level. Ecological responses to general trends in climate and to irregular extreme events are complex and difficult to assess. Despite this, and despite the importance of climatic factors to ecology and crop production, there is a surprising lack of systematic monitoring and analysis of species and system response to climate change. The Rothamsted Insect Survey is one comprehensive study in which population changes can be related to climatic variation, and there are some relevant species studies on individual sites (Taylor, 1989). However, it would be sensible to establish, now, situations in which to detect degrees and types of biological and ecological response. The questions asked in these situations are likely to focus on changes in the distribution of species and in the composition of communities of nature conservation or of amenity interest, in the production of agricultural and forest crops, in the occurrence of crop pests, and in land-use patterns. While the current concern is the consequences of long-term projected trends in climate resulting from the greenhouse effect, the biota will respond to the frequency and intensity of irregular extreme events (for example, dry summers or mild winters) as much as to general trends in mean values of temperature or precipitation.

In all these questions, effects are likely to be greatest, and most detectable, at the climatically determined margins of distribution of a species or production system. Thus the questions can be targeted quite specifically by selection of species and sites according to their biology and distribution. For example, repeated annual sampling showed that population regulation of the homopteran *Neophilaenus lineatus* was density dependent at Wytham Woods, a low-altitude site, but density independent and climatically determined at Moor House, a more northerly high-altitude site (Figure 3.1) (Whittaker, 1971). Population response of *N. lineatus* to climate change would, therefore, be most readily detected at Moor House, and it is a suitable indicator species because its density can be monitored by counting numbers of readily observed spittle. Similarly, the annual distribution and success of insects

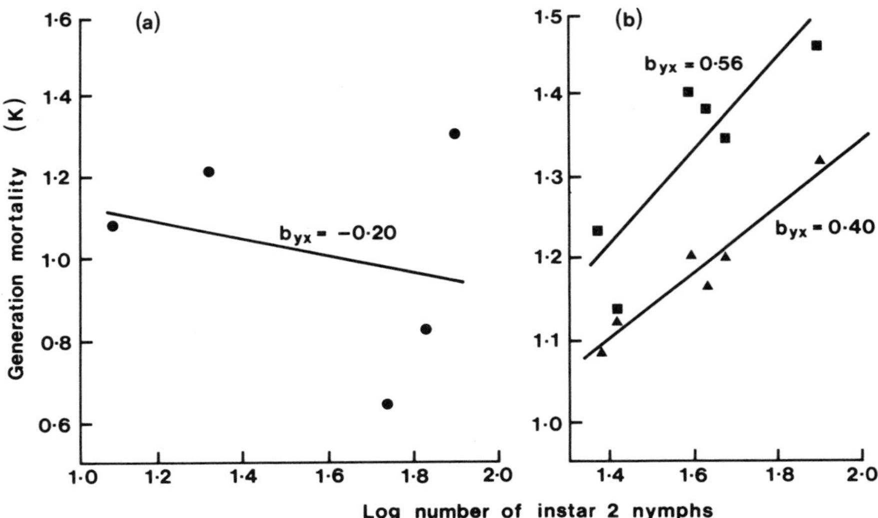

Figure 3.1 The relation between the logarithm of numbers of instar 2 in each generation and the subsequent 'generation' mortality (K) in *Neophilaenus lineatus* (from Whittaker, 1971), (a) at the *Juncus* site, Moore House, (b) at Upper Seeds, Wytham Woods. ■: including parasitoid; ▲ : excluding parasitoid (Coulson and Whittaker, 1978)

along altitudinal rather than geographical gradients of climate can be directly related to climatic variation (Coulson and Whittaker, 1978; Whittaker, 1985). The same principles apply for plants in the selection of species, sites, and parameters to detect response to climate change. For example, in the uplands of Britain, *Calluna vulgaris* and *Rubus chamaemorus* are at the cold and warm ends, respectively, of their climatically determined distribution. The upper and lower distribution of these species can be readily identified and marked to define the extent to which hypothesized or observed future changes in climate affect vegetation. Such basic monitoring can be enhanced by associated research, as in the case of *Calluna* and *Rubus*, in which physiological studies have defined their photosynthetic responses to climatic variables (Figure 3.2). From these response surfaces, Grace and Marks (1978) showed a differential response of the two species to warm years, and how growth of *Calluna* could affect *Rubus* through light interception. This type of information allows prediction of both short- and long-term responses to particular climatic variables, with potential validation by field measurement.

The examples from insects and plants indicate how studies can be focused on questions of which species will be affected, and where and why. When expanded to broader questions of land use, the analysis of Parry and Carter (1985) is particularly relevant. They showed that historically, the frequency of crop failure was related to the occurrence of climatic extremes. During periods of climatic change the distribution of agricultural use expanded or contracted depending on the frequency of failure. As a result, change in land use in upland areas was concentrated in a

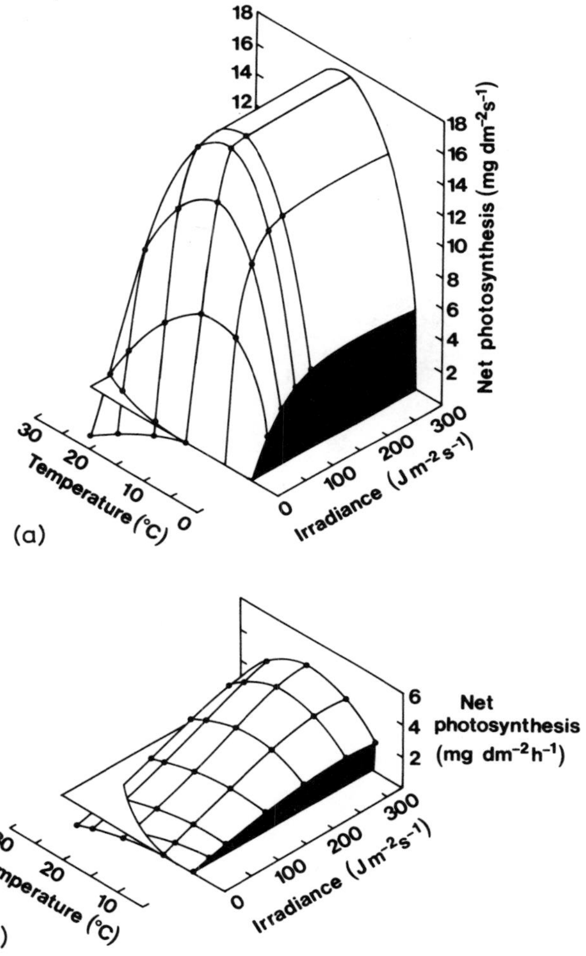

Figure 3.2 (a) Photosynthesis of *Rubus chamaemorus*, aged 11 weeks, as a function of temperature and irradiance (380–720 nm) (Grace and Marks, 1978); (b) photosynthesis of *Calluna vulgaris* shoots, 35 days after budbreak, with no flowers, and with a temperature treatment of 10°C (Grace and Marks, 1978)

distinct geographical zone at the margin of particular crop production systems. This is illustrated in Figure 3.3 for a particular valley in which improvement of moorland for agriculture, and abandonment of farmland to moor, has been restricted for over 200 years to a narrow fringe between stable areas of moorland and farmland.

The growth of species or crops within the main part of their range will be affected by climate, but other factors will also limit their success. By definition, climate will have its greatest impact where it is the main controlling variable, i.e.

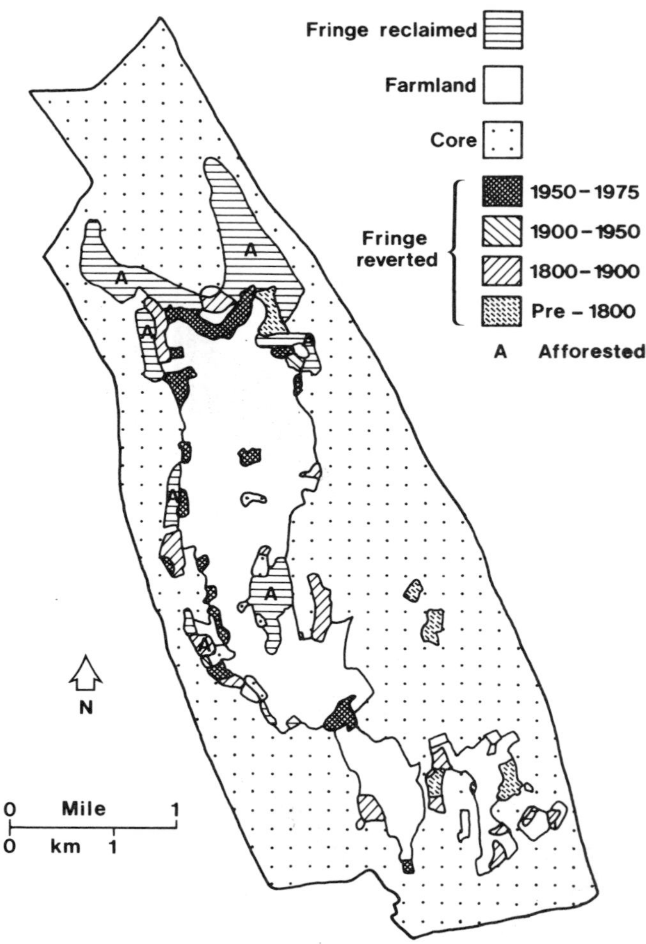

Figure 3.3 The distribution of land which has been reclaimed from or reverted to moorland in Bransdale in the North York Moors

at the climatically determined margins of distribution. The implications for long-term monitoring and research of the examples quoted are (1) sites should be large enough to include climatic gradients or be selected to cover a gradient, (2) species sensitive to climatic change can be identified through their distribution patterns and population dynamics, and (3) relatively simple measures, such as phenological observations, distribution limits, the current year's shoot increment, and population success, can, given a knowledge of the species based on detailed research, provide measures of short- and long-term response.

3.2.2 CHEMICAL POLLUTANTS

Experience has shown that the range of pollutants and their effects are extremely variable and complex, yet there is frequent pressure to define cause and effect at short notice. However, there are general lessons to be learned, particularly in relation to the selection of sites for long-term monitoring and/or research. In considering pollutant problems such as acid deposition (N, S, O_3), heavy metals (Pb, Zn, Cd), fertilizers (N, P), pesticides (DDT, 2,4-D), or radionuclides (^{134}Cs, ^{137}Cs, ^{239}Pu, ^{240}Pu), the common questions concern the rates of deposition, the transformation and retention of the element or compound within an ecosystem, its transfer through biological pathways, and its toxicology. In other words, questions of element dynamics which require detailed understanding of ecosystem processes. Unless there are overriding site-specific issues, it is sensible, therefore, to associate studies on new pollutants with sites for which the dynamics of other elements are known.

In assessing the consequences of pollutants, there are three main recurrent questions concerned with spatial variation, infrequent events, and interaction effects:

(1) Which areas can be identified as 'hotspots' associated with particular deposition or retention characteristics? For example, concentrations of ^{137}Cs following the Chernobyl accident showed high within and between site variation in the UK, resulting not only from the rainfall pattern but also from small-scale variation in surface water movement and varying ion exchange characteristics or organic and mineral soils (Figure 3.4). The decline of ^{137}Cs in vegetation predicted from studies of predominantly mineral soils did not correspond well with observed time trends on peaty soils, emphasizing again the effect of varying site conditions (Figure 3.5). Experience in acid deposition research has also shown the importance of climatic and topographic conditions, vegetation cover, and soil properties in determining variation within and between sites in cause and effect interactions. In particular, the combination of factors has been important in answering the question of which areas constitute hotspots which are most sensitive to acid deposition.

(2) Which climatic conditions are particularly important in causing pollution problems? The climatically induced episodes of deposition and of release of acidity in snow melt, and of periods of high O_3 concentrations and flushes of NO_3 associated with drought conditions (Roberts *et al.*, 1989), all emphasize the significance of short-term fluctuations within the context of long-term measurements.

(3) To what extent can observed effects be attributed to specific pollutants? Even the immediate direct effects of a pollutant on tree growth are confused by the interaction or combined effects of a pollution cocktail or by interaction with other environmental factors, as in the case of increased sensitivity of red Appalachian spruce to frost associated with atmospheric deposition (Fowler

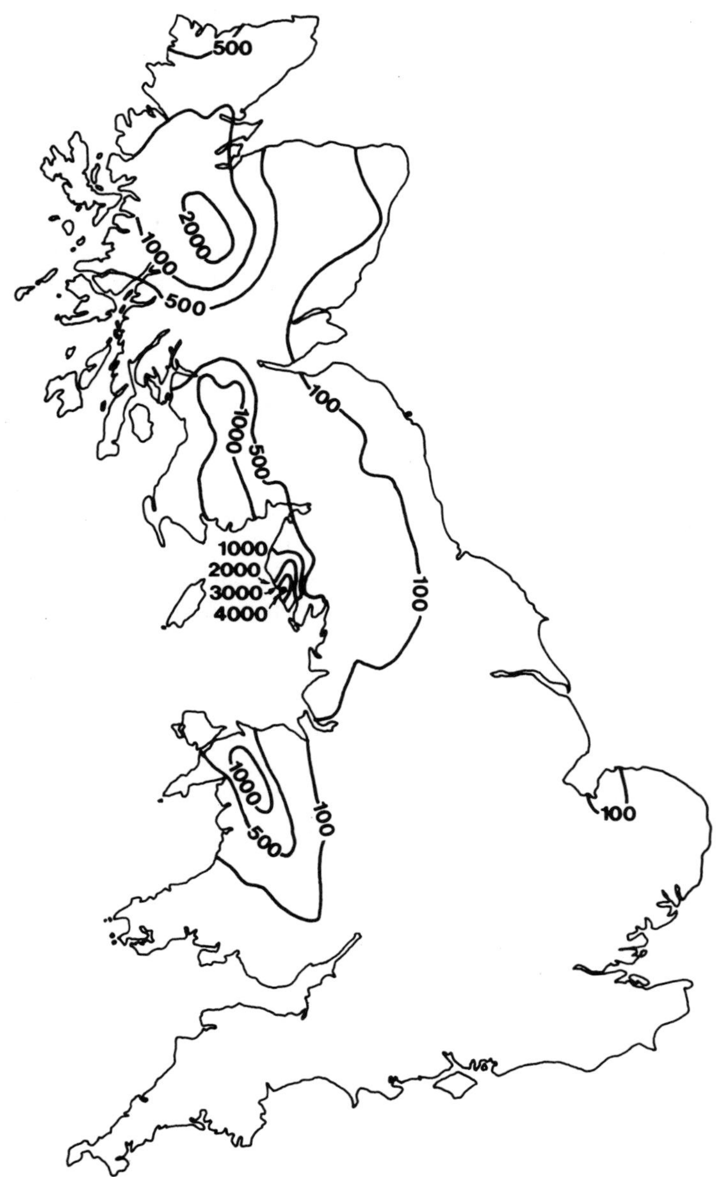

Figure 3.4 ^{137}Cs in UK grassland vegetation following the Chernobyl accident (Bq/m^2)

et al., 1989). More indirect effects are well known, such as the release of toxic Al from weakly buffered acid soils as a result of shifts in ionic balance following acid deposition (Ulrich, 1987), or the concentration and transfer of some pesticides, heavy metals, and radionuclides along food chains.

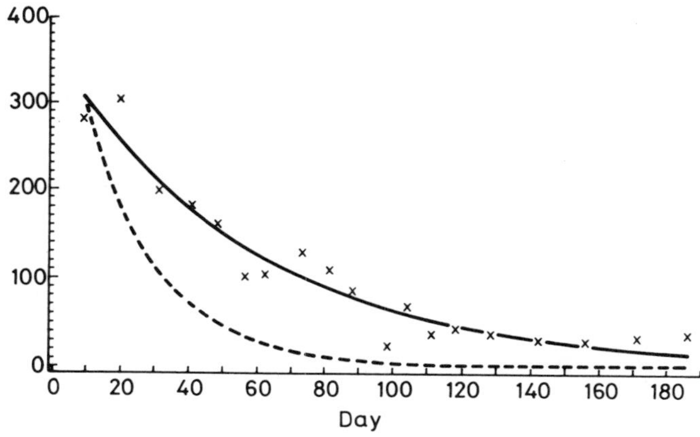

Figure 3.5 Predicted (- - -) and observed(—) concentration of ^{137}Cs following the Chernobyl accident in UK upland vegetation on peaty soils

The implications of these questions for further site monitoring and the assessment of pollutant cause and effect are that sampling must enable:

(1) Detection of small within-site hotspots, particularly of aerial pollutants;
(2) Sites to cover a range of environmental variation to allow identification of areas sensitive to mobilization or concentration of pollutants;
(3) Measurement of a range of variables other than the pollutant necessary to detect interactions and indirect effects; and
(4) Measurements on time scales frequent enough to detect climatic episodes, which are at least as important as long-term mean characteristics.

3.2.3 MANAGEMENT AND LAND USE

Some questions concerned with such specific management practices as fertilizer or pesticide application are considered under the heading of pollution, but there are also repeated questions concerning the amount and distribution of changes in management systems and land use, and of the consequential alteration in flora and fauna. The more obvious changes in land cover (for example, hedgerow removal or loss of moorland through agricultural improvement (Figure 3.3)), are readily determined given adequate sampling design and consistency of recording. In contrast, detection of the ecological consequences of such changes, and of more extensive management changes such as modified grazing type and intensity, are more challenging (Usher and Thompson, 1988). These consequences are particularly relevant to long-term research because the response times are often measured in decades. Three topics, not altogether new, on which questions are focused are succession, soil processes, and land cover mosaics. These are discussed below.

The limited number of detailed long-term studies of the successional sequence of vegetation following changes in management is remarkable (Miles, 1979). Those detailed long-term studies that exist, classically Watt (1960) and the Park Grass Experiment (Taylor, 1989; Thurston *et al.*, 1976), show marked short-term variations in species composition which could be misinterpreted as long-term trends if the studies were restricted in time (Figure 3.6). In these studies, as in others on vegetation and fauna (Marrs *et al.*, 1988; Willis, 1988; Peterkin and Jones, 1989) which constitute long-term measurement of responses to changes in land management, site specificity limits the ability to make predictions of the consequences of land-use change and to assess management options. This limitation can be overcome by integrating results from isolated studies into descriptive general hypotheses of succession (see Figure 3.7) (Miles, 1988), or

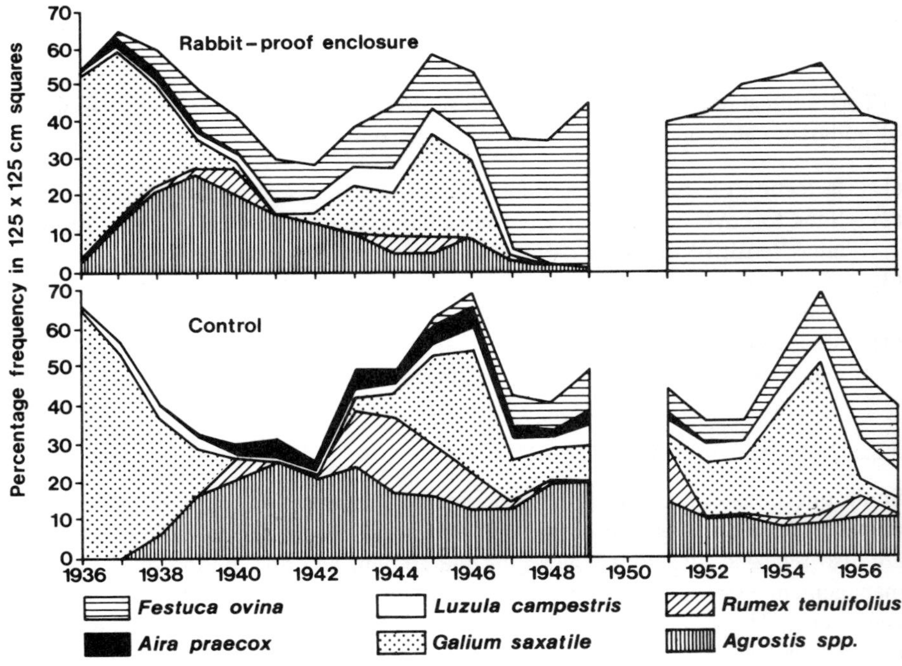

Figure 3.6 Changes in species composition of a grassland inside and outside a rabbit-proof enclosure in the English Breckland (from Watt, 1960, p. 220, courtesy of *Journal of Ecology*)

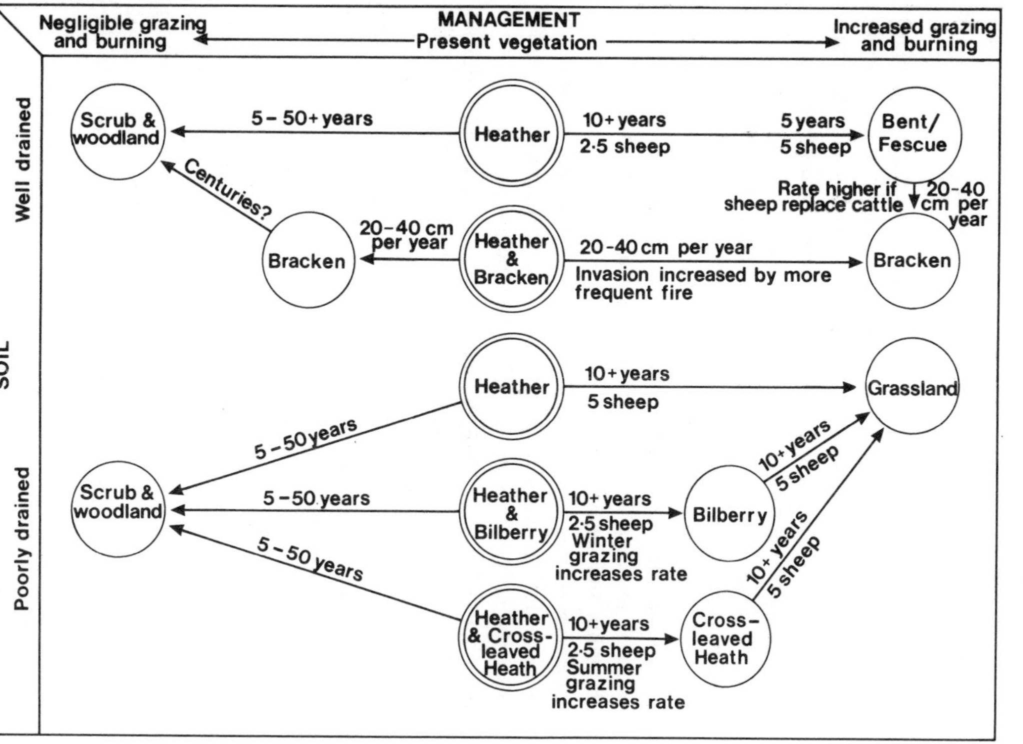

Figure 3.7 Changes in heather-dominated dwarf shrub heaths as a result of variations in intensity of grazing and burning. Present vegetation is shown in double circles: probable and possible stages in the vegetation succession are shown by single and broken circles respectively. Rate of change where known is shown above the arrows: sheep numbers are per hectare. Differing soil conditions are noted in the left-hand margin (after ITE, 1978)

by more mechanistic hypotheses based on growth strategies (Grime, 1987; Noble and Slatyer, 1980). These site-specific and theoretical approaches need to be complemented by extensive surveys undertaken at infrequent intervals, as in the National Ecological Survey of Britain carried out by the Institute of Terrestrial Ecology in 1978, which involved a stratified sample of 256 sites. The survey provides a baseline definition of species composition of open areas, and linear features. It was repeated in 1990 to determine the degree of vegetation change associated with measured changes in land use.

Studies of the response of soil processes to land management are essentially long term, given the nature of pedogenetic processes. There are many pragmatic principles of soil protection, particularly those to minimize erosion and maximize fertilizer use efficiency, which are amenable to short-term studies. However, recurrent questions of sustained fertility, of soil improvement or degradation by trees, and of the effects of pesticides demand long-term field observations. The Rothamsted Experiments, described by Johnston in Chapter 6 of this volume, are unique in their time scale. They are also important in demonstrating the value of including management variations, even with inadequate statistical design (Taylor, 1989). One key feature in the determination of long-term trends in soil properties is the definition of within-site heterogeneity; geostatistics are a valuable aid in this respect. Retrospective studies (Davis, 1989) and space-for-time substitution (Pickett, 1989) provide valuable complementary approaches to long-term studies on sites which provide the only means of controlled manipulation.

Questions of the effects of loss or gain of habitats are being expressed with increasing frequency in terms of fragmentation, isolation, connectivity, or unit size and shape, i.e. in the terms of theories of island biogeography. This emphasis is related to recent legislation which provides financial support for management practices to sustain or enhance wildlife in environmentally sensitive areas, and to support farm woodlands and set-aside of agricultural land. The response of invertebrates and flora, as mentioned above, is long term, involving questions of dispersal, colonization, and establishment. Such research is appropriate in scale to most study sites, but observation of existing patterns is unlikely to provide sufficient variation to give satisfactory answers. The subject is very amenable to experimental manipulation yet, surprisingly, is virtually unexplored. On a much larger scale, studies of the effect of structural characteristics of woodlands or of habitat mosaics on large mammals and avian predators have very particular site requirements which are not compatible with most other topics.

3.2.4 SPECIES INTRODUCTIONS

In *The Ecology of Invasions by Animals and Plants*, Elton (1958) documented and analyzed many cases of the population expansion of an introduced species. Interest in the subject has been sustained in the UK by population expansion of introductions such as coypu, mink, sika and muntjak deer, and collared dove. More recently, the subject has been stimulated by questions concerning the environmental consequence of the accidental or intentional introduction of genetically manipulated organisms (Kornberg and Williamson, 1987).

Following the course of accidental introductions, whether genetically manipulated or not, may require long-term studies, but obviously these will have to be done at the sites of occurrence. The role of established sites will be to provide an opportunity to examine experimentally the response of the introduced species to a range of environmental conditions and communities. The potential role of a network of study sites for systematic testing of introductions has apparently not been considered.

The above discussion of the four general subjects to which environmental questions are addressed has identified a number of factors which must be considered when planning long-term research and defining the role of study or reference sites. The main considerations are that:

(1) Monitoring is likely to be most effective when it is integrated with field experiments and with research which can help to assess causal and consequential relationships. Study sites must, therefore, be large enough to allow a range of activities.
(2) Research and monitoring of different subjects may require very different approaches in terms of site selection. For example, some questions on effects of climate change or species introductions require analysis of population distributions which are not defined by sites. However, selection of study sites along major environmental gradients will provide a suitable focus for many such analyses.
(3) Most long-term studies are site-specific. This restricts the ability to define the limits of the problem, and is, therefore, a danger in concentration of effort. The need to define both within- and between-site variation argues for the provision of a gradient of study sites.
(4) Occasional extremes in climate, or other short-term episodes, are important determinants of long-term change. The need for long-term observations to distinguish short-term fluctuations from longer trends has been emphasized in the work of Likens (1985) and Tilman (1989). The case is well illustrated in Figure 3.6. In the rabbit-proof enclosure and in the Park Grass plot receiving mineral fertilizer there were periods in which changes in species composition over a number of years moved contrary to the long-term trend. In both cases the successional trend in response to an abrupt change in management took decades to emerge.
(5) Short-term 'natural' variation (for example, year-to-year changes in species composition) needs to be separated from long-term trends. This is discussed more fully in the next section.

3.3 SEPARATION OF 'NOISE' FROM 'SIGNAL'

To identify environmental change and its consequences we have to determine the existing or normal fluctuations (short-term noise) and trends (long-term noise) in order to distinguish new variations (signal) induced by extrinsic factors.

Monitoring by itself can be of value in some circumstances, particularly following the introduction of a new element whether it be a chemical (e.g. pesticide or ^{137}Cs), a species (e.g. coypu), a crop (e.g. Sitka spruce), or even a gene. However, in detecting changes in existing elements (e.g. nitrate concentration in water, atmospheric SO_2, or sward species composition) and in system responses (e.g. plant growth or soil organic matter accumulation or frequency of flooding), monitoring is of limited value. In these cases monitoring data do not allow distinction between the observed and the expected. It is not possible to determine whether the nitrate concentration is high because of increased fertilizer use or because of particular climate conditions; whether the decline in the bird population is part of a cyclical change or caused by increased predation; or whether the expansion of regeneration at the tree line is due to increased temperature, reduced grazing pressure, or a good seed-production year. These questions can only be answered if monitoring is an integral part of research, and particularly where it is used to test predictions, i.e. to separate noise from signal.

In selecting sites for long-term monitoring and research it is important to understand the fluctuations and trends that might be expected. Trends, whether they are determined by management or occur naturally, are often induced in response to events occurring years or decades previously, and knowledge of site history is particularly important. The state of the system will influence, for example, the deposition and retention of pollutants or nutrients through variation in vegetation cover and soil organic matter content; the plant growth response to climate through variation in species composition or stage of development; and the success of introduced species through variation in sites for colonization or in competition.

Although there are many variations on the theme, five general patterns of long-term trends are recognized—constant, cyclical, directional, episodic, and catastrophic. These are illustrated hypothetically in Figure 3.8, with short-term fluctuations superimposed. The y axis may represent many different parameters (p) for example, plant production, soil nitrogen or organic matter concentration, species diversity, and soil water table or moisture content. The trend is determined by the type of system. Thus, an arable system with consistent short-term management would be expected to show only annual variations around constant values of p (Figure 3.8(a)) in contrast to repeated management or natural cycles such as heather burning and regeneration (Figure 3.8(b)) and natural succession or afforestation with a long-term directional trend (Figure 3.8(c)). The pattern induced by episodes such as drought or defoliation which allow recovery (Figure 3.8(d)) contrasts with catastrophic situations (Figure 3.8(e)) which result in a long-term change from one state to another, as in the conversion of grassland to arable. It can be argued that the constant, cyclical, and directional patterns simply represent expansion of time-scale with annual arable cropping, 20-year burning, and 50-year forest rotation. Similarly, the episodic and catastrophic patterns simply represent extreme variations

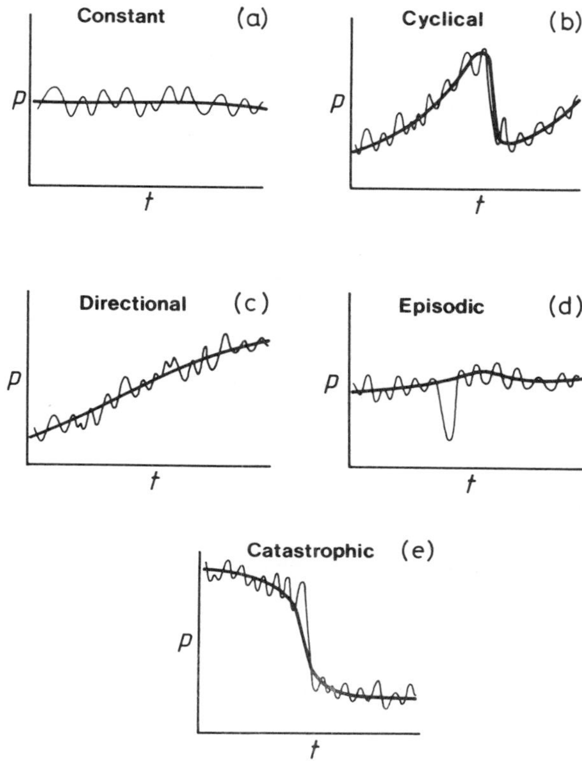

Figure 3.8 Main types of long-term trends in ecosystems in response to management or to natural succession, with short-term fluctuations superimposed (see text)

in the rate or degree of recovery (e.g. from defoliation). Note also that in episodic or catastrophic situations, while some attributes such as standing crop or species diversity will change rapidly, others such as soil organic matter may continue to respond over decades: the echo from the past.

The real world is much more complex than the simple representations presented here. However, the key points arising from this discussion are that our ability to detect changes caused by particular human activities can be confused by the uncertainty of whether the changes result from previous actions or from natural trends such as pedogenesis or vegetation succession. In monitoring change, the greatest success is likely to be achieved when the monitoring is combined with research which includes prediction of fluctuations and trends against which data from monitoring can be tested. There are advantages in the selection of sites with a known history of management and which are sensitive to the targeted change. In the latter case, the site may be marginal to a species or crop production system, and change will be readily apparent, although the fact that fluctuations will also be greatest at the margin may cause a problem.

3.4 SELECTION OF SITES

Environmental problems are complex, and understanding the cause–effect relationships requires in-depth study involving a number of disciplines which must focus, sooner or later, on the field situation. For this, there must be a stage of site selection. The normal process is to identify sites which are well known, then see how they fit together to cover environmental variation which at best has been broadly defined. During the International Biological Program sites were selected by participant countries for research on production processes and then combined to form international Biome networks. Analysis of the Tundra Biome, constituted in this way, showed that the sites covered a wide range of variation along environmental gradients of temperature, moisture, soil organic matter, acidity, and nutrient availability (Figure 3.9), against which biological processes could be related (French, 1981). This approach of selecting then classifying sites resulted in important gaps, although flexibility in the Canadian program did allow the late selection of a missing high arctic site at Devon Island. The classification also showed that local site conditions (for example, the wet and dry subsites on Devon Island) had more in common with geographically distant subsites (Finland, Antarctica, Ireland) than with their immediate neighbors. This 'accidental' inclusion of local variation greatly enhanced the potential of between-site comparison and synthesis for determination of global controls of production and decomposition processes (Bliss *et al.*, 1981).

The method of site selection in the International Biological Program had certain drawbacks. An alternative would have been to determine the environmental variation to be covered, stratify, and then select sites within the strata. An example of this approach was an ecological survey of the UK which was subsequently used in assessing change in land cover and use (Bunce and Heal, 1984). From a hierarchical classification based on climatic, topographic, and geological attributes, 32 land classes were used as strata from which 256 sample sites (1 km^2) were randomly selected (eight squares in each stratum) for survey and monitoring. This approach allows ease of regional and national quantification and of extension of sampling for particular purposes. It does not select sites for which information is readily available, but it provides a framework within which these sites can be placed. Thus, the two approaches of select and classify or classify and select can be complementary rather than exclusive.

Figure 3.9 (*opposite*) Abiotic analysis of tundra sites (from French, 1981), showing the distribution of sites along components I and II, indicating primary clusters. The analysis used seven climatic variables, four soil temperature variables, two soil moisture variables and eight chemical variables. Arrows show the nearest linkages of 'outlier' sites. Main site codes: G: Glenamoy, Ireland; MH: Moor House, UK; H: Hardangervidda, Norway; K: Kevo, Finland; A: Abisko, Sweden; D: Devon Island, Canada; B: Point Barrow, Alaska, USA; T: Tareya, Taimyr, USSR; M: Macquarie Island, Australia; SG: South Georgia, Antarctica; S: Signy Island, Antarctica; DK: Disko Island, Greenland; N: Niwot Ridge, Colorado, USA. Other letters and numbers in codes refer to sub-sites

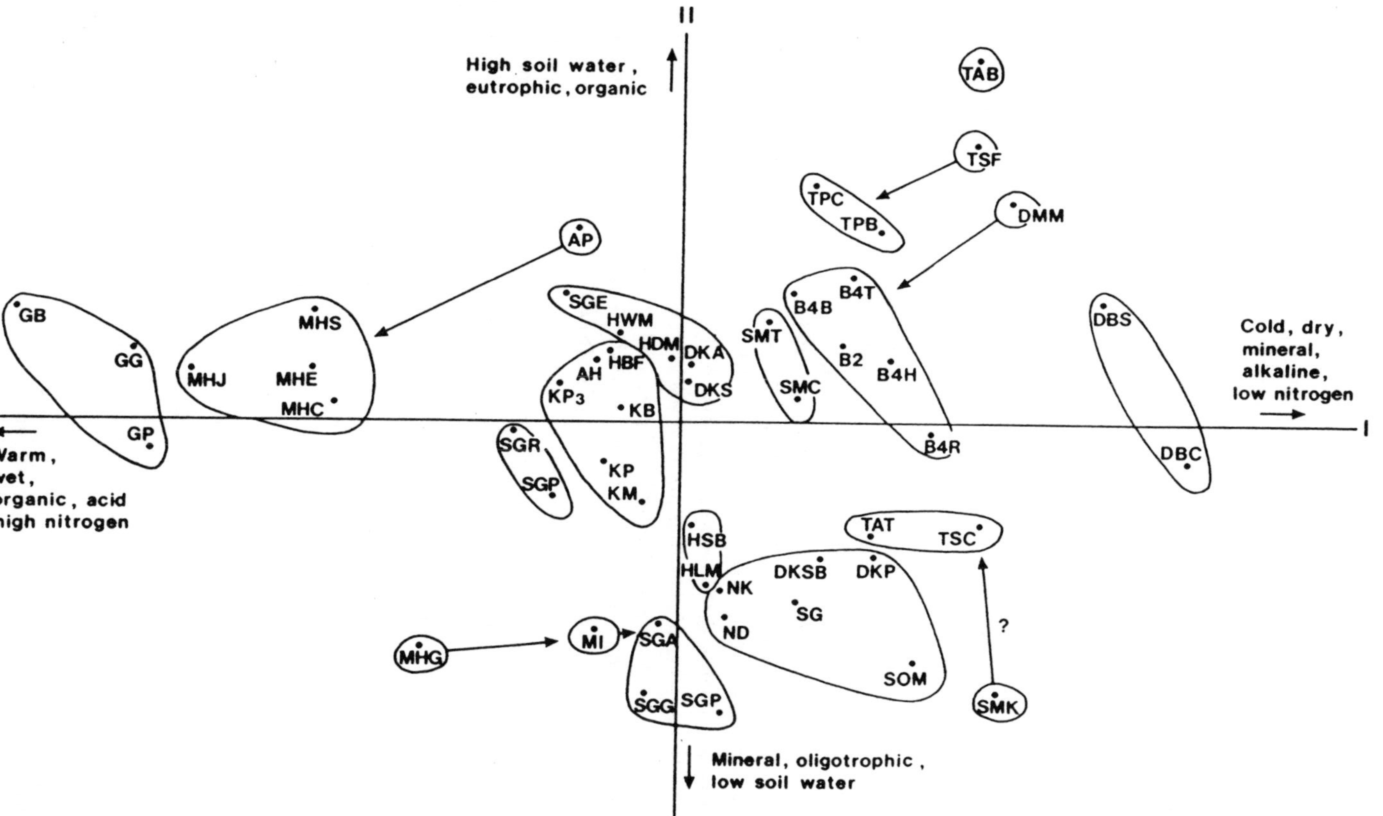

Figure 3.9 (*caption on facing page*)

As already indicated, sites which are at the margin of the distribution of a species or land use will be more sensitive to change in environmental conditions than those towards the center of the range. Thus, these latter sites, which are considered to be typical or representative and are selected for characteristics close to the mean values will, for some purposes, be of limited value. The tree line at high latitudes or elevation provides an obvious example, and one which corresponds with the ecotone concept (di Castri *et al.*, 1988). However, an important component of the ecotone concept is zones of transition in time. In such situations a small change in a particular variable such as temperature or element concentration can cause a relatively large change in the state of the species or system, i.e. it is at or near a threshold. di Castri *et al.* (1988) tended to focus on spatially defined boundaries or zones of transition, but some boundaries are temporally defined, particularly by the state of soil conditions. This is best shown by the following three soil examples.

Miles and Young (1980) showed that soil improvement by birch tended to occur on brown podzolic soils in Scotland and northern England. They selected sites with different ages of birch (chronosequences) and found that the ability of birch to reduce acidity and increase nutrient availability was determined by the initial state of the soil; some sites changed while others did not.

In a second example, J.D. Ovington, W.H. Pearsall, and others hypothesized that some soils were particularly susceptible to the effect of coniferous planting, i.e. they were on a threshold between soil types. They selected a site at Gisburn in north-west England on which *Pinus sylvestris*, *Picea abies*, *Alnus glutinosa*, and *Quercus petraea* were planted in both pure and mixed stands. After 30 years, distinction between the effects of conifer and deciduous trees on soil conditions were apparent, with some unexpected responses (Brown and Harrison, 1983).

In the third example, Hornung *et al.* (1989) formalized the theoretical and practical information on the sensitivity of waters to acid deposition and afforestation. Based on climatic, geological, and pedological data, they mapped the distribution of areas in Wales in which water acidification (and Al toxicity) would be most likely to occur through afforestation because of the soil buffering characteristics and the increased deposition in forests (Fowler *et al.*, 1989) (Figure 3.10). The acid soils with Al buffering sensitive in the short term, and those which have little residual Ca buffering were found to be sensitive in the medium term.

These three examples emphasize the threshold or poised nature of some soils that will determine the response of the plant community and chemical condition to changes in management or pollution. Thus, selection of sites will determine whether or not a response is detectable. The first two examples also illustrate alternative approaches in research design, i.e. selection of a series of sites which represent stages in time as distinct from long-term analysis of an experimental manipulation. The third example indicates the possibility of selecting sites to test defined hypotheses. In all three examples, the concept of a threshold of response is explicit (though not quantified) in the selection of sites. Reversibility is another question, but it is one that Miles (1988) is examining by following the observational chronosequence approach with experimental planting of *Calluna* on cleared birch plots to complement validation experimental planting of birch on *Calluna* sites.

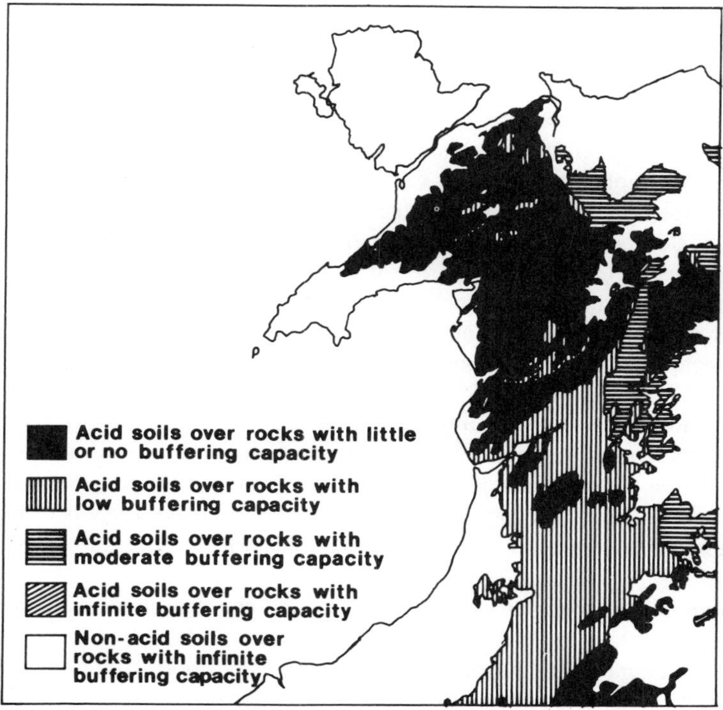

Figure 3.10 Probable occurrence of acid waters in Wales predicted from soils, geology and land use

The identification of constant, successional, or recurrent management as an important criterion in site selection has already been made, and is illustrated in Figure 3.8. Having identified the management and environmental conditions, the question then is the selection of intensive sites which combine into a network. However, an important variation on the theme is the organization of a network combining intensive and extensive sites. The concept was clearly planned in the US Grassland Biome of IBP (van Dyne, 1972), with detailed process-related research at the intensive site designed to predict responses over a wider range of conditions, the predictions being tested at the extensive series of sites. Variations on this approach, using a number of sites on which limited observations are made to complement an intensive site, are distinct from a network of many sites with the same measurements, and can provide a powerful tool in combining detailed understanding with widespread application.

3.5 CONCLUSIONS

Many of the points made in this chapter are obvious, but the experience of the current discussion in the UK indicates that they are worth re-examining. The main lessons seem to be:

(1) Defining the problem is critically important in selection of sites for long-term ecological research and monitoring. The problems of the future may be unforeseen, but are likely to belong to four generic types which have particular requirements (climate change, pollution, land use, species introductions).
(2) Definition of inherent short-term fluctuations and longer-term trends will be important in detecting changes attributable to more recent activities of man. Because of this, monitoring is likely to be most useful when combined with research, and when based on sites which have a good knowledge of management history and understanding of processes.
(3) For many problems the influence of man is most likely to be important and detectable under conditions which are marginal to the species or system, or which are transitional between states. Thus the selection of ecotone or threshold sites can be as important as selection of sites which represent more average conditions, depending on the question.
(4) Selection of sites should include careful analysis of the range of variation to be covered and can usefully incorporate and build on a planned network of intensive and extensive sites.

Throughout this chapter questions of spatial and temporal scale have been implicit rather than explicit. The questions are appropriate to site selection and to the measurements to be made on the sites, from soil crumb to catchment. Many of the issues are identified in Risser (1986) with the clear message that the scales adopted, and the interrelationships between the hierarchy of scales, must be carefully identified in relation to the problem being considered. The challenge in long-term studies is in economy of effort to answer the problem. As identified by participants in the International Biological Program: 'We underestimated our ability to collect data and overestimated our ability to do something with them.'

3.6 REFERENCES

Ball, D.F., Dale, J., Sheail, J. and Heal, O.W. (1982). *Vegetation Change in Upland Landscapes*. Institute of Terrestrial Ecology, Cambridge, UK.

Bliss, L.C., Heal, O.W. and Moore, J.J. (1981). *Tundra Ecosystems: a Comparative Analysis*. Cambridge University Press, Cambridge, UK.

Brown, A.H.F. and Harrison, A.F. (1983). Effects of tree mixtures on earthworm populations and nitrogen and phosphorus status in Norway spruce (*Picea abies*) stands. In Lebrun *et al.* (Eds) *New Trends in Soil Biology*. Dieu-Brichart, Louvain-la-Neuve, 101–108.

Bunce, R.G.H. and Heal, O.W. (1984). Landscape evaluation and the impact of changing land use on the rural environments: the problem and an approach. In Roberts, R.D. and Roberts, T.M. (Eds) *Planning and Ecology*, Chapman & Hall, London, 164–188.

Coulson, J.C. and Whittaker, J.B. (1978). Ecology of moorland animals. In Heal, O.W. and Perkins, D.F. (Eds) *Production Ecology of British Moors and Montane Grasslands*. Springer-Verlag, Berlin, 52–93.

Davis, M.B. (1989). Retrospective studies. In Likens, G.E. (Ed.) *Long-term Studies in Ecology*. Springer-Verlag, New York, 71–89.

di Castri, F., Hansen, A.J. and Holland, M.M. (1988). A new look at ecotones. Emerging International Projects on Landscape Boundaries. *Biology International*, Special Issue 17.

Elton, C. (1958). *The Ecology of Invasions by Animals and Plants*. Methuen, London.

Fowler, D., Cape, J.N., Deans, J.D., Leith, I.D., Murray, M.B., Smith, R.I., Sheppard, L.J. and Unsworth, M.H. (1989). Effects of acid mist on the frost hardiness of red spruce seedlings. *New Phytologist*, **113**, 321–335.

Fowler, D., Cape, J.N. and Unsworth, M.H. (1989). Deposition of atmospheric pollutants on forests. *Phil. Trans. R. Soc. Lond. B.*, **324**, 247–265.

French, D.D. (1981). Multivariate comparisons of IBP Tundra Biome site characteristics. In Bliss, L.C., Heal, O.W. and Moore, J.J. (Eds) *Tundra Ecosystems, a Comparative Analysis*. Cambridge University Press, Cambridge, UK, 47–75.

Grace, J. and Marks, T.C. (1978). Physiological aspects of bog production at Moor House. In Heal, O.W. and Perkins, D.F. (Eds) *Production Ecology of British Moors and Montane Grasslands*. Springer-Verlag, Berlin, 38–51.

Grime, J.P. (1987). Dominant and subordinate components of plant communities: implications for succession, stability and diversity. In Gray, A.J., Crawley, M.J. and Edwards, P. J. (Eds) *Colonization, Succession and Stability*, Blackwell Scientific, Oxford, 413–428.

Hornung, M., Le Grice, S., Brown, N. and Norris, D. (1989). The role of geology and soils in controlling surface water acidity in Wales. In Stoner, J. and Edwards, R.W. (Eds) *Acid Waters in Wales*. Kluwer Academic Publishers, Dordrecht, The Netherlands, 55–66.

Kornberg, H. and Williamson, M.H. (1987). *Quantitative Aspects of the Ecology of Biological Invasions*. The Royal Society, London.

Likens, G.E. (1985). An experimental approach for the study of ecosystems. The fifth Tansley Lecture. *J. Anim. Ecol.*, **73**, 381–396.

Likens, G.E. (Ed.) (1989) *Long-Term Studies in Ecology*. Springer-Verlag, New York.

Marrs, R.H., Bravington, M. and Rawes, M. (1988). Long-term vegetation change in the *Juncus sqarrosus* grassland at Moor House NNR in northern England. *Vegetatio*, **76**, 179–187.

Miles, J. (1979). *Vegetation Dynamics*. Chapman & Hall, London.

Miles, J. (1988). Vegetation and soil change in the uplands. In Usher, M.B. and Thompson, D.B.A. (Eds) *Ecological Change in the Uplands*. Blackwell Scientific, Oxford, 57–70.

Miles, J. and Young, W. (1980). The effects on heathland and moorland soils in Scotland and northern England following colonization by birch (*Betula* spp). *Bulletin d'Ecologie*, **11**, 233–244.

Noble, I.R. and Slatyer, R.O. (1980). The use of vital attributes to predict successional changes in plant communities subject to recurrent disturbances. *Vegetatio*, **43**, 5–21.

Parry, M.L. and Carter, T.R. (1985). The effects of climatic variations on agricultural risk. *Climatic Change*, **7**, 95–110.

Peterkin, G.F. and Jones, E.W. (1989). Forty years of change in Lady Park Wood: the young-growth stands. *J. Ecol.*, **77**, 401–429.

Pickett, S. (1989). Space-for-time substitution as an alternative to long-term studies. In Likens, G.E. (Ed.) *Long-Term Studies in Ecology*. Springer-Verlag, New York, 110–135.

Risser, P.G. (1986). Spatial and temporal variability of biospheric and geospheric processes. *Report of a Workshop*, ICSU Press, Paris.

Roberts, T.M., Skeffington, R.A. and Blank, L.W. (1989). Causes of Type 1 spruce decline in Europe. *Forestry*, **62**, 180–222.

Taylor, L.R. (1989). Objective and experiment in long-term research. In Likens, G.E. (Ed.) *Long-Term Studies in Ecology*. Springer-Verlag, New York, 20–70.

Thurston, J.M., Williams, E.D. and Johnson, A.E. (1976). Modern developments in an experiment on permanent grassland started in 1856: effects of fertilisers and lime on botanical composition and crop and soil analyses. *Annales agronomiques*, **27**, 1043–1082.

Tilman, D. (1989). Ecological experimentation: strengths and conceptual problems. In Likens, G.E. (Ed.) *Long-Term Studies in Ecology*. Springer-Verlag, New York, 136–157.

Usher, M.B. and Thompson, D.B.A. (Eds) (1988). *Ecological Change in the Uplands*. Blackwell Scientific, Oxford.

Van Dyne, G.M. (1972). Organization and management of an integrated ecological research programme—with special reference on systems analysis, universities and scientific co-operation. In Jeffers, J.N.R. (Ed.) *Mathematical Models in Ecology*. Blackwell Scientific, Oxford, 111–172.

Watt, A. (1960). Population changes in acidiphilous grass-heath in Breckland, 1936-57. *J. Ecol.*, **48**, 605–629.

Whittaker, J.B. (1971). Population changes in *Neophilaenus lineatus* (Homoptera: *Cercopidae*) in different parts of its range. *J. Anim. Ecol.*, **40**, 425–443.

Whittaker, J.B. (1985). Population cycles over a 16-year period on an upland race of *Strophingia ericae* (Homoptera: *Psylloidea*) on *Calluna vulgaris*. *J. Anim. Ecol.*, **54**, 311–321.

4 Expanding the Temporal and Spatial Scales of Ecological Research and Comparison of Divergent Ecosystems: Roles for LTER in the United States

JOHN J. MAGNUSON, TIMOTHY K. KRATZ, THOMAS M. FROST, CARL J. BOWSER, BARBARA J. BENSON, and REDWOOD NERO
North Temporate Lake LTER Site, Center for Limnology, University of Wisconsin-Madison, Madison, WI 53706, USA

4.1 INTRODUCTION

Ecological processes operate at a broader range of temporal and spatial scales than is typically addressed in ecological studies. The Long-Term Ecological Research Program funded by the US National Science Foundation was established to allow ecologists to work over this broader range (Callahan, 1984; Swanson and Franklin, 1988). The expanded temporal and spatial scale of Long-Term Ecological Research (LTER) reveals processes and events that have often been invisible. Our

Long-term Ecological Research. Edited by Paul G. Risser

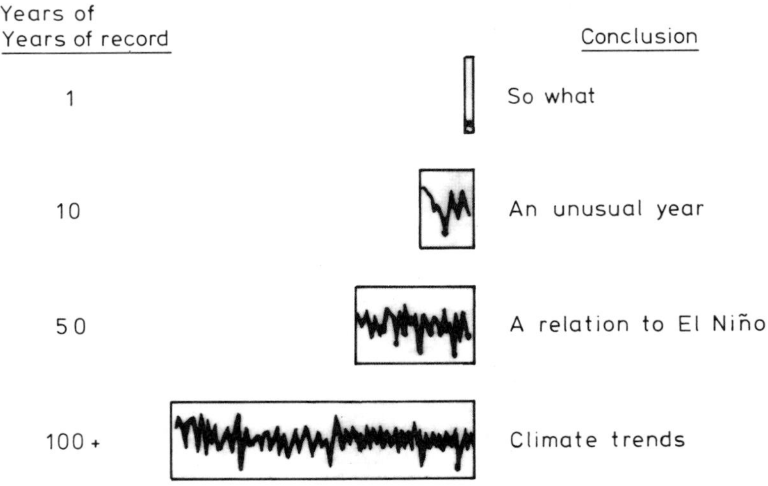

Figure 4.1 Observations of the ice cover records of Lake Mendota demonstrate the importance of a broad temporal scale in rendering the present visible

understanding of these processes and events is further augmented by a functioning network of research sites which allows the contrast of similarities and differences among systems which differ greatly from each other, such as deserts, grasslands, forests, and tundra, or even lakes, islands, oceans, and continents.

In this chapter we point out the advantages of studying ecology at longer temporal scales than common in most ecological research, make a case for expanding spatial scales of ecological research through explicit treatment of scale, provide examples of the use of LTER sites for comparisons of divergent ecological systems, and describe one LTER site, the North Temperate Lakes Site at the University of Wisconsin-Madison, which was designed to make possible long-term ecological research at both the site and intersite levels.

4.2 EXPANDING TEMPORAL SCALES

Short-term studies do not reveal slow changes which occur over years and decades, or allow the interpretation of the cause and effect relations of these slow changes. Processes which act over decades are hidden in an 'invisible present' (Magnuson *et al.*, 1983; Magnuson, 1990). Ecologists often must work in this 'invisible present' in which the significance of current conditions cannot be assessed without additional years of information to place the present into perspective.

Observations of the ice cover records of Lake Mendota demonstrate the importance of a broad temporal scale in rendering the present visible (Robertson, 1989; Magnuson, 1990). The duration of ice cover in a particular year, the winter of 1982–1983 (Figure 4.1), is relatively uninteresting, uninterpretable, and, for that matter, soon forgotten. Yet, when the time series of annual records is extended to 10 years, to 50 years, or to the length of Lake Mendota's record of 132 years, the present is put into the context of time and can be better understood. With 10 years of records it is apparent that the duration of ice cover in 1983 was about 40 days shorter than in any of the other nine years and far exceeded the typical range of variation. It also becomes apparent that the duration of ice cover varies considerably from year to year. Records from 50 years show that 1983 and other El Niño years all tended to have shorter durations of ice cover (Robertson, 1989), and the phenology of ice cover can be seen to be linked to a major feature of global climate: the southern ocean oscillation index (Quinn *et al.*, 1978; Mysak, 1986). With 132 years of records, a general warming trend becomes visible that had been invisible with the 10- and 50-year records (Robertson, 1989, 1990). The little ice age ended about 1890 (Wahl and Lawson, 1970; Lamb, 1977) and its end is reflected in the decrease in the duration of ice cover on Lake Mendota. In the most recent years there is a hint that another warming has begun, which perhaps, along with the change in the 1890s, signals CO_2-induced global climate warming (Robertson, 1990). When the entire time series is viewed, the 1983 ice cover duration is seen to be the shortest of the entire 132 years. Thus, each increase during the period of record reveals new insights that make the condition in 1983 more understandable and more interesting.

Time lags of longer than a year between cause and effect or response to a disturbance permeate natural systems, and contribute to our inability to interpret short-term records in observational research. We frequently observe the response of an ecological system to a cause that occurred before our observations began, or, in manipulative ecological studies, see a transition of the system rather than the new state of the system after manipulation (Magnuson, 1990).

Time lags in ecological systems occur for many reasons. Some processes, such as the accumulation of biomass, simply take time. The time lag between the year when a cohort of plants or animals is formed and when it reaches maximum biomass depends on the mortality and growth rates of organisms in that particular population. For many fishes in freshwater lakes the time lag is about 2 to 4 years, for long-lived trees, it may be 200 years or longer.

The spatial dimensions of landscapes in which ecosystems occur also induce time lags; movements of water, materials, and organisms take time. Classic examples of this include the time for dispersal of exotic plants and animals across the landscape. Time lags can also occur because a process requires the coincidence of two or more low-probability events; in some cases many years may pass before the events occur together. Successful reproduction of plant and animal species is often dependent on such events. Time lags are also generated when a cause-and-effect chain of events accumulates the lags from each link in the chain reaction: for example, the effects of an important consumer on the rest of a food web can step through a series of interactions, each of which takes time to unfold (Hrbacek, 1958; Hrbacek *et al.*, 1961; Henrikson *et al.*, 1980; Carpenter *et al.*, 1985; Northcote, 1988).

Examples of each of the above time lags can be found in Magnuson (1990) for LTER sites in the United States. Another type of time lag is presented here, with an example from alpine tundra (Magnuson, 1990). Relict biota persist in an ecosystem even after conditions change. *Kobresia* is a sedge which, on the Niwot Ridge LTER Site in the Rocky Mountains, prefers sites free of snow in winter and rarely reproduces sexually. Kobresia effectively maintains itself through vegetative tillering. Pat Webber and his students (Emerick, 1976; Webber *et al.*, 1976; Keigley, 1987) assessed the resistance of the *Kobresia* community to environmental change by increasing snowpack with snowfences and following the subsequent performance of *Kobresia* (Figure 4.2). In the first 2 years after greater snowpack, the plant produced more leaves and appeared to be vigorous. But the snowpack prevented the generation of new shoots and within 5 years *Kobresia* was sparse; by 10 years it had completely disappeared. A time lag, linked causally to the biology of this species, is evident in the disappearance of *Kobresia*. A conclusion made in the first two years after the manipulation would have been misleading.

Long-term ecological research, by putting the present into the context of temporal change and by allowing for the analysis of time lags greater than a year between cause and effect, provides an approach which can help to penetrate the invisible present. The long-term dynamics of ice cover on Lake Mendota and the time-lagged influence of snowpack on tundra vegetation illustrate the case.

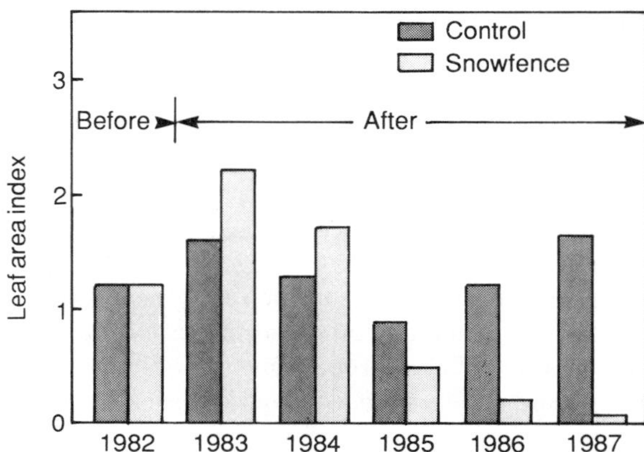

Figure 4.2 Leaf area indices of *Kobresia* in the alpine tundra at the NIWOT Ridge LTER site in Colorado in a treatment area where snowpack was increased with a snowfence and in a reference area where it was not. Data and its interpretation are from Emerick, 1976; Keigley, 1987; Webber *et al.*, 1976

4.3 EXPANDING THE SPATIAL SCALE OF ECOLOGICAL RESEARCH

LTER uses not only longer time scales than most ecological research but also a broader range of spatial scales (Figure 4.3). The analogy of an 'invisible place' as with the 'invisible present' is an appropriate metaphor because ecological processes and responses depend on the spatial context of an observation as well as on its temporal context (Swanson and Sparks, 1990). The following examples point out some of the reasons why spatial scale should and can be treated explicitly in ecological studies.

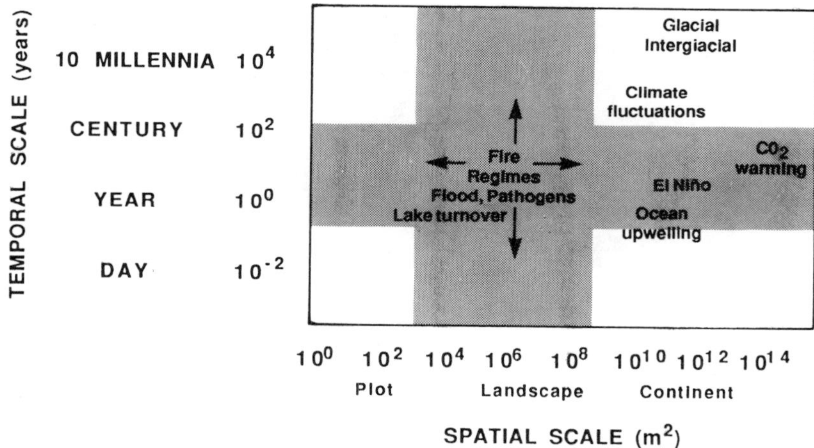

Figure 4.3 A diagram of the temporal and spatial scales used in the long-term ecological research

Remote sensing tools that observe at known spatial scales appear to provide opportunities for ecologists to address scale explicitly in ecosystem- and landscape-level research. Landsat and SPOT produce high-quality images of both land and water ecological systems at the same spatial resolution, $30m^2$ for Landsat, or as small as $10m^2$ for SPOT. The data also can be spatially aggregated to produce coarse-grained images. In aquatic systems, remotely sensed data from the water column, obtained by using sonar, have features similar to remotely sensed views of the earth's surface from space. Sonar data (Figure 4.4, top) are depicted in a two-dimensional raster data file displaying echo strengths which are proportional to the biomass of organisms at that depth along a ship's transect. For remotely sensed acoustic data in a lake or ocean, we see a profile view (length and depth) of animal biomass rather than a plane (length and width) view of chlorophyll as seen from remote sensing from space. Another analogy is that different sound frequencies are reflected differentially from small and large animals, and different wave lengths of light are reflected differentially from soil, green vegetation, and water.

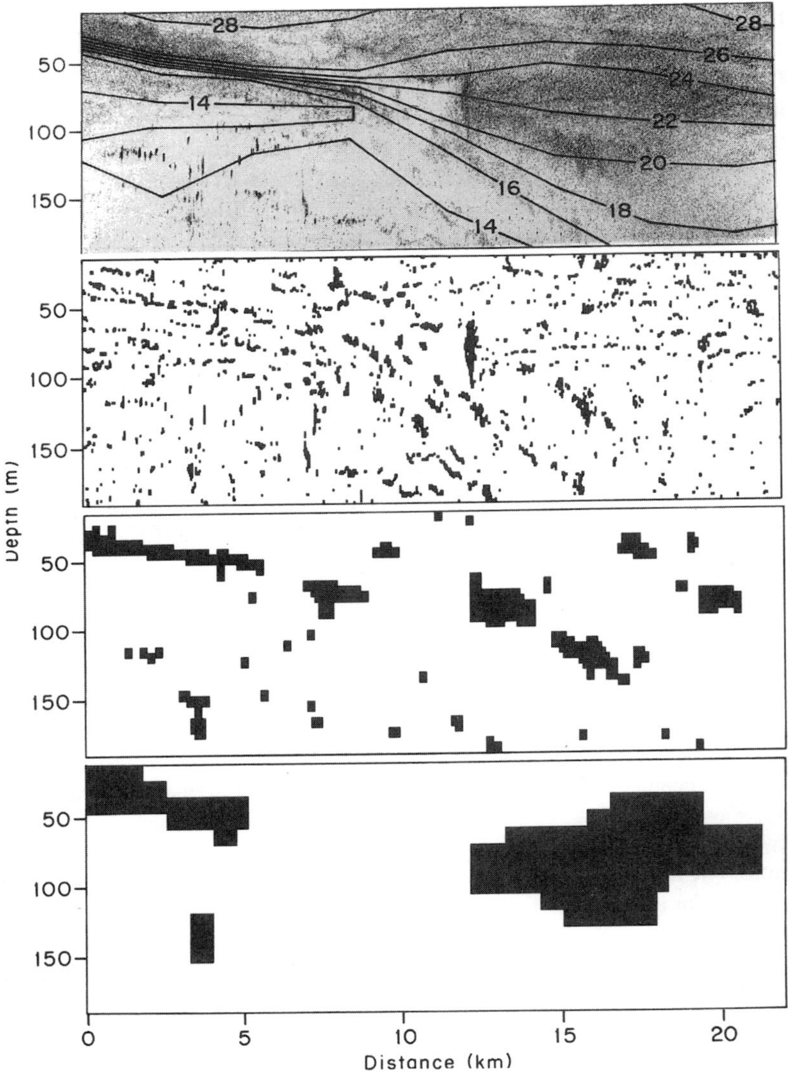

Figure 4.4 The influence of changing the scale of observation on the patchiness of scattering from a 70 kHz sonar crossing the north wall of the Gulf Stream in the Atlantic Ocean off Cape Hatteras. This plane view is 23 km in length with data selected from 9 to 188 m depths below the surface. The transect runs from the north-west on the left to the south-east on the right and is centered at 35°50′N latitude and 74°36′W longitude. The size of pixels grades from 1×25 (m height by width) on the top to 12×300 on the bottom. The isotherms were drawn by eye from computer-generated isotherm maps. (Nero and Magnuson, in press)

We would like to compare landscape and lakescape structure at the same scale and with nested sets of spatial scales, because differences between landscape and lakescape structure generate variations in the movements of materials, organisms, and energy across lake or land systems. These differences, in turn, should influence the local and regional dynamics in species diversity, population dispersal, population temporal dynamics, and biogeochemistry. Steele (1985) argued that paying explicit attention to the time and space characteristics of variability may allow comparisons between terrestrial and ocean systems to better test 'general hypotheses about resilience, food web dynamics and other ecological concepts'. Species turnover, the rate of change in species composition, would be especially interesting to relate to the scale of patchiness in lake versus ocean environments (Magnuson, 1988) and ocean versus terrestrial environments (Steele, 1985).

Being able to observe different systems at the same spatial scales is important because when the same system is observed at several spatial scales, completely different characteristics in the spatial distribution of organisms are revealed (O'Neill, 1989; Turner *et al.*, 1989b). For example, two of us, Nero and Magnuson, analyzed the patchiness of macrozooplankton and small nekton near the north wall of the Gulf Stream off Cape Hatteras and observed a multitude of patches associated with social or feeding aggregations at a fine spatial scale (Figure 4.4, second panel). Using the same data, but increasing the grain, revealed patchiness associated with the Gulf Stream front, with a major concentration of biota at the sharp thermal and salinity gradient between the Gulf Stream on the right and slope water on the left (Figure 4.4, third panel). At an even greater grain, two large patches of organisms were apparent, one where the isotherms approach the surface, and the other in the Gulf Stream (Figure 4.4, bottom panel). When we compare ecosystems using tools that observe different systems at different unique scales, apparent variations among the systems may be attributable to the spatial scaling used, rather than to any difference between the systems themselves. Another interesting alternative is to compare different systems at particular spatial scales relevant to the process or structure under study; the appropriate scale for comparison may differ among ecosystems.

The remote sensing technologies are available to contrast the patchiness of chlorophyll in terrestrial and aquatic systems in addition to animal biomass in lakes and marine environments at nested sets of spatial scales. There should be basic differences in the patterns of plant distribution; terrestrial vegetation patchiness is related closely to relatively permanent landforms, whereas surface chlorophyll distributions of lakes and oceans are tied closely to dynamic currents, eddies, upwellings, convergences, and fronts. A landscape view of the earth's surface is incomplete without the structural features of patchiness in aquatic systems.

Analogies and contrasts between land and water systems have been made (Barbour and Brown, 1974; Magnuson, 1976, 1988). Land or continental systems are continuous and highly connected, with rapid dispersal of organisms, whereas lakes are discontinuous, isolated, and patchy environments in a matrix of land where dispersal of organisms is slow (Magnuson, 1988). Spatial analyses of land and water by analogy often have been in the context of island biogeography, where the isolation and connectedness of individual islands, lakes or habitat islands have been related to species richness and composition in individual lakes (Barbour and Brown, 1974; Tonn and Magnuson, 1982; Rahel, 1984, 1986; Eadie *et al.*, 1986) or islands (MacArthur and Wilson, 1967; Brown, 1971; Diamond, 1975). It should be possible to increase the scale of the comparisons to lake districts and island archipelagos. Such scaling up to larger regions is essential as ecologists attempt to address global change issues (Turner *et al.*, 1989a). However, the processes and models with explanatory power at the level of lake or island may not be appropriate at larger spatial scales. A first step is to quantify the spatial features of lakescapes and landscapes.

The structure of land patches imbedded in a matrix of water and water patches within a matrix of land can be compared by using measures derived from landscape ecology (Naveh and Libermann, 1984; Risser *et al.*, 1984; Forman and Gudron, 1986; Franklin and Forman, 1987; Turner and Ruscher, 1988). As the proportion of land to water grades from 100% water through island archipelagos and lake districts to 100% land, the properties of the two systems might become transposed. A null hypothesis would be that the transposition is symmetrical, i.e. water patches at 25% water would have the same character as land patches at 25% land, or at 50% land and water the structure of the two systems would be the same. This transposition of character can be examined at several scales using routines available in geographic information systems. Measures to examine this hypothesis include size and shape of patches, the proportion of interior versus edge habitat, and fractal dimensions, in addition to relational measures such as connectedness, texture, interpatch distances, fragmentation, general impedance to movement, and anisotrophy.

Such a quantitative first step could assist in the generation of hypotheses about the differences in the temporal dynamics of chemicals, populations, and biotic diversity across this continuum of ocean to islands to lakes to continents. Interpretation also can be made regarding the influence of humans. For example, are human effects symmetrical on land and water habitats with respect to whether they contribute to or reduce fragmentation, impedance, etc.? We think they are not, and suggest that human activity acts very differently on lakescapes and landscapes by contributing to increased fragmentation of landscapes and to increased connectedness of lakescapes. Cutting forests and clearing land have reduced the size of forest patches and favored edge species over interior species (Franklin and Forman, 1987; Britingham and Temple, 1983). Patches of remaining forest also become too small to sustain minimum viable populations; as a consequence local extinctions result (Soule, 1987). On the other hand, for lakes, canal building provides access for exotic species to previously isolated lakes, such as permitting the movement of sea lamprey around Niagara Falls into the upper Great Lakes (Pearse *et al.*, 1980); ballast water transports organisms such as zooplankters and larval molluscs from continent to continent and contributes to invasions of exotics (Carlton, 1985; Evans,

1988; Hebert *et al.*, 1989); and intentional stocking of sport fish moves predaceous fish across traditional land barriers (Courtenay and Stauffer, 1984).

These increases in connectivity of previously isolated freshwater environments are expected to increase the rate of local extinctions and species turnover (Magnuson, 1976). The differences between lakes and continents are not as well explained by the fact of one being land and the other being water, as by the differences in the spatial structure of the two landscapes. In other aquatic environments, such as rivers and streams, humans have reduced connectedness by dam construction (Magnuson, 1978), and have thus contributed to increased fragmentation characteristic of terrestrial landscapes; terrestrial islands, on the other hand, can have the same problems of community destablization from invasions by exotics as those suffered by lakes.

4.4 COMPARISON OF DIVERGENT ECOLOGICAL SYSTEMS

At the intersite level, the LTER network can provide opportunities to understand ecosystem properties and functions by comparing divergent ecological systems. Each of us is most knowledgeable about a particular kind of ecosystem and a particular set of organisms; many technical constraints prevent us from routinely considering systems outside our own expertise. Yet there are benefits to be derived from comparing systems as different as lakes and islands, and oceans and continents. In long-term ecological research, and perhaps in all of science, comparisons of divergent systems can help to assess the generality of the processes and events being studied. Comparisons can break down disciplinary boundaries, and can help us to achieve new levels of understanding.

The importance of any finding increases with its generality. For example, the food web in Cedar Bog Lake, Minnesota, would not be all that exciting to a terrestrial ecologist in Bavaria if Raymond Lindeman (1942) had not generalized his results to a trophic dynamic concept that went beyond specific habitat and taxa. The generality of his conceptualization made Lindeman's paper a cornerstone in ecosystem ecology (Cook, 1977). Seeking generality among ecological systems often requires a translation to a terminology which is not ecosystem- or taxon-specific. In the case of trophic ecology, units of energy simplified the diverse and forbidding taxonomy of trillions of food webs.

Breaking down disciplinary boundaries among ecosystem investigators allows ideas and tools which are developed and common in one area to spread rapidly through the ecological sciences. Time lags are often long in transfers between fields; for example, Johannes Muller first used a net to sample plankton in the North Sea in 1845, but it was not until 1868 that P.E. Muller first used a net to verify the existence of plankton in a Swiss lake, a lag of 23 years in the transfer of a simple technology from oceanography to limnology. Such time lags are common in all fields of research.

Perhaps the most exciting reason to compare diverse systems is the prospect of an entirely new level of understanding of ecological processes and events. We provide examples of two approaches to enhance comparisons of divergent ecosystems: (1) using the same theory or hypothesis and (2) employing dimensionless metrics of temporal and spatial variability.

4.4.1 SAME THEORY: LAKES AS ISLANDS

By applying the ideas of island biogeography (MacArthur and Wilson, 1963, 1967; MacArthur, 1972) to lake community ecology, limnologists at the University of Wisconsin (Magnuson, 1976; Tonn and Magnuson, 1982; Rahel, 1984; Tonn, 1985; Magnuson *et al.*, 1989; Tonn *et al.*, 1990) have been able to predict and explain the fish assemblage structure in small forest lakes of the north temperate zone. The approach was robust enough to apply in detail across the faunas of Wisconsin in North America and Finland in northern Europe. The same extinction factors—low pH, winter anoxia, small lake size, and low productivity—and the same invasion factors—absence of land barriers or steep stream gradients—determined the type of fish assemblages in both regions. Of the two processes of extinction and invasion that are responsible for assemblage structure in small, relatively isolated lakes, extinction is by far the more important in determining the assemblage at a given time. This is apparently due to the fact that invasions are rare and the extinction factors act rapidly after an invasion, making the fish assemblage most likely to be observed at any one time the one present after extinctions rather than the one present after invasions. The point is that the use of a concept from terrestrial ecology greatly increased our ability to understand community ecology of small lakes at the temporal scale of invasion and local extinctions.

4.4.2 DIMENSIONLESS METRICS OF TEMPORAL AND SPATIAL VARIABILITY

Ecologists do not have many metrics to compare diverse ecosystems. Many of the things we study are ecosystem-specific. For example, forests do not have thermocline depths or hypolimnetic oxygen deficits, lakes do not have fire frequencies or soil moisture. Metrics which reflect important properties of systems and are not ecosystem-specific are needed to facilitate meaningful comparisons in any search for generality. Here we describe an approach to evaluating temporal and spatial variability of diverse ecosystems with dimensionless metrics.

Variability is common to all ecological systems and is a feature thought to be important by ecologists. We expect that a serious analysis of ecosystem variability can provide insight into the forces affecting ecosystem processes and be a meaningful measure with which to compare and characterize divergent ecological systems. Initially, we analyzed the variability of various limnological parameters from several lakes at the North Temperate Lakes LTER site (Kratz *et al.*, 1987b). We classified these parameters into three groups: 'location-specific' (varying among locations, but consistent at each location among years), 'year-specific' (varying among years at a location, but consistent among locations), or 'complex' (varying both among locations and years). We reasoned that location-specific parameters would be those affected by the individual details of a specific location, such as the morphometry of the lake, watershed area, hydrology, etc.; whereas year-specific parameters would be those regionally affected by year-to-year differences in the weather. Complex parameters would indicate interaction by location-specific details and weather. Therefore, the type of variability that parameters exhibit can be used to infer the processes controlling the parameters' behavior.

We are presently applying the analysis to contrast similarities and differences among the divergent set of ecosystems in the US LTER network. LTER sites constitute a broad array of ecosystems with which to test the robustness of the theory of using variability to compare ecosystems. Twelve LTER sites chose to participate in the study: H.J. Andrews Experimental Forest LTER Site (Arthur McKee), Bonanza Creek Experimental Forest LTER Site (John Yarie), Cedar Creek Natural History Area LTER Site (Richard Inouye), Central Plains Experimental Range LTER Site (Jon Hanson), Coweeta Hydrologic Laboratory LTER Site (Cory Berish), Hubbard Brook Experimental Forest LTER Site (Randy Dahlgren), Illinois River and Mississippi Rivers LTER Site (Peter Bayley), Jornada LTER Site (Gary Cunningham), Konza Prairie LTER Site (Donald Kaufman), Niwot Ridge/Green Lakes Valley LTER Site (James Halfpenny), North Inlet Marsh-Estuarine System LTER Site (Elizabeth Blood), North Temperate Lakes LTER Site (Barbara Benson, Carl Bowser, Steve Carpenter, Thomas Frost, Dennis Heisey, Timothy Kratz, John J. Magnuson), and NSF (Caroline Bledsoe) (see Brenneman and Blinn, 1987 and Brenneman *et al.*, 1987a,b, 1988, for general information on the LTER sites).

4.4.3 THE DATA SET

Each LTER site provided five years or more of annual data (*ca* 1981-1986), usually for five locations or more from their site and for a variety of parameters. At the North Temperate Lakes LTER Site, locations are defined as different lakes along an altitudinal gradient; at the Jornada Desert, locations are different geomorphic features along a catina, again along an altitudinal gradient. A number of other LTER sites (Hubbard Brook, Illinois River, Andrews Forest, and North Inlet Estuary) also chose locations along altitudinal gradients. In contrast, at the Konza Prairie and Coweeta LTER Sites, locations are different watersheds; on the NIWOT Alpine Tundra LTER Site they are different plant communities, and at the Alaska Taiga they are successional ages.

Four types of ecosystem parameters were used in this analysis of variability: climate data, such as precipitation and temperature; edaphic data, such as nutrient (N, P) concentrations; and biological data on selected plants and animals, such as abundance, species richness, reproductive success, or production.

Variability estimates for each parameter were computed from a standard two-way analysis of variance where 'location' and 'year' were the two treatment effects

(Figure 4.5). Because there was only one value of a parameter for each location–year combination, there was no replication and the effects of interaction and error could not be separated for individual LTER sites. We lumped interaction and error together into a category called 'other'. The two-way analysis of variance was conducted on both the raw data as supplied by each LTER site, and also on relativized data. Data were relativized by dividing each value in a location–year matrix by the grand mean for that matrix. This relativization dampened the effects of measuring different parameters with different units, and thus made comparisons among different parameter types easier. We used only matrices that had no missing data.

The magnitude of variation owing to 'location', 'year', and 'other' were computed for both the raw and relativized data using the expected values for a two-way analysis of variance. We computed the proportion of variance (r^2) explained by 'location', 'year', and 'other' from a Model II ANOVA (Sokal and Rohlf, 1981). We chose this model because we viewed 'years' and 'locations' as a sample to infer the properties of a large population of years and locations.

On 18 to 22 April 1988, an intersite workshop of participating scientists was convened at the Trout Lake Station (Center for Limnology, University of Wisconsin-Madison) to examine and compare variability in North American ecosystems. The data, standardized for all sites, were available on pairs of Apple Macintosh and IBM AT machines so that each subgroup of two to three persons could conduct immediate analyses on questions generated at the workshop using familiar spreadsheets and statistical packages. The data (VARiation in North

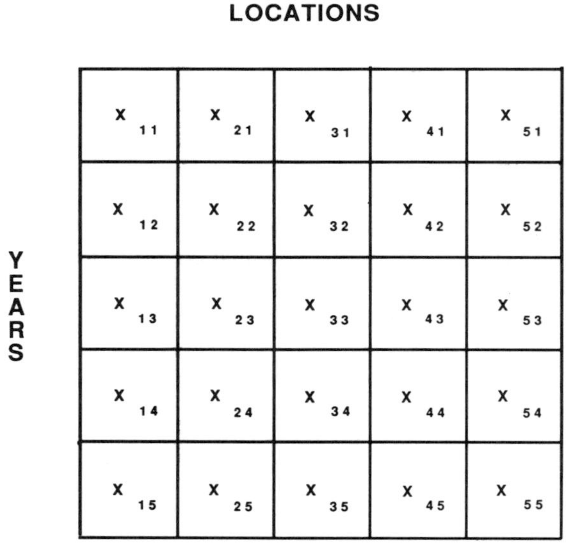

LOCATIONS

YEARS					
	X_{11}	X_{21}	X_{31}	X_{41}	X_{51}
	X_{12}	X_{22}	X_{32}	X_{42}	X_{52}
	X_{13}	X_{23}	X_{33}	X_{43}	X_{53}
	X_{14}	X_{24}	X_{34}	X_{44}	X_{54}
	X_{15}	X_{25}	X_{35}	X_{45}	X_{55}

Figure 4.5 Diagram of a two-way analysis of variance which provides estimates of variation associated with year, location and other (interaction) from 12 LTER sites

American Ecosystems or VARNAE) constitute one of the first integrated data sets created by the LTER network. The workshop which catalyzed the analysis of variability in these ecosystems continues as an interactive analysis and writing project among 12 LTER sites.

4.4.4 RESULTS

In each of the 12 ecosystems, variation in edaphic parameters among locations (within an ecosystem) was larger than that among years. This result was not particularly surprising but it did emphasize the importance of encompassing spatial variability of a research site in long-term studies. In our analysis, however, it was surprising that the ecosystem most similar to lakes in patterns of edaphic variability among locations was the desert, not the more aquatic stream, river, or estuary. It seems that, in patterns of variability, systems do not sort along traditional disciplinary lines. This interesting result and other differences and similarities among divergent ecosystems are stimulating us to generate new hypotheses about the systems we study. For example, how does variability among locations contribute to temporal dynamics of populations, communities, and ecosystem processes? Or more specifically, do deserts and lakes, by having common spatial variability in edaphic features, also have common dynamic properties?

Analyzing data from the Northern Temperate Lakes site, Kratz *et al.* (in press) noted a relationship between the among-year variability exhibited by a lake and that lake's topographic position in the landscape. We asked whether the same pattern held true for estuaries, deserts, forests, etc. The patterns for lakes and the estuary were opposites; lakes with lower landscape position had less temporal variability than those high in the landscape, whereas in the estuary, lower locations were more variable. These patterns suggest that even though there is a large amount of information in landscape position, the mechanisms by which landscapes interact with ecosystems are likely to be ecosystem dependent.

To discern general properties of variability in the ecosystems, we compared variability of climatic, edaphic, plant, and animal components across the 12 ecosystems. By analogy, each ecosystem served as would a fish, mammal, bird, reptile, or amphibian in a generalization about the properties of vertebrates. Each is very different, yet they share common properties. In our analysis of ecosystem variability, we looked for the common properties of temporal and spatial variability of ecosystems. One example follows.

Before the workshop we had noted that signals are passed up through ecosystem components from climatic to edaphic to plant and to animal components, on the one hand, but also are passed down through ecosystem components from carnivores to herbivores to plants and even to chemical concentrations and microclimates. As these signals move through an ecosystem it is interesting to speculate on whether they are amplified or attenuated. For example, does variation in precipitation from year to year cause a higher year-to-year variability in nutrient availability, which in turn causes still higher variability in animals? Or does variability in precipitation cause less variability in nutrient availability, and sequentially, in plants and in animals?

With the data from all of the sites, we evaluated this general question using relative variance, which is analogous to the coefficient of variation, and r^2 or the proportion of variance owing to years, locations, or other. Relative variance tended to increase from nonbiological to biological measures for all three sources of variability—'year', 'location', and 'other' (Figure 4.6). The increase appeared to be greatest for 'location'. One interpretation of these results is that the signal is amplified through the system from climate to animals, especially with regard to the relative variability associated with 'location' and 'other' and the proportion of variability owing to 'other'. Year-to-year variability, however, does not appear to be amplified.

Other explanations, in addition to the amplification hypothesis, are also possible. The first is that measurement error increases from climatic to animal measures and that there is no signal in the data. Climate parameters, for example, may be measured more accurately and also may represent more aggregated measures than the animal data.

However, variability ascribed to 'other' includes not only measurement error but also interaction between the main effects of 'year' and 'location'. Thus, an additional possibility is that the increasing relative variance for 'other' represents interaction between temporal and spatial dynamics of ecosystems. The increasing variability owing to 'other' and 'location' (Figure 4.6) indicated that distribution of organisms was more patchy than that of climatic and edaphic features, and that their distribution changed in an interactive way with 'location' and 'year'; different locations may be more suitable in different years. The proportion of variance owing to interaction between 'year' and 'location' would be expected to be greater for more mobile organisms that could respond to the changing environment during a particular year. For example, small mammals at the desert, alpine tundra, and long-grass prairie sites could move among adjacent plant communities or watersheds to seek out the most favorable locations, which may differ among years. Alternatively, interaction would be expected to be greater for plants and animals with large differences in annual reproductive success or production if different locations were more favorable in different years. For example, interlake movement is unlikely in the isolated basins at the North Temperate Lakes LTER Site and the abundances of plants and animals there are largely a response to annual conditions in each lake. Regardless of cause, the observation that variability changes systematically among ecosystem components stimulates inquiry into the causes of ecosystem variability in time and space. For example, what are the characteristics of sites that deviate from the pattern observed here (variation increasing from nonbiological to biological components)?

Comparisons of the dynamics of divergent ecosystems in a coherent manner depends upon the design of research at individual sites. The LTER sites, used in the above analysis of temporal and spatial variability, incorporated measurements of climatic, edaphic, plant, and animal components at one-year intervals and at a

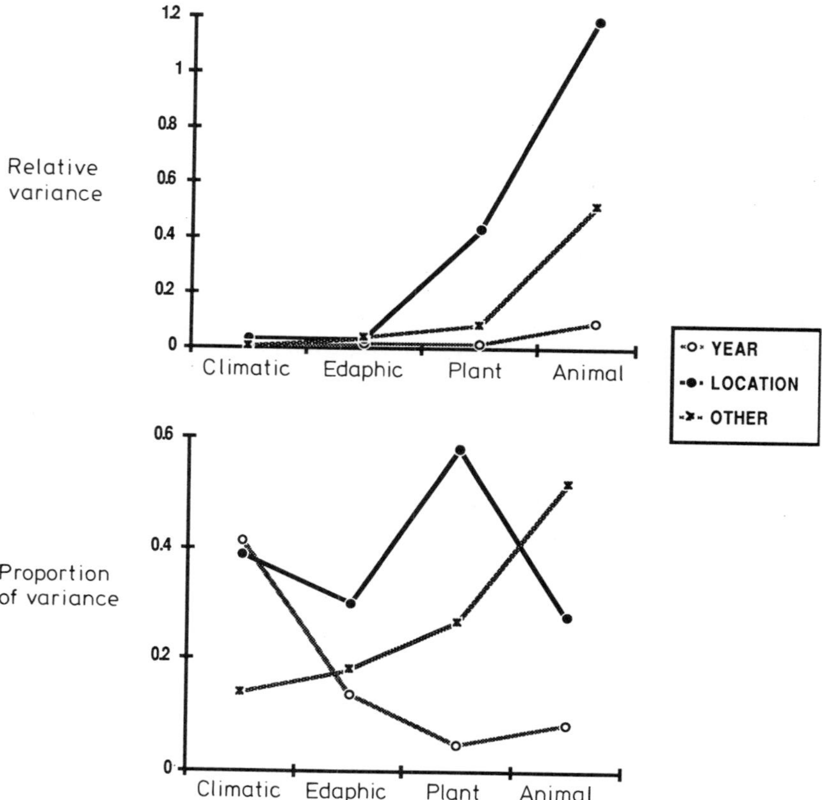

Figure 4.6 The influence of ecosystem component (climatic, edaphic, plant and animal) on the relative variances and r^2s, the proportion of variance, associated with time (among years), location (among locations at an LTER site) and other (interaction and measurement error) at 12 LTER sites

number of locations in each ecosystem. In the following pages we describe the characteristics and design of one of these sites, the North Temperate Lakes LTER Site, with which we are most familiar.

4.5 NORTH TEMPERATE LAKES LONG-TERM ECOLOGICAL RESEARCH PROJECT

The North Temperate Lakes LTER project (Magnuson *et al.*, 1984; Kratz *et al.*, 1986) was designed to take advantage of the benefits of conducting ecological research on broader and more varied temporal and spatial scales than are characteristic of most ecological research.

Lakes are excellent systems for long-term ecological research. Their boundaries are distinct, and adjacent lakes, although similar to each other in one sense, may differ greatly in physical, chemical, and biological properties and processes. Thus, a set of lakes can be used in natural experiments to isolate, at least partially, important control factors external to and within the lake. Clarity of the boundaries makes lake sites especially useful for analysis of the dynamics of landscape scale processes. Important ecological variables have been identified for lakes, and information is available to suggest measurement frequencies that minimize interferences from short-term variations.

The north temperate lakes provide an ideal system from which to explore, develop, and test ecological theory at the intersite level. Lake ecosystems superficially appear quite different from the principal ecosystems at the other LTER sites, i.e. forest, prairie, desert, tundra, saltmarsh, and river. To include lakes in comparative ecosystem research, we must operate at a very general level. We think ideas generated or tested in this way will apply to a more diverse group of ecosystems than would theory developed for a single, more narrow range of ecosystems. The previous section on the comparison of temporal and spatial variability among divergent ecological systems describes one of our approaches to comparing diverse systems.

The LTER project on north temperate lakes has the following scientific objectives:

(1) To perceive and describe long-term trends and patterns in physical, chemical, and biological properties of lake ecosystems;
(2) To understand the dynamics of internal and external processes affecting lake ecosystems;
(3) To analyze the temporal responses of lake ecosystems to disturbance and stress;
(4) To evaluate the interaction between spatial heterogeneity and temporal variability of lake ecosystems; and
(5) To expand our understanding of lake-ecosystem properties to a broader, regional context.

The LTER project is question driven rather than descriptive, but new questions continue to emerge from examination of earlier and developing data sets. Some questions have involved major manipulations such as the experimental acidification of Little Rock Lake, a project funded primarily by the United States Environmental Protection Agency. Other questions have evolved from natural disturbances to lakes such as the invasion of an exotic fish or crayfish. In addition, the LTER site is designed to be attractive to a wide variety of ecologists, who bring their specific research to the North Temperate Lakes LTER site partly because of the LTER project, but also because of the facilities provided by the University of Wisconsin, the natural history and resources of the region, and the ongoing research activity.

In the first five years of the project (1981 to 1985), we recognized that until long-term data sets were available we would have to develop approaches in initiating long-term ecological research on lake ecosystems using basic concepts and short-term data. We compared recent LTER data with similar historic data collected on northern Wisconsin lakes during the 1925 to 1942 Birge and Juday era (Bowser, 1986; Rudstam, 1984). We evaluated interyear variability within and among lakes with historic and recent data (Kratz *et al.*, 1987a; Robertson, 1989; McLain and Magnuson, 1989; Magnuson *et al.*, 1990) and we began to develop an understanding of the principal external and internal linkages affecting the dynamics of LTER lakes (Hurley *et al.*, 1985; Rudstam and Magnuson, 1985; Anderson and Bowser, 1986; Kratz *et al.*, 1987; Lodge *et al.*, 1989; Kratz and Medland, 1989).

At the same time, we initiated long-term measurements on specific lakes so that investigators could, as the time series evolved, observe and analyze quantitative patterns of long-term changes in the physical, chemical, and biological features of lake ecosystems. Our choices of lakes and measurements were guided by a desire to identify and answer important ecological and natural resource questions about lakes with respect to long-term (Likens, 1983; LeCren, 1984; Strayer *et al.*, 1986) and landscape level (Naveh and Lieberman 1984; Risser *et al.*, 1984) phenomena. Questions specific to lakes include the importance of external versus internal cycling of nutrients, the interactions between nutrient inputs and internal consumer controls of plankton, and the influence of landscape position on variability and coherence of lake ecosystems. Our long-term measurements at various ecological levels and on several temporal and spatial scales are intended to capture the essential structure and functioning of lake ecosystems and enable analyses of interactions among principal ecosystem components. We also have extended our analyses in time using sediment cores.

We chose to study a suite of lakes (Figure 4.7, and Table 4.1) linked through a common groundwater and surface flow system and common climatic, edaphic, and biogeographic systems. The seven primary lakes are located within 5.3 km of the Trout Lake Station in north central Wisconsin. By this choice of primary lakes, groundwater, based on its importance in regulating differences in the chemical composition of lakes and in linking terrestrial and lake ecosystems (Likens *et al.*, 1977; Winter, 1978; Crowe and Schwartz, 1981a,b; Frape *et al.*, 1984,), becomes one central focus of the North Temperate Lakes LTER project. Even with these commonalties, the suite of lakes includes oligotrophic, dystrophic, and mesotrophic lakes with marked differences in size, morphometry and habitat diversity, thermal and chemical features, species richness and assemblies, and biological productivity (Table 4.1).

We also have secondary lakes on which less complete data are collected in most cases. The choice of secondary lakes and measurements on them may change with time, but these lakes are studied with long-term research goals in mind. They serve for comparison with the primary lakes on specific questions. For example, Lake

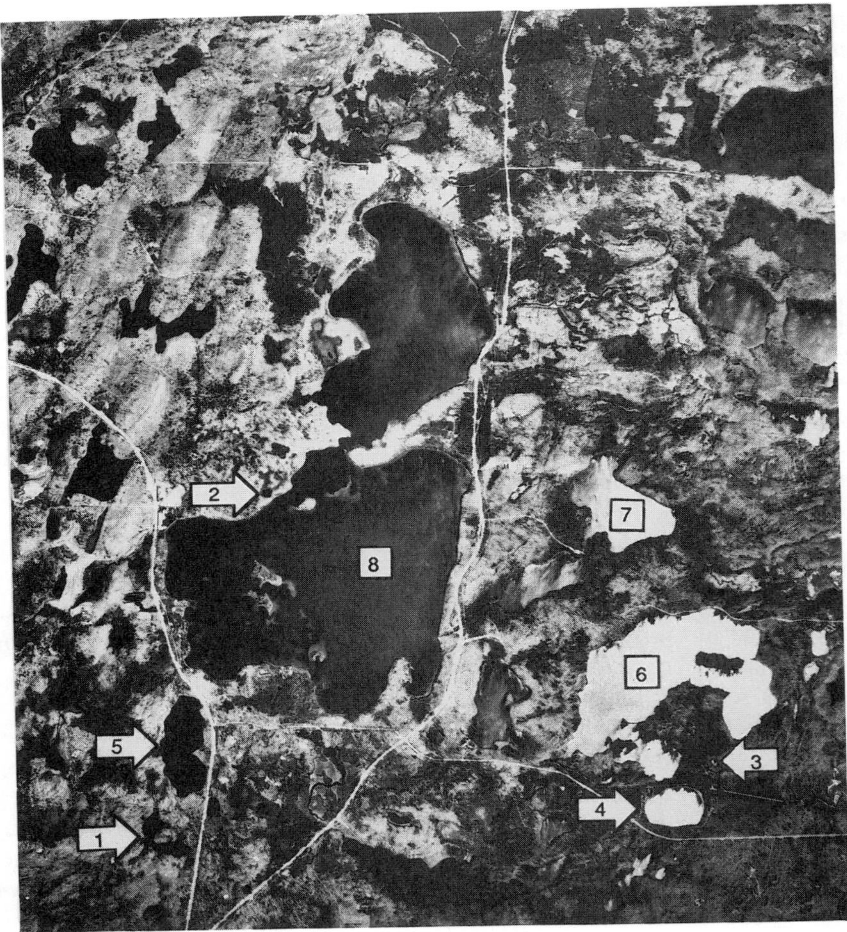

Figure 4.7 Air photo of the North Temperate Lakes LTER site in Northern Wisconsin identifying the LTER lakes. The center of the photo is near 46°2′30″ N latitude and 89°37′40″ W longitude. For scale, north and south Trout Lake combined are *ca* 7.2 km long. Photo from NHAP (National High Altitude Photography Program). LTER lakes are 1 = Little Rock Lake, 2 = Trout Bog (12–15), 3 = Crystal Bog (27–2), 4 = Crystal Lake, 5 = Sparkling Lake, 6 = Big Muskellunge Lake, 7 = Allequash Lake, and 8 = Trout Lake. Little Rock Lake is a secondary LTER lake divided by a rubber curtain and used in an experimental acidification project

Table 4.1 Selected characteristics of the lakes at the North Temperate Lakes LTER Site, the seven primary lakes plus the unacidified side of one secondary LTER lake, Little Rock Lake

Characteristics	Little Rock Lake	Trout Bog (12–15)	Crystal Bog (27–2)	Crystal Lake	Sparkling Lake	Big Muskellunge Lake	Allequash Lake	Trout Lake
Location in Fig. 4.7	1	2	3	4	5	6	7	8
Area (ha)	8.1	2.0	1.0	36.654	64.3	396.34	168.45	1607.9
Max. depth (m)[a]	6.5	7.5	2.5	20.4	20.0	21.4	7.6	35.7
Ave. length of ice cover (days)[b]	151	160	155	143	145	145	152	143
Mean winter temperature (°C)[c]	—	2.6	3.1	2.9	2.9	2.2	2.1	1.9
Mean summer epilimnion temperature (°C)[d]	22.7	20.8	21.5	21.1	21.2	21.2	21.6	19.7
Mean summer secchi (m)[e]5.3	1.2	1.4	8.3	6.6	7.2	3.0	4.6	
pH[f]	6.1	4.55	4.82	5.72	7.14	7.07	7.29	7.41
Conductivity (spring mixis) (μmhos/cm)[g]	12.6	20	12	14	72	43	82	85
Total P (spring mixis) (filtered μg/l)[h]9.0	4.0	5.8	2.3	3.9	2.8	8.4	2.0	
Mean summer chlorophyll (μg/l)[i]2.2	8.6	4.5	1.1	1.1	1.7	3.8	1.6	
Number fish species[i]	10	3	1	8	17	27	25	33

a Fall 1981–Spring 1984/Little Rock—1983–85
b The volumetric mean average temperature from 1 January to 30 March 1982–83 (three temperature profiles per year)/Little Rock—not available
c The average temperature at 1 meter from 1 June to 31 August 1981–84 (six temperature profiles per year)/Little Rock—June to Sept. 1984
d Average value 1982–83 from 1 June to 31 August/Little Rock—May to October 1984
e Weighted average of spring and fall mixis values 1981–83/Little Rock 1988
f Average surface—most value 1982–83 at spring mixis/Little Rock 1986
g 1983 surface—most value at spring mixis/Little Rock 1986
h Average surface value 1982–84 from 1 June to 31 August Little Rock 1986
i Cumulative number present 1981–83/Little Rock—May to October 1984

Mary, a meromictic lake, has bottom waters that have been extremely stable over time, and is useful in comparisons with the other lakes which are more influenced by year-to-year changes in weather. Little Rock Lake is a secondary lake used for an intense experiment on lake acidification (Brezonik *et al.*, 1986; Watras and Frost, 1989) and for which the LTER primary lakes serve as reference lakes. Another secondary lake, Lake Mendota, is our only study lake located outside the Northern Highlands area. We have included it because there are extensive and historic data available for Mendota and because it is so accessible to the Laboratory of Limnology.

We spent considerable effort designing and implementing a balanced and integrated data collection program, which is described by Kratz *et al.* (1986). Our sampling program on the primary lakes allows comparisons of parameters among seasons, years, and lakes. We sample most major physical and chemical parameters. Likewise, information is collected on all trophic levels from primary producers to top consumers. We sample most intensively at four key times of the year; spring overturn, maximum stratification in summer, fall overturn, and winter stratification. These are the four seasons for a dimictic lake at temperate latitudes.

Frequency of sampling is matched as closely as practically possible with the appropriate time scale for each parameter's dynamics. Between the intensive quarterly sampling periods, phytoplankton and zooplankton samples are usually collected biweekly, although the sampling frequency is reduced to monthly when water temperatures are low. Temperature, dissolved oxygen, chlorophyll *a*, and nutrient profiles are measured in each lake every 2 or 3 weeks during the open water season and every 5 weeks under ice cover. Parameters varying over longer time scales, such as macrophytes, fishes (abundance, biomass, and community structure), and benthic invertebrate abundance, are measured annually in August. These last items are measured at a fixed set of locations to capture the within-lake distributions and spatial heterogeneity. All other parameters are measured at a central station in each lake.

Early in the project we produced a detailed inventory of the available historic data from the Trout Lake site. The earliest and most complete aquatic data were collected by E.A. Birge, C. Juday, and their colleagues from 1924 through 1942 (Frey, 1963). The extent and scope of their work are impressive, spanning more than 500 lakes in the Northern Highlands Lake District and covering numerous physical, chemical, and biological parameters. Major portions of the water chemistry, phytoplankton, and zooplankton data have been computerized by various agencies. In addition, we gathered historic maps and documents relevant to the site and are in the process of developing a Geographic Information System (GIS) with satellite and ground-based spatial data.

We developed a database system using a mixed environment of microcomputers and campus mainframe systems. We have used microcomputers for data entry, data reduction, data transmission, and statistical graphical analysis. Storage and handling of full-record files are accomplished using a hierarchical database named 'SIR' (Scientific Information Retrieval) maintained on the campus VAX-780 computer cluster. Records of data at various stages are archived, and are available to users in forms ranging from unverified raw data to error-checked, finalized data files available from the SIR data bases.

The North Temperate Lakes LTER project is operated by the Center for Limnology at the University of Wisconsin-Madison. Interests of individual researchers range from geology to meteorology to ecology. The Trout Lake Station provides access to lakes and facilities at the principal field site in the Northern Highlands Lake District. The Northern Highlands and American Legion State Forests protect about two-thirds of the lake frontage. Many lakes have totally forested watersheds and no private frontage. The area was intensely logged and burnt early in the twentieth century. Today, outdoor recreation and forestry form the economic base for local communities. The site has one of the highest concentrations of lakes in the world. The range of limnological conditions is remarkable, and includes oligotrophic, mesotrophic, eutrophic, and dystrophic lakes; dimictic, meromictic and monomictic lakes; atmosphere-dominated, groundwater-dominated, and drainage lakes; lakes with varved sediments, winterkill lakes, temporary and permanent forest ponds, beaver ponds, small reservoirs, and wetlands, in addition to a variety of warm- and cold-water streams (Brenneman and Blinn, 1987). The climate is temperate, and lakes are usually ice covered from late November until late April.

4.6 CONCLUSIONS

The LTER network provides an opportunity to conduct ecological research at temporal and spatial scales broader than has been possible with conventional research funding. Design features of an LTER project include (1) focusing on long-term phenomena unveiled over years, decades, or even centuries, (2) operating with a suite of locations within an ecosystem or with sets of ecosystems, (3) observing at a nested set of spatial scales, (4) collaborating within an active, interdisciplinary group of question-driven scientists, (5) implementing a dependable and sustainable system of data management, and (6) creating an institutional setting sufficient to sustain a long-term effort without a suppression of creativity. Individual long-term ecological research sites and projects can gain greatly by being part of a network of structurally similar projects conducted at ecologically or geographically divergent sites. The latter feature challenges ecologists to generate knowledge and understanding which can be generalized to all natural systems rather than to only a specific ecosystem.

4.7 ACKNOWLEDGMENTS

We thank M.L. Smith, C.M. Hughes and D.E. Chandler for preparation of the text and figures. The acoustic transect at the edge of the Gulf Stream was obtained and analyzed through the support of Office of Naval Research grant N00014-79-C-0703. The North Temperate Lakes LTER site is funded by National Science Foundation grant BSR8514330.

4.8 REFERENCES

Anderson, M.P. and Bowser, C.J. (1986). The role of groundwater in delaying lake acidification. *Water Resources Research*, **22**, 1101–1108.

Barbour, C.D. and Brown, J.H. (1974). Fish species diversity in lakes. *American Naturalist*, **108** (962), 473–489.

Bowser, C.J. (1986). Historic data sets: lessons from the past, lessons for the future. In Michener, W.K. (Ed.) *Research Data Management in the Ecological Sciences*. University of South Carolina Press, Columbia, South Carolina, 155–179.

Brenneman, J. and Blinn, T. (Eds) (1987). Long-term ecological research in the United States, a network of sites. *Long-Term Ecological Research Network*, fourth edition, revised. Department of Forest Science, Oregon State University, Corvallis, Oregon.

Brenneman, J., Franklin, J.F. and Magnuson, J.J. (1987a). *LTER Network News*. **1** (Spring). Department of Forest Science, Oregon State University, Corvallis, Oregon.

Brenneman, J., Franklin, J.F. and Magnuson, J.J. (1987b). *LTER Network News*. **1** (Fall). Department of Forest Science, Oregon State University, Corvallis, Oregon.

Brenneman, J., Franklin, J.F. and Magnuson, J.J. (1988). *LTER Network News*. **1** (Spring). Department of Forest Science, Oregon State University, Corvallis, Oregon.

Brezonik, P.L., Baker, L.A., Eaton, J., Frost, T.M., Garrison, P., Kratz, T.K., Magnuson, J.J., McCormick, J.H., Perry, J.A., Rose, W.J., Shephard, B.K., Swenson, W.A., Watras, C.J. and Webster, K.E. (1986). Experimental acidification of Little Rock Lake, Wisconsin. *Water, Air, and Soil Pollution*, **31**, 115–122.

Britingham, M.C. and Temple, S.A. (1983). How cowbirds caused forest songbirds to decline. *BioScience*, **33**, 31–35.

Brown, J.H. (1971). Mammals on mountaintops: non-equilibrium insular biogeography. *American Naturalist*, **105**, 467–478.

Callahan, J.T. (1984). Long-term ecological research. *BioScience*, **34**, 363–367.

Carlton, J. (1985). Transoceanic and interoceanic dispersal of coastal marine organisms, the biology of ballast water. *Oceanography and Marine Biology Annual Review*, **23**, 313–371.

Carpenter, S.C., Kitchell, J.F. and Hodgson, J.R. (1985). Cascading trophic interactions and lake productivity. *BioScience*, **35**, 634–639.

Cook, R.E. (1977). Raymond Lindeman and the trophic-dynamic concept in ecology. *Science*, **198**, 22–26.

Courtenay, W.R. Jr and Stauffer, J.R. Jr (Eds) (1984). *Distribution, Biology, and Management of Exotic Fishes*. Johns Hopkins University Press, Baltimore, Maryland.

Crowe, A.S. and Schwartz, F.W. (1981a). Simulation of lake-watershed systems: I Description and sensitivity analysis of the model. *Journal of Hydrology*, **52**, 71–105.

Crowe, A.S. and Schwartz, F.W. (1981b). Simulation of lake-watershed systems: II Application to Baptiste Lake, Alberta, Canada. *Journal of Hydrology*, **52**, 107–125.

Diamond, J.M. (1975). Assembly of species communities. In Cody, M.L. and Diamond, J.M. (Eds) *Ecology and Evolution of Communities*. Harvard University Press, Cambridge, Massachusetts.

Eadie, J.McA., Hurly, T.A., Montgomerie, R.D. and Teather, K.L. (1986). Lakes and rivers as islands: species–area relationships in the fish faunas of Ontario. *Environmental Biology of Fishes*, **15**(2), 81–89.

Emerick, J. C. (1976). Effects of artificially increased winter snow cover on plant canopy architecture and primary production in selected areas of the Colorado alpine tundra. PhD Dissertation, University of Colorado at Boulder, Colorado.

Evans, M.S. (1988). *Bythotrephes cederstroemi*: its new appearance in Lake Michigan. *Journal of Great Lakes Research*, **14**(2), 234–240.

Forman, R.T.T. and Gudron, M. (1986). *Landscape ecology*. John Wiley, New York.

Franklin, J.F. and Forman, R.T.T. (1987). Creating landscape patterns by forest cutting: ecological consequences and principles. *Landscape Ecology*, **1**, 5–18.

Frape, S.K., Fritz, P. and McNutt, R.H. (1984). Water–rock interaction and chemistry of groundwaters from the Canadian Shield. *Geochimica et Cosmochinica Acta*, **48**, 1617–1627.

Frey, D. G. (1963). *Limnology in North America*. University of Wisconsin Press, Madison, Wisconsin.

Hebert, P. D. N., Muncaster, B.W. and MacKie, G.L. (1989). Ecological and genetic studies on *Dreissena polymorpha* (Pallas): a new mollusc in the Great Lakes. *Canadian Journal of Fisheries and Aquatic Sciences*, **46**, 1587–1591.

Henrikson, L.H., Nyman, G., Oscarson, H.G. and Stenson, J.E. (1980). Trophic changes without changes in external nutrient loading. *Hydrobiologia*, **68**, 257–263.

Hrbacek, J. (1958). Density of the fish population as a factor influencing the distribution and speciation of the species in the genus *Daphnia*. *XVth International Congress of Zoology*, Section X, Paper 27, 794–795.

Hrbacek, J., Dvorakova, M., Korinek, V. and Prochazkova, L. (1961). Demonstration of the effect of the fish stock on the species composition of zooplankton and the intensity of metabolism of the whole plankton association. *International Association of Theoretical and Applied Limnology Proceedings*, **14**, 192–195.

Hurley, J.P., Armstrong, D.E., Kenoyer, G.J. and Bowser, C.J. (1985). Groundwater as a silica source for diatom production in a precipitation-dominated lake. *Science*, **227**, 1576–1579.

Keigley, R.B. (1987). Effect of experimental treatments on *Kobresia myosuroides* with implications for the potential effect of acid deposition. PhD Dissertation, University of Colorado at Boulder, Colorado.

Kratz, T.K., Benson, B.J., Blood, E.R., Cunningham, G.L. and Dahlgren, R.A. The influence of landscape position on temporal variability of four North American ecosystems. *The American Naturalist* (in press).

Kratz, T.K., Cook, R., Bowser, C.J. and Brezonik, P.L. (1987). Winter and spring pH depressions in northern Wisconsin lakes caused by increases in pCO_2. *Canadian Journal of Fisheries and Aquatic Sciences*, **44**, 1082–1088.

Kratz, T.K., Frost, T.M. and Magnuson, J.J. (1987). Inferences from spatial and temporal variability in ecosystems: long-term zooplankton data from lakes. *The American Naturalist*, **129** (6), 830–846.

Kratz, T.K. and Medland, V.M. (1989). Relationship of landscape position and groundwater input in northern Wisconsin Kettle-hole peatlands. In Sharitz, R. and Gibbons, J. (Eds) *Freshwater Wetlands and Wildlife*. CONF-8603101, DoE Symposium Series No. 61. US DoE Office of Scientific and Technical Information, Oak Ridge, Tennessee.

Kratz, T.K., Magnuson, J.J., Bowser, C.J. and Frost, T.M. (1986). Rationale for data collection and interpretation in the Northern Lakes Long-Term Ecological Research program. In *American Society for Testing and Materials (ASTM). Proceedings of Symposium on Rationale for Sampling and Interpretation of Ecological Data in the Assessment of Freshwater Systems*, Philadelphia, Pennsylvania, 22–23.

Lamb, H.H. (1977). *Climate: Present, Past, and Future*. Vol.II. *Climatic History and the Future*. Methuen, New York.

LeCren, D. (1984). Letters to the editor. Long-term ecological research. *British Ecological Society Bulletin*, **15**(4), 185–187.

Likens, G.E. (1983). A priority for ecological research—address of the past president. *Bulletin of the Ecological Society of America*, **64**(4), 234–243.

Likens, G.E., Bormann, F.H., Pierce, R.S., Eaton, J.S. and Johnson, N.M. (1977). *Bio-geochemistry of a Forested Ecosystem*. Springer-Verlag, New York.

Lindeman, R.L. (1942). The trophic-dynamic aspect of ecology. *Ecology*, **23**, 399–418.

Lodge, D.M., Krabbenhoft, D.P. and Striegl, R.G. (1989). A positive relationship between groundwater velocity and submerged macrophyte biomass in Sparkling Lake, Wisconsin. *Limnology and Oceanography*, **34**(1), 235–239.

MacArthur, R.H. (1972). *Geographical Ecology—patterns in the distribution of species*. Harper and Row, New York.

MacArthur, R.H. and Wilson, E.O. (1963). An equilibrium theory of insular biogeography. *Evolution*, **17**, 373–387.

MacArthur, R.H. and Wilson, E.O. (1967). *The Theory of Island Biogeography*. Princeton University Press, Princeton, New Jersey.

Magnuson, J.J. (1976). Managing with exotics—A game of chance. *Transactions of the American Fisheries Society*, **105**(1), 1–9.

Magnuson, J.J. (1978). Ecological approaches to hydraulic structures. In Driver, E.E. and Wunderlich, W.O. (Eds) *Environmental Effects of Hydraulic Engineering Works*. Tennessee Valley Authority, Knoxville, Tennessee, 11–28.

Magnuson, J.J. (1988). Two worlds for fish recruitment; lakes and oceans. Early Life History Series Publication (AFS), *Proceedings of the 11th Annual Larval Fish Conference*, **5**, 1–6.

Magnuson, J.J. (1990). The invisible present. *BioScience*, **40**(7), 495–501.

Magnuson, J.J., Benson B.J. and Kratz, T.M. Temporal coherence in the limnology of a suite of lakes in Wisconsin, USA. *Freshwater Biology*, **23**, 145–159.

Magnuson, J.J., Bowser, C.J. and Beckel, A.L. (1983). The invisible present, long-term ecological research on lakes. *L&S Magazine*, College of Letters and Sciences, University of Wisconsin-Madison, Madison, Wisconsin. Premier Issue, Fall, 3–6.

Magnuson, J.J., Bowser, C.J. and Kratz, T.K. (1984). Long-term ecological research on north temperate lakes (LTER). *Verhandlungen Internationale Vereinigung Limnologie*, **22**, 533–535.

Magnuson, J.J., Paszkowski, C.A., Rahel, F.J. and Tonn, W.M. (1989). Fish ecology in severe environments of small isolated lakes in northern Wisconsin. In Scharitz, R. and Gibbons, J. (Eds) *Freshwater Wetlands and Wildlife*. CONF-8603101, DoE Symposium Series No. 61, US DoE Office of Scientific and Technical Information, Oak Ridge, Tennessee.

McLain, A.S. and Magnuson, J.J. (1989). Analysis of recent declines in cisco (*Coregonus artedii*) populations in several northern Wisconsin Lakes. *Finnish Fisheries Research*, **9**, 155–164.

Mysak, L.A. (1986). El Niño, interannual variability and fisheries in the Northeast Pacific Ocean. *Canadian Journal of Fisheries and Aquatic Sciences*, **43**, 464–497.

Naveh, Z. and Lieberman, A.S. (1984). *Landscape Ecology, theory and application*. Springer-Verlag, New York.

Nero, R.W. and Magnuson, J.J., Effects of changing spatial scale on acoustic observations of patchiness in the Gulf Stream. *Landscape Ecology* (in press).

Northcote, T.G. (1988). Fish in the structure and function of freshwater ecosystems: a 'top-down' view. *Canadian Journal of Fisheries and Aquatic Sciences*, **45**, 361–379.

O'Neill, R.V. (1989). Perspectives in hierarchy and scale. In Roughgarden, J., May, R.M.

and Levin, S.A. (Eds) *Perspectives in Theoretical Ecology*. Princeton University Press, Princeton, New Jersey, 140–156.

Pearse, W.A., Braem, R.A., Dustin, S.M. and Tibbles, J.J. (1980). Sea lamprey (*Petromyzon marinus*) in the lower Great Lakes. *Canadian Journal of Fisheries and Aquatic Sciences*, **37**, 1802–1810.

Quinn, W.H., Zopf, D.O., Short, K.S. and Kuo Yang, R.T.W. (1978). Historical trends and the statistics of the southern oscillation, El Niño, and Indonesian droughts. *Fisheries Bulletin*, **76**(3), 663–677.

Rahel, F.J. (1984). Factors structuring fish assemblages along a bog lake successional gradient. *Ecology*, **65**(4), 1276–1289.

Rahel, F.J. (1986). Biogeographic influences on fish species composition of northern Wisconsin lakes with applications for lake acidification studies. *Canadian Journal of Fisheries and Aquatic Sciences*, **43**, 124–134.

Risser, P.G., Karr, J.R. and Forman, R.T.T. (1984). Landscape ecology: directions and approaches. Special Publication 2, Illinois Natural History Survey, Urbana, Illinois.

Robertson, D.M. (1989). The use of lake water temperature and ice cover as climatic indicators. PhD Dissertation, University of Wisconsin-Madison, Wisconsin.

Robertson, D.M. (1990). Lakes as indicators of and responders to climatic change. *Proceedings of the LTER Intersite Workshop on Climatic Variability and Ecosystem Response*. In press.

Rudstam, L.G. (1984). Long-term comparison of the population structure of the cisco (*Coregonus artedii LeSueur*) in smaller lakes. *Wisconsin Academy of Science, Arts and Letters*, **72**, 185–200.

Rudstam, L.G. and Magnuson, J.J. (1985). Predicting the vertical distribution of fish populations: analysis of cisco, *Coregonus artedii*, and yellow perch, *Perca flavescens*. *Canadian Journal of Fisheries and Aquatic Sciences*, **42**, 1178–1188.

Sokal, R.R. and Rohlf, F.J. (1981). *Biometry: the principles and practice of statistics in biological research*. Second edition, W.H. Freeman and Company, San Francisco, California.

Soule, M.E. (1987). *Viable Populations for Conservation*. Cambridge University Press, New York.

Steele, J.H. (1985). A comparison of terrestrial and marine ecological systems. *Nature*, **313**, 355–358.

Strayer, D., Glitzenstein, J.S., Jones, C.G., Kolasa, J., Likens, G.E., McDonnell, M.J., Parker, G.G. and Pickett, S. T.A. (1986). Long-term ecological studies: an illustrated account of their design, operation, and importance to ecology. Occasional Publication of the Institute of Ecosystem Studies, Number 2, August. New York Botanical Garden, Millbrook, New York.

Swanson, F.J. and Sparks, R.E. (1990). Long-term ecological research and the invisible place. *BioScience*, **40**(7), 502–508.

Swanson, F.J. and Franklin, J.F. (1988). The long-term ecological research program. *Eos*, **69**(3), 34, 36, 46.

Tonn, W.M. (1985). Density compensation in *Umbra-Perca* fish assemblages of northern Wisconsin lakes. *Ecology*, **66**(2), 415–429.

Tonn, W.M. and Magnuson, J.J. (1982). Patterns in the species assemblages in northern Wisconsin lakes. *Ecology*, **63**(4), 1149–1166.

Tonn, W.M., Magnuson, J.J., Rask, M. and Toivonen, J. (1990) Intercontinental comparison of small-lake fish assemblages: the balance between local and regional processes. *The American Naturalist*, **136**(3), 345–375.

Turner, M.G. and Ruscher, C.L. (1988). Changes in landscape patterns in Georgia, USA. *Landscape Ecology*, **1**, 241–251.

Turner, M.C., Dale, V.H. and Gardner, R.H. (1989a). Predicting across scales: theory development and testing. *Landscape Ecology*, **3**(3,4), 245–252.

Turner, M.G., O'Neill, R.V., Gardner, R.A. and Milne, B.T. (1989b). Effects of changing spatial scale on the analysis of landscape pattern. *Landscape Ecology*, **3**(3,4), 153–162.

Wahl, E.W. and Lawson, T.L. (1970). The climate of the midnineteenth century United States compared to the current normals. *Monthly Weather Review*, **98**(4), 259–265.

Watras, C.J. and Frost, T.M. (1989). Little Rock Lake (Wisconsin): perspectives on an experimental ecosystem approach to seepage lake acidification. *Archives of Environmental Contamination and Toxicology*, **18**, 157–165.

Webber, P.J., Emerick, J.C., Ebert May, D.C. and Komarkova, V. (1976). The impact of increased snowfall on alpine vegetation. In Steinhof, Y.H.W. and Ives, J.D. (Eds) *Ecological Impacts of Snowpack Augmentation in the San Juan Mountains, Colorado*. Final Report, San Juan Ecology Project, Colorado State University Publication, Fort Collins, Colorado, 201–264.

Winter, R.C. (1978). Ground-water component of lake water and nutrient budgets. *International Association of Theoretical and Applied Limnology Proceedings*, **20**, 438–444.

5 Long-term Studies: Past Experience and Recommendations for the Future

STEWARD T.A. PICKETT
Institute of Ecosystem Studies, The New York Botanical Garden, Mary Flagler Cary Arboretum, Box AB, Millbrook, NY 12545, USA

5.1 INTRODUCTION

The mandate of the Institute of Ecosystem Studies (IES) is to study the disturbance and recovery of north temperate ecosystems. Although 'ecosystem' is the overarching label, our interests include populations, communities, and landscapes as well as resources and stress factors. In this chapter, the term 'system' is used to encompass this multiplicity of subjects. One of the strategies for fulfilling

Long-term Ecological Research. Edited by Paul G. Risser

the mandate of the IES is to employ long-term studies (LTS). Because LTS can be a major commitment of personnel, land, facilities, and money, the IES chose first to study the characteristics of long-term studies. We hope this effort will help us to design and conduct better LTS than we otherwise might. This chapter shares the experience we have gained. Much of the background comes from the IES second Occasional Publication (Strayer *et al.*, 1986), which laid out the kinds of LTS, and sought to determine what contributed to their success. The chapter also examines alternatives to LTS which were explored in much greater depth in the 1987 Cary Conference, entitled *Long-Term Studies in Ecology: Approaches and Alternatives*. That conference, which drew on all aspects of ecology from populations through communities to ecosystems, had an international scope, permitting a thorough analysis of the features of long-term studies (Likens, 1989). In addition, my comments will be based to some extent on my experience in contributing to a 31-year study of plant succession.

This chapter also illustrates the nature of successful LTS, and explores the motivations of such studies. Whether the success of LTS can be unambiguously founded on rigorous *a priori* arguments, or whether it is an empirical matter, will also be addressed. The kinds of important questions remaining for LTS, especially those that are only apparent from LTS, will be presented. Finally, the advice of the participants of the 1987 Cary Conference on methods will be summarized.

Strayer *et al.* (1986) sought input on LTS from practicing ecologists for a survey that was initially intended to be a rigorous, perhaps statistical, study of LTS. However, it became clear that the difficulty of ferreting out a representative sample of LTS, including those that had failed, would make the original intent impossible. Therefore, the study evolved into a survey of existing LTS, using questionnaires and interviews with the scientists conducting them. The first insight that emerged was a catalog of the kinds of phenomena amenable to LTS. These included slow, rare, subtle, and complex phenomena. The nature of LTS that emerged from the survey is summarized in this chapter.

5.2 THE SUBJECTS OF LONG-TERM STUDIES

Long-term studies are designed to capture the effects of the environment and biotic interactions in ongoing ecological processes. Such conditions and interactions are likely to change through time. The past is composed of the actual prior environmental conditions affecting a system. A documented past is labelled 'history', but even when it is undocumented, the past may still have an influence on present and future states and trajectories of ecological systems. The environmental conditions that affect a system through time can be divided exhaustively into the following three categories (Pickett, 1989). The first two apply to all systems, while the third applies to systems having a clear beginning.

(1) Constraining or enabling conditions, external to the system, can affect its performance. These can be called boundary conditions.
(2) The particular order of events over time can influence the subsequent trajectory of a system.
(3) Conditions that hold at the beginning of a system, e.g. a succession or anthropogenic impact, can affect the subsequent trajectory of the system. Such conditions can be labeled 'initial conditions'. Initial conditions apply to systems that have a demarcated start.

The different ways that conditions can affect a system over time can be summarized diagrammatically (Figure 5.1). To emphasize that the effects of a past environment may not be obvious in a study of conditions that are part of the current environment of a system, the concept of 'echoes of the past' serves as a useful flag (Garcia Novo, personal communication).

The goal of LTS is to document changing environmental influences and system states before they become lost to the historical record. The following four kinds of phenomena are profitably addressed by LTS.

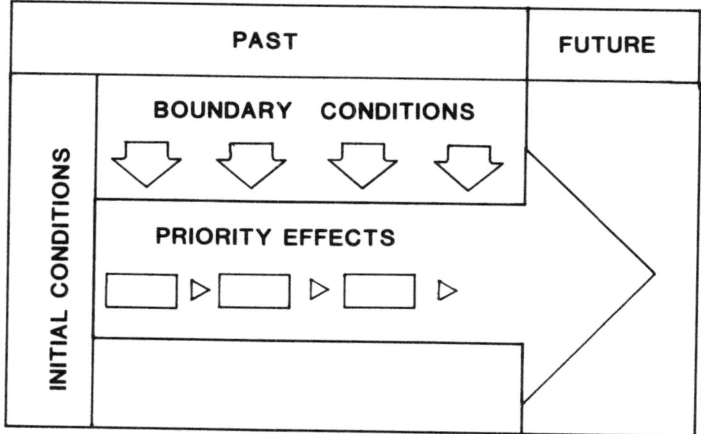

Figure 5.1 Schematic representation of the kinds of influences on an ecological process or system through time. Initial and boundary conditions and priority effects are defined in the text. Initial and boundary conditions are separated in cases where a process has a demarcated start. The large horizontal arrow represents the course of a system through time, and the small boxes enclosed represent specific system states whose order can influence the outcome of the entire process. The vertical line separating the head from the body of the arrow represents the present, with the past and future to the left and right respectively. Long-term studies can capture the environmental conditions and system states determining the trajectory of a process before they are lost to an unrecorded past

5.2.1 SLOW PHENOMENA

Slow phenomena are those that take a long time to occur. The practical definition of long-term will be persistence beyond the usual limits of funding cycles, completion of a graduate degree, or the length of time 'hot' ideas remain fashionable. LTS are defined as those lasting more than 5 years. Community and ecosystem succession is the paradigm of a slow process. Space-for-time substitution, or chronosequence, is the commonest tactic for studying succession. However, chronosequences have often proven to be misleading, and long-term studies were recognized early to be necessary for the understanding of succession (Clements, 1916). In salt marsh succession, permanent plots and longer-term sediment cores have documented that the expected autogenic and deterministic patterns were not found (Clark, 1986; Niering, 1987). On stripmined sites in Pennsylvania, the long-accepted chronosequence was found to reflect, in part, different initial conditions. Sites used to construct the chronosequence were found to differ in initial acidity of the soil as a result of the recent mining of deeper, less weathered strata (R.S. Hedin, in Pickett, 1989).

In spite of the failings of chronosequences, their value is clear if certain limits are kept in mind. In the case of oldfield succession, space-for-time substitution has documented general trends. For instance, trends in life history types, the order of dominant species, convenient 'stages', and regional differences have emerged from chronosequences. Many of these insights could just as well have come from LTS, of course. However, the understanding that has emerged from the few LTS of oldfield succession is of a different sort. LTS are beginning to expose the nature of transitions, the role of year-specific conditions, the problems with end points, and the role of newly invading exotic species in succession (Pickett, 1989). Too few cases of yearly variation have been studied at this time to produce empirical principles to understand them.

Functional parameters (e.g. nutrient availability, plant–animal interaction) have been conspicuously underrepresented in studies employing space-for-time substitution (Vitousek, 1977; Thorne and Hamburg, 1985). Chronosequences have been most often employed to assess structural or compositional aspects of ecological systems. Perhaps functional parameters are difficult to assess other than by LTS. The role of debris dams and organic storage in streams is a novel concern that has emerged only as long-term records of disturbance and sediment trapping have accumulated. The cascading effects of trophic interactions on productivity in lakes (Carpenter *et al.*, 1985; Carpenter, 1988; Mills and Forney, 1988) have appeared in LTS. The mechanism has been elucidated using models and retrospective studies. Other indirect effects (Pace and Cole, 1989) have appeared as a result of LTS. The persistent effect of salt marsh wrack, potentially including biogeochemistry (Valiela, 1989), is another example. Because indirect effects often involve organisms with different longevities, long-lasting ecosystem components, or persistent site conditions, LTS have been important in their elucidation. A potentially important, although neglected, area is the long-term changes in the genetics of populations. Such changes may well affect other ecological processes (Pace and Cole, 1989). The reliable aging of many long-lived animals requires LTS (Taylor, 1989; McAninch and Strayer, 1989). Here the slow dynamics of cohorts is the limiting factor.

Transient phenomena can be missed or misinterpreted by short-term studies. Tilman (1989) found that 70% of ecological experiments last less than 1 year. Such limited duration might miss important results, or worse, cause the results to be misinterpreted (Carpenter, 1988). Tilman (1989) discusses the results of Toumey and Keinholz's (1931) famous trenching experiments in forest understory with a study of the plots 21 years later (Lutz, 1945). While the basic conclusion of the importance of root competition held, the role of light and the 'winning' species had to be re-evaluated. Transient effects are likely to appear in many systems as they equilibrate with new experimentally imposed environmental regimes.

Interactions between two species or components of a system that are mediated by a third party or component are called 'indirect effects'. Indirect effects are those incidental to a direct ecological interaction. Such indirect effects might not appear in the short term. For example, herbivores may have effects on an ecosystem other than the consumption of plants (Shachak *et al.*, 1987; Naiman, 1988). Brown *et al.* (1986) discovered unsuspected indirect interactions among unrelated taxa in the seed-based food web of Arizona deserts using experimental LTS. It took approximately a decade to expose the indirect effects (Brown *et al.*, 1986) because they are slow to resonate through complex trophic webs. As more information becomes available and indirect effects are more widely appreciated, perhaps they will be easier to address in short-term studies.

A more obvious problem than slow and transient phenomena in the short duration of most ecological experiments is the variation in outcome resulting from environmental variability. Experiments in different years often yield different results (Tilman, 1989). Differences in initial conditions which prevail when the experiment is established and changing boundary conditions during the experiment (Figure 5.1) (Pickett, 1989) can be important factors. Some variability may be accommodated by repeating studies at different times if no persistent priority effects (Paine, 1987) operate in the system. It is clear from the specific cases presented, and the classification of kinds of histories and indirect effects that might impinge on ecological systems, that LTS are the only sure way to determine slow processes.

5.2.2 RARE EVENTS

Rare events are the second major type of process amenable to LTS. Such events are those that occur infrequently, and may be either periodic and predictable or unpredictable. An extreme type of rare event is one that is unique, that is, unprecedented and unrepeated. If one is lucky, rare events that are predictable to some degree can be captured in extensive spatial networks of short-term studies. Phenomena such as periodic recruitment, periodic mortality, or small gap disturbance are amenable to extensive short-term sampling (Franklin, 1989).

Unique events, such as the invasion of an exotic species or a disease, are only available to direct long-term study (Taylor, 1989). The anthropogenic displacement of ecological systems to new states is a unique event. It is not likely that concepts of resistance and resilience will be completely successful in anticipating unique events, although measures of resistance and resilience may help to explain the response of different systems to extreme or unique events. Unfortunately, only LTS can capture unique events.

Other forms of rare events have been uniquely illustrated by LTS. Description of disturbance regimes provides an example of the relationship of LTS to rare events. The documentation of single- or few-treefall gaps by extensive short-term surveys has been successful (Runkle, 1982), and assuming equilibrium, the small-gap aspect of the disturbance regime can be characterized. However, testing the assumption of equilibrium in the disturbance regime, and determination of the spatial and temporal distribution of large gaps or catastrophes, has yielded only to direct LTS (Falinski, 1977; Runkle and Yetter, 1987).

Issues concerning which species fill gaps, and whether advanced regeneration of new invaders, sprouts, or seedlings succeed in different kinds of gaps, have been addressed definitively by LTS, but only hypothetically by short-term studies. The question of whether entire landscapes are in equilibrium under the influence of large-patch disturbances is unanswerable by chronosequence. One may assume equilibrium in a sufficiently large area, but the assumption must be tested. If catastrophic or large disturbances are clustered in time or space, only direct LTS or, in some cases, historical or paleoecological reconstruction is productive (Romme, 1982). The question of whether (or what) ecological systems are in equilibrium is a crucial one (Remmert, 1985; Shugart, 1989), and the prevalence of climate change over centuries and millennia (Davis, 1983) casts doubt on the efficacy of short-term alternatives to LTS to answer the question.

5.2.3 SUBTLE PROCESSES

Subtle processes are embedded in a variable matrix and, therefore, cannot be extracted without a long record (Strayer *et al.*, 1986). While an overall or net trend may exist, the temporal variance will obscure the trend in a short-term study. The long-term record of CO_2 in Hawaii is a case in point. Systems or processes strongly influenced by climate are likely to exhibit subtle behavior (Franklin, 1989). For instance, what constitutes a normal deviation in a successional trajectory? What is a catastrophic yearly decline in fish stocks in a lake? Or, relevant to public concern over the drought of 1988 in North America, what differentiates a normal extreme year from an entirely new trajectory? If such questions about climate are troubling, how much worse is the uncertainty over normal (but periodic) tree mortality as opposed to irretrievable forest dieback? As in the case of slow processes, a sound underlying mechanistic understanding of the process can help extract trends in subtle processes from the temporal variance. However, LTS are still likely to be critical in generating or testing such understanding. The ability to discriminate normal from unusual ecological events, as meteorologists do, would be a powerful tool for management (Pace and Cole, 1989).

5.2.4 COMPLEX PROCESSES

Complex processes are those that have multiple causes. How can causality be evaluated in large systems or areas that cannot be manipulated? This question can be answered by observing the system over a time long enough to encompass periods when different causes dominate its structure or function. For systems that are replicated adequately in space, a comparative approach may answer the question of control by different causes. Population regulation is perhaps the paradigm of a multivariate problem. Populations can be controlled by production, predation and herbivory, competition, or dispersal, among other factors. The abiding controversies about population regulation (Andrewartha and Birch, 1984) attest to the complexity of the situation. Taylor (1989) notes that long runs of animal population data have been very helpful in disentangling the roles of various controlling factors. The interaction between wolves and moose on Isle Royale in Lake Superior is a sterling example. Likewise, the widespread declines or extirpations of migrant birds from their breeding ranges has been exposed by LTS, the spatial extent of the samples being an important feature of the data (Leck *et al.*, 1981). The reliable aging of many long-lived animals would also be impossible without LTS (Taylor, 1989; McAninch and Strayer, 1989). It is of interest to note that many such studies of animal populations were begun out of curiosity rather than any theoretical motivations (Taylor, 1989).

5.3 CORRELATES OF SUCCESS

Strayer *et al.* (1986) extracted two clear administrative correlates of success from their survey of LTS. First, all successful LTS were associated with at least one dedicated leader who apparently felt personally responsible for the project. Second, the successful studies were simple and accommodating in design, which meant they were relatively easy to run, their output could be adapted to various uses, and ancillary studies could be associated with them.

Long-term studies also exhibit a variety of other features. However, not all of these are associated with all successful studies, nor were they mentioned as critical features by all the scientists who were contacted. These secondary features include experimentation as a part of the LTS, clearly defined objectives, protected sites, archiving of samples for future analysis, short-term justification by published productivity or societal import, and synthesis and modeling during the course of the study.

The third aspect of success, identification of particular ecological processes successfully addressed by LTS, is a difficult matter. In order to determine what processes are most amenable to LTS, the universe of comparison would have to include failed studies. However, only the successful studies are available for consideration, and these cover a broad range which includes population dynamics, community dynamics, and ecosystem processes. The survey conducted by Strayer *et al.* (1986) indicates that substantial effort has been expended on LTS in all three of these broad areas, and examples of unexpected and important results can be cited for each area (Strayer *et al.*, 1986; Likens, 1989). Specific LTS in each of the broad areas usually address, in order, population regulation, patterns and causes of community dynamics, and regulation of nutrient flow. Emerging interests in LTS include third-party mediation of interactions (i.e. indirect effects), herbivory, and disturbance. Notably, the study of several levels of organization and potentially interacting phenomena at the same site are new trends in LTS that are quite promising. Many insights have emerged from individual studies conceived and conducted by one or a few ecologists, and no productive and fundamental area of ecology has failed to benefit from LTS.

5.4 MOTIVATIONS OF LONG-TERM STUDIES

One of the mandates of the workshop which generated this book was a rigorous examination of the motivations of LTS. The question that must be asked is, 'Are LTS unambiguously necessary on scientific grounds?' This question might be answered unequivocally by falsifying the conditions which are recognized as conducive to LTS. The strategy of attempting to falsify propositions about LTS was proposed by the meeting organizers. In attempting such a strategy, I have accepted the classification by Strayer *et al.* (1986) of phenomena appropriate to LTS, as it covers most of the six conditions for LTS used in this workshop, and as it has been used by others (Franklin, 1989). Obviously, ecologists have been motivated to perform LTS by the notion that the phenomenon of interest falls into one of the categories of slow, rare, subtle, or complex. However, there are three other ways to conceive a motivation for LTS, and these are theoretical, empirical, and political motivations.

5.4.1 THEORETICAL MOTIVATIONS

In a discussion of the theories that have demanded LTS, succession is a topic that is proposed as having clear theoretical motivation (Odum, 1969; Jackson, 1981; Pickett *et al.*, 1987). Long-term studies have been instrumental in rejecting tenets of the earlier versions of succession theory and in refining the theory. Some specific, theoretically motivated questions have addressed the dominance of facilitation as a mode of species turnover, the existence of equilibrium communities terminating succession (Remmert, 1985), and the role of seed rain versus seed banks. Indeed, the approximately 30-year-old Buell Succession Study was motivated in part by the desire to test Egler's (1954) idea of initial floristic composition (H. Buell, personal communication). The classic experiments at Rothamsted were motivated by a controversy over fertilization (Johnston, 1989) that may be considered theoretical in a broad sense. LTS on population regulation have been stimulated by the persistent debate over density-dependent versus density-independent population regulation (Taylor, 1989). The 1987 Cary Conference did not identify other theories that stimulated initiation of LTS. Perhaps this is because most ecological theories are explicitly equilibrial (Valiela, 1989) and do not address slow processes or lags, indirect effects, or rare events.

5.4.2 EMPIRICAL MOTIVATIONS

The strongest empirical motivation for LTS is the accumulating experience with rare events or continuous change (Pickett and White, 1985; Weatherhead, 1986). Such experience is contrary to much classical thinking in ecology (Valiela, 1989), and has helped to shape a revolution in the discipline. One of the most impressive bodies of data demonstrating the ubiquity and magnitude of long-term dynamics in ecology is the paleoecological record. Species have migrated individualistically since deglaciation (Davis, 1983; Jacobson *et al.*, 1987), and climatic fluctuation has been common for even longer intervals (COHMAP Members, 1988). The various experiments at Rothamsted, including both the classic and modern LTS, illustrate the empirical value of LTS in general. The original goals of several of the Rothamsted experiments have long since been fulfilled (Taylor, 1989; Johnston, 1989); however, they continue to be of value as demonstrations and as sources of continuing insight into agricultural practice (Johnston, 1989). Unexpected environmental factors were found to be important during the course of the studies. Such insight is purely empirical, since it was not anticipated by theory or *a priori* hypothesis.

The unexpected empirical discovery of major anthropogenic changes is so striking that it continues to motivate ecologists and governments (Franklin, 1989) to pursue LTS. The long-term weather and stream gauging records are clear cases. Networks must be established or improved to capture anthropogenic change and to relate it explicitly to potential change in ecological interactions and systems. Networks should include both modified and still unmodified systems, for even in modified systems synergisms and new stresses may appear, and systems as yet unmodified are unlikely to remain so as population pressures grow and global climate change proceeds. The strength of the empirical motivation of LTS is reflected in the finding of Strayer *et al.* (1986) that no LTS has been abandoned voluntarily by any of the approximately 100 scientists they surveyed.

5.4.3 POLITICAL MOTIVATION

The need to convince policy or decision makers that some environmental changes such as abatement, mitigation, and prevention are worth attention is a potentially powerful justification of LTS. This motive is, however, difficult for an ecologist to evaluate, and political scientists or historians are perhaps in a better position to determine whether data from LTS have in fact motivated legislative or executive decisions. The reaction of the Reagan administration against the 'let burn' policy of the US Park Service, as a result of the large number of intense fires in the severe drought year of 1988, is an indication that factors other than ecological knowledge (some of which is based on LTS) may interfere with evaluation of any sort of ecological study (Findlay and Jones, 1989).

5.4.4 EXISTING LONG-TERM STUDIES AS A DATA BASE

The desire to evaluate the motivations and conditions under which LTS are unquestionably justified assumes that the data base on LTS is unbiased. Are the LTS surveyed by Strayer *et al.* (1986) and the contributors to Likens (1989) indeed an unbiased sample? Possibly not. Failed LTS have not been adequately sampled (Strayer *et al.*, 1986). More fundamentally, in spite of any perceived or documented values of LTS, concerns with career 'fitness' and the nature of funding schedules may have limited the number and nature of LTS in the past. For instance, Strayer *et al.* (1986) note that most LTS are less than 20 years old, and their number has been increasing since the late 1960s.

A second major problem in using the existing body of LTS to guide design of new studies is the assumption that the strategies that worked in the past are valid now and for the future. The social and political environment for research may have changed. The correlation of simplicity and a dedicated single investigator with existing studies may reflect the limits of funding and the past reward system of the ecological community as much as desirable design features. Funding and scientific recognition of LTS, and appreciation of multi-investigator projects, have all increased. Furthermore, society may recognize the benefits of ecological LTS and decide to institutionalize them. Under such circumstances, the attributes of simplicity and a single dedicated leader may not be required for success. While the formulas for success based on existing LTS clearly embody much wisdom, they should not be followed blindly. Rather, the extent to which features of existing LTS reflect a vanished research environment, and the environment that new studies will be likely to encounter, must be evaluated.

5.5 THE 'NATURAL HISTORY' OF LONG-TERM STUDIES

The value of LTS seems unarguable when based on the sources I have abstracted. The phenomena and motivations listed earlier are all represented among the ranks of successful LTS. Important insights have resulted from LTS where short-term studies had been equivocal, silent, or even wrong (Pickett, 1989; Tilman, 1989; Taylor, 1989).

The existing body of LTS seems to have much in common with the products of biological evolution. Its products are diverse, the features of specific studies are contingent upon the situation from which they arose, and their structure is opportunistic. Diversity of LTS is illustrated by the fact that some are manipulative, while others are observational; some are focused on populations, while others focus on ecosystems; and some document patterns, while others examine mechanisms. Long-term studies are contingent upon context in that some are performed by 'lone wolves' on shoe-string budgets, measuring a modest number of parameters, while others are multi-investigator and multi-site studies which have cost 'megabucks'. Long-term studies are opportunistic in that they take advantage of various combinations of research strategies, including retrospective studies, short-term experiments, and modeling; they may also incorporate historical data and data from other disciplines when it is appropriate and profitable.

This richness of long-term studies was the most striking feature of the survey conducted by Strayer *et al.* (1986) and it was confirmed by the analyses at the 1987 Cary Conference (Likens, 1989). Indeed, it would appear that the past success of LTS as a class has something to do with its richness and flexibility. Some philosophers and historians argue that the success of science as a whole results from similar flexibility and opportunism (Cohen, 1985; Lloyd, 1987). The flexibility of LTS permits new concerns to be addressed, and the simplicity enhances comparability in time and space. But it is important to recognize the evolving needs for LTS. Geographical breadth and comparability are important features which will be required in the future. A network of planned LTS may differ in important ways from existing LTS, most of which (Strayer *et al.*, 1986) were not originally intended to be long term. For this reason, while still retaining the benefits of the evolved nature of most extant LTS, networks of LTS may require a more systematic approach. It is possible that the opportunistic, *ad hoc* nature of most existing LTS might, for example, hinder cross-site comparison.

5.6 RECOMMENDATIONS

The participants at the 1987 Cary Conference met in various discussion groups to evaluate LTS and make recommendations for planning and executing future LTS. The recommendations are abstracted here, with citations of the authors of the meeting group reports. The number of participants in the discussions was large, and attempts were made to reach consensus or to note the more important alternatives. The names of all contributors to each discussion are found in the papers cited below (Likens, 1989).

5.6.1 CONCEPTUAL MODELS

Begin with a conceptual model. Although many studies surveyed by Strayer *et al.* (1986) violated this requirement, some of the most productive and ultimately largest LTS (e.g. those at Hubbard Brook and Coweeta Hydrologic Laboratory) had a motivating conceptual model (Caraco and Lovett, 1989). Eaton and McDonnell (1989) suggested that any experimental LTS must have a conceptual model. I emphasize 'conceptual model' rather than simple hypothesis because it is important to spell out the assumptions and logical claims of argument in order to evaluate the tests of hypotheses. If it is impossible to construct a tentative conceptual model, then experimentation is premature. In considering the establishment of a network of LTS, it would seem even more critical to have a conceptual model. Such a model will guide the selection of sites and will suggest the manipulations to be performed (if it suggests any at all) and the parameters to be measured. A model will also assist in determining what pre-treatment parameters should be measured (Berkowitz *et al.*, 1989). Some situations may suggest alternative models; the rationale for specifying them is the same as that outlined above.

5.6.2 DATA HANDLING

One striking problem with LTS is the overwhelming body of data that can be accumulated. There is already a wealth of relatively unexploited data from existing LTS. Franklin (1989) suggested that resources for analysis and synthesis were needed. At the very least, more funding should be made available to permit productive use of existing, high-quality data sets. Even where there may be some problems with the data from existing studies, timely analysis may be the best way to learn what those problems are and to reduce the likelihood of repeating or perpetuating the errors. Most of the LTS on the list accumulated via questionnaire in the *Permanent Plotter Newsletter* (G.G. Parker, unpublished manuscript) have yet to produce available output. Without timely analysis of the potentially rich resource of existing long-term data, we may be reinventing wheels.

5.6.3 APPLICATION TO EXPERIMENTS

Long-term work is required even in field experimental work (Tilman, 1989). To avoid the conclusion that transient dynamics are the ultimate outcome of an experiment, they must be followed for longer periods; to this end, ecological research that is not necessarily designed to address one of the traditional areas of concern of LTS (e.g. succession, population dynamics, stress effects) should be conducted for longer periods than is now common. This may require protection and commitment of sites, and funding for longer periods than was common in the past. To summarize this point, ecology as a whole can benefit from extending its temporal scope.

5.6.4 SPATIAL EXTENT

A broader spatial extent is required in LTS as it is in all ecological research. A high degree of generality of some patterns and phenomena is now emerging, for example, for acid rain and natural disturbance. However, most ecological research, whether short term, long term, observational, or experimental, is conducted at single sites.

The lack of comparison saps the ability to evaluate the degree of generality of the results (Bradshaw, 1987). The power of the large networks of dated paleoecological cores is a convincing argument for comparison among LTS (COHMAP Members, 1988). National parks would make ideal components for such networks (Parsons, 1989).

5.6.5 QUESTIONS MANDATING LONG-TERM STUDIES

Discussion at the 1987 Cary Conference attempted to identify classes of questions that absolutely require LTS. Two conditions were suggested as logically demanding LTS (Pace and Cole, 1989): unanticipated changes, and situations where no surrogate method is available. Transient dynamics or indirect effects are likely to be common among unanticipated changes. Unexpected changes that catastrophically shift systems to new domains are also possible, and are important motivations for LTS (Garcia Novo, 1977). Because such catastrophic shifts have been neglected in ecology, there is neither theory nor empirical generalization to guide the search for these phenomena. Theory and generalization may better develop to incorporate transient and indirect effects, but some effects are simply unprecedented and cannot be anticipated. Unanticipated changes are common among the catalog of anthropogenic effects, and are often startlingly novel. Witness the question of ozone depletion in the stratosphere. The most widespread and significant changes now facing ecology and society are anthropogenic ones. The spread of the megalopolis throughout the world suggests a strategy for establishing networks to capture the changes that are almost certain to result in natural populations, biotic communities, ecosystems, and landscapes. Both terrestrial and aquatic realms are involved, and networks will be needed to answer questions about unexpected anthropogenic changes in all ecological systems. A hub and spoke design to assess gradients of anthropogenic effects on forests centered around a major city is presently being planned (McDonnell and Roundtree, unpublished manuscript).

Changes in global climate are the other major class of environmental changes that are likely to be unanticipated. The magnitude and significance of such changes are only now beginning to be appreciated. The role of LTS in documenting such changes and the ecological responses to them is clear (Pace and Cole, 1989). LTS are also required to validate mechanistic and predictive models of global change and ecological results. LTS are mandated when no surrogate method is available (Pace and Cole, 1989). Various contributors to the 1987 Cary Conference (Likens, 1989) evaluated surrogate methods. The evaluations suggest that chronosequences are not uniformly good substitutes for LTS (Pickett, 1989), and retrospective studies cannot forecast unforeseen future changes, although they can give necessary perspective to the type and distribution of rare or slow events in the past (Davis, 1989). Modeling can give insight into suspected trends that may result from future periodic or unique events (Shugart, 1989), but not all societally impactful changes are likely to be suspected by ecologists. Likewise, the applicability of microcosms may be limited by problems of scaling (Shugart, 1989). It would seem unwise to rely upon surrogate methods entirely in the future, since they have not compensated completely for LTS in the past.

Having considered the circumstances in which LTS are called for, several broad questions that require LTS can be presented. Although the number of specific questions that might arise in various specialties is too large to list here, it is possible to indicate the kind of pure ecological question that might be most profitably addressed through LTS. Such questions will deal, in all likelihood, with the degree to which various ecological systems are organized and the role of internal versus external organizing influences. This issue is stated here in an admittedly abstract form, but it can be made operational in various specific ecological contexts (Pickett *et al.*, 1989). The degree of organization will be reflected in the persistence of functions and their attendant structures, in resistance to change, and in resilience after disturbance. Because the temporal dimension is implicit in each of these questions, the use of LTS will be critical.

5.6.6 METHODS

Methods were addressed by many contributors to the 1987 Cary Conference. The resulting recommendations should be especially useful in planning networks of LTS. They should apply to both large multi-investigator projects and the work of a single investigator. Samples should be archived for future analyses not now possible or considered important (Lewis *et al.*, 1984), and for calibration with new techniques that may become available (Strayer *et al.*, 1986; Caraco and Lovett, 1989). LTS should be combined with other methods (Caraco and Lovett, 1989). While direct LTS will usually be the backbone of the effort, new questions may constantly arise (Johnston, 1989). Such questions may often be about mechanisms or details of pattern, and can be profitably addressed with shorter-term experiments associated with a monitoring effort (Pace and Cole, 1989). In general, a combination of methods has been used in LTS studies addressing processes at all levels of organization (Caraco and Lovett, 1989). Variance of forecasts from LTS may be reduced if they are based on a variety of methods (Parker, 1989). Likewise, extrapolation of one LTS to other sites requires a variety of safeguards not yet adequately explored by ecologists (Berkowitz *et al.*, 1989).

Long-term studies must accommodate different scales of variability (Eaton and McDonnell, 1989). Because the scales of events likely to be encountered over a long time span probably differ greatly, the design of LTS should be capable of capturing events on different scales. Attention should be paid to sampling regimes that can be aggregated or disaggregated to fit the scale of a process or structure of interest. The problem of scale deserves increased attention, especially since unanticipated changes may involve scale shifts. For example, not even long-term succession studies have effectively accommodated the changing scales of various growth forms. The parameters measured throughout a study must be selected to be relevant over the range of phenomena expected to impinge on the system. For example, 'standard' plant community parameters and plot sizes are not designed to address non-equilibrium processes and patchiness; nor are they sensitive to architectural changes in the community. The value of LTS will be increased when they overlap paleoecological records (Davis, 1989).

These various recommendations result from the combined experience of many ecologists responsible for the design, execution, and interpretation of existing LTS. Because of the diversity of approaches, levels of focus, and questions addressed by these LTS (Strayer *et al.*, 1986), it may be that the recommendations encapsulate the vast majority of problems likely to exist in LTS. The recommendations should be of great value in avoiding problems in work that is explicitly designed to contribute over the long term.

5.7 ACKNOWLEDGMENTS

I appreciate the comments of Dave Strayer, Gene Likens, Johnny Johnston, Paul Risser, and Mark McDonnell. This paper is a contribution to the program of the Institute of Ecosystem Studies with financial support from the Mary Flagler Cary Charitable Trust.

5.8 REFERENCES

Andrewartha, H.G. and Birch, L.C. (1984). *The Ecological Web: More on the Distribution and Abundance of Animals*. University of Chicago Press, Chicago.

Berkowitz, A.R., Kolasa, J., Peters, R.H. and Pickett, S T.A.. (1989). How far in space and time can the results from a single long-term study be extrapolated? In Likens, G.E. (Ed.) *Long-Term Studies in Ecology: Approaches and Alternatives*. Springer-Verlag, New York, 192–198.

Bradshaw, A.D. (1987). Comparison—its scope and limits. *New Phytologist*, **106** (supplement), 3–21.

Caraco, N.M. and Lovett, G.M. (1989). How can the various approaches to studying long-term ecological phenomena be integrated to maximize understanding? In Likens, G.E. (Ed.) *Long-Term Studies in Ecology: Approaches and Alternatives*. Springer-Verlag, New York, 186–188.

Carpenter, S.R., Kitchell, J.F. and Hodgson, J.R. (1985). Cascading trophic interactions and lake productivity. *BioScience*, **35**, 634–639.

Carpenter, S.R. 1988. Transmission of variance through lake food webs. In Carpenter, S.R. (Ed.) *Complex Interactions in Lake Communities*. Springer-Verlag, New York, 119–135.

Christensen, N.L. and Peet, R.K. (1981). Secondary forest succession on the North Carolina Piedmont. In West, D.C., Shugart, H.H. and Botkin, D.B. (Eds) *Forest Succession: Concept and Application*. Springer-Verlag, New York, 230–245.

Clark, J.S. (1986). Late-Holocene vegetation and coastal processes at a Long Island tidal marsh. *Journal of Ecology*, **74**, 561–578.

Clements, F.E. (1916). *Plant Succession*. Carnegie Institution of Washington, Publication No. 242. Washington, DC.

Cohen, I.B. (1985). *Revolution in Science*. Harvard University Press, Cambridge, Massachusetts.

COHMAP Members (1988). Climatic changes of the last 18,000 years: observations and model simulation. *Science*, **241**, 1043–1052.

Davis, M.B. (1983). Quaternary history of deciduous forest of eastern North America and Europe. *Annals of the Missouri Botanical Garden*, **70**, 550–563.

Davis, M.B. (1989). Retrospective studies. In Likens, G.E. (Ed.) *Long-Term Studies in Ecology: Approaches and Alternatives*. Springer-Verlag, New York, 71–89.

Eaton, J.S. and McDonnell, M.J. (1989). What are the difficulties in establishing and interpreting the results from a long-term manipulation? In Likens, G.E. (Ed.) *Long-Term Studies in Ecology: Approaches and Alternatives*. Springer-Verlag, New York, 189–191.

Egler, F.E. (1954). Vegetation science concepts. I. Initial floristic composition, a factor in old field vegetational development. *Vegetatio*, **4**, 412–417.

Elliott, J.M. (1985). The choice of a stock-recruitment model for migratory trout, *Salmo trutta*, in an English Lake District stream. *Arch. Hydrobiol.*, **104**, 145–168.

Falinski, J.B. (1977). Research conducted in vegetation and plant population dynamics conducted in Bialowieza Geobotanical Station of Warsaw University in the Bielowieza Primaeval Forest. *Phytocoenosis*, **6**, 5–148.

Findlay, S.E.G. and Jones, C.G. (1989). How can we improve the reception of long-term studies in ecology? In Likens, G.E. (Ed.) *Long-Term Studies in Ecology: Approaches and Alternatives*. Springer-Verlag, New York, 201–202.

Franklin, J.F. (1989). Importance and justification of long-term studies in ecology. In Likens, G.E. (Ed.) *Long-Term Studies in Ecology: Approaches and Alternatives*. Springer-Verlag, New York, 3–19.

Garcia Novo, F. (1977). The ecology of vegetation of the dunes in Doñana National Park (South-west Spain). In Jefferies, R.L. and Davy, A.J. (Eds) *Ecological Processes in Coastal Environments*. Blackwell, Oxford, 571–592.

Hamburg, S.P. and Cogbill, C.V. (1988). Historical decline of red spruce populations and climatic warming. *Nature*, **331**, 428–431.

Jackson, J.B.C. (1981). Interspecific competition and species distribution: the ghosts of theories and data past. *American Zoologist*, **21**, 889–901.

Jacobson, G.L., Webb, III, T. and Grimm, E.C. (1987). Patterns and rates of vegetation change during the deglaciation of eastern North America. In Ruddiman, W.F. and Wright, E.H. (Eds) *North America and Adjacent Oceans During the Last Deglaciation*. Geological Society of America, Boulder, Colorado, 277–288.

Johnston, A.E. (1989). The value of long-term experiments—a personal view. In Likens, G.E. (Ed.) *Long-Term Studies in Ecology: Approaches and Alternatives*. Springer-Verlag, New York, 175–179.

Leck, C.F., Murray, B.G. Jr. and Swinebroad, J (1981). Changes in breeding bird populations at Hutcheson Memorial Forest since 1958. *William L. Hutcheson Memorial Forest Bulletin*, **6**, 8–15.

Lewis, R.A., Stein, N. and Lewis, C.W. (Eds) (1984). *Environmental Specimen Banking and Monitoring as Related to Banking*. Nijhof, Boston.

Likens, G.E. (Ed.) (1989). *Long-Term Studies in Ecology: Approaches and Alternatives*. Springer-Verlag, New York.

Lloyd, E.A. (1987) Confirmation of ecological and evolutionary models. *Biology and Philosophy*, **2**, 227–293.

McAninch, J.B. and Strayer, D.L. (1989). What are the tradeoffs between immediacy of management needs and the longer process of scientific discovery? In Likens, G.E. (Ed.) *Long-Term Studies in Ecology: Approaches and Alternatives*. Springer-Verlag, New York, 203–205.

Mills, E.L. and Forney, J.L. (1988). Trophic dynamics and development of freshwater pelagic foodwebs. In Carpenter, S.R. (Ed.) *Complex Interactions in Lake Communities*. Springer-Verlag, New York, 11–30.

Naiman, R.J. (1988). Animal influences on ecosystem dynamics. *BioScience*, **38**, 750–752.
Niering, W.A. (1987). Vegetation dynamics (succession and climax) in relation to plant community management. *Conservation Biology*, **4**, 287–295.
Odum, E.P. (1969). The strategy of ecosystem development. *Science*, **164**, 262–270.
Pace, M.L. and Cole, J.J. (1989). What questions, systems, or phenomena warrant long-term ecological study? In Likens, G.E. (Ed.) *Long-Term Studies in Ecology: Approaches and Alternatives*. Springer-Verlag, New York, 183–185.
Paine, R.T. (1987). Controlled manipulations in the marine intertidal zone, and their contributions to ecological theory. Special Publication of the Academy of Natural Sciences of Philadelphia, **12**, 245–270.
Parker, G.G. (1989). Are currently available statistical methods adequate for long-term studies? In Likens, G.E. (Ed.) *Long-Term Studies in Ecology: Approaches and Alternatives*. Springer-Verlag, New York, 199–200.
Parsons, D.J. (1989). Evaluating national parks as sites for long-term studies. In Likens, G.E. (Ed.) *Long-Term Studies in Ecology: Approaches and Alternatives*. Springer-Verlag, New York, 171–173.
Pickett, S.T.A. (1989). Space-for-time substitution as an alternative to long-term studies. In Likens, G.E. (Ed.) *Long-Term Studies in Ecology: Approaches and Alternatives*. Springer-Verlag, New York, 110–135.
Pickett, S.T.A., Collins, S.L. and Armesto, J.J. (1987). Models, mechanisms and pathways of succession. *Botanical Review*, **53**, 335–371.
Pickett, S.T.A. and White, P.S. (Ed.) (1985). *The Ecology of Natural Disturbance and Patch Dynamics*. Academic Press, Orlando.
Pickett, S.T.A., Kolasa, J. Armesto, J.J. and Collins, S.L. (1989). The ecological concept of disturbance and its expression at various hierarchical levels. *Oikos*, **54**, 129–136.
Remmert, H. (1985). Was geschiet im Klimax-Stadium? *Naturwissenschaften*, **72**, 505–512.
Romme, W.H. (1982). Fire and landscape diversity in subalpine forests of Yellowstone National Park. *Ecol. Monogr.*, **52**, 199–221.
Runkle, J.R. (1982). Patterns of disturbance in some old-growth, mesic forests of eastern North America. *Ecology*, **63**, 1533–1546.
Runkle, J.R. and Yetter, T.C. (1987). Treefalls revisited: gap dynamics in the southern Appalachian Mountains. *Ecology*, **68**, 417–424.
Shachak, M., Jones, C.G. and Granot, Y. (1987). Herbivory in rocks and the weathering of a desert. *Science*, **236**, 1098–1099.
Shugart, H.H. (1989). The role of ecological models in long-term ecological studies. In Likens, G.E. (Ed.) *Long-Term Studies in Ecology: Approaches and Alternatives*. Springer-Verlag, New York, 90–109.
Strayer, D., Glitzenstein, J.S., Jones, C.G., Kolasa, J., Likens, G.E., McDonnell, M.J., Parker, G.G. and Pickett, S.T.A. (1986). Long-term ecological studies: an illustrated account of their design, operation, and importance to ecology. Occasional Publication of the Institute of Ecosystem Studies, Number 2. Millbrook, NY.
Taylor, L.R. (1989). Objective and experiment in long-term research. In Likens, G.E. (Ed.) *Long-Term Studies in Ecology: Approaches and Alternatives*. Springer-Verlag, New York, 20–70.
Thorne, J.F. and Hamburg, S.P. (1985). Nitrification potentials of an old-field chronosequence in Compton, New Hampshire. *Ecology*, **66**, 1333–1338.
Tilman, D. (1989). Ecological experimentation: strengths and conceptual problems. In Likens, G.E. (Ed.) *Long-Term Studies in Ecology: Approaches and Alternatives*. Springer-Verlag, New York, 136–157.
Valiela, I. (1989). Conditions and motivations for long-term ecological research: some notions from studies on salt marshes and elsewhere. In Likens, G.E. (Ed.) *Long-Term Studies in Ecology: Approaches and Alternatives*. Springer-Verlag, New York, 158–169.

Vitousek, P.M. (1977). The regulation of element concentrations in mountain streams in the northeastern United States. *Ecological Monographs*, **47**, 65–87.
Weatherhead, P.J. (1986). How unusual are unusual events? *American Naturalist*, **128**, 150–154.

6 Benefits from Long-term Ecosystem Research: Some Examples from Rothamsted

A. E. JOHNSTON
Soils Department, AFRC Institute of Arable Crops Research, Rothamsted Experimental Station, Harpenden, Herts AL5 2JQ, UK

6.1 INTRODUCTION

In many parts of the world terrestrial ecosystems are more or less man-managed, and these management inputs interact with climatic change, the physico-chemical properties of the soil, and various other inputs. In the absence of enlightened

Long-term Ecological Research. Edited by Paul G. Risser

management, these other inputs can have large effects on soils, plants, and animals; even to the extent of affecting the soil microbial biomass, which acts as a source and sink for nutrients in many natural ecosystems. An example of such an input is acid rain, which has been the subject of much publicity in recent years, and its obviously adverse effect on sensitive ecosystems. The effects of acid rain can be mitigated, however, as the extent of soil acidification depends not only on the amount of acidifying inputs, but also upon the soil properties and the vegetative cover, both of which can be controlled to some extent by correct management. With changing land-use patterns, a changing environment, and increased pressures on the landscape, it is essential to try to assess what the effects of some of these changes may be if irreparable damage to the ecosystems is to be avoided.

Field experiments started in the 1840s to 1860s by Lawes and Gilbert at Rothamsted were designed to find which of the elements present in farmyard manure (FYM) are essential for plant growth, and in what quantity they are essential. The effects of nitrogen, phosphorus, potassium, magnesium, and sodium were tested singly and in various combinations, using simple chemical salts, and were compared with those of FYM. Eight of the original experiments still continue, some with little change and others appreciably modified. Because each of these experiments also has soils which have received no amendments, it is possible to assess the effects of man's anthropogenic activities on their fertility.

Data from these and other long-term experiments are used here to assess, where possible, effects of non-agricultural as well as agricultural inputs on the physico-chemical properties of soils and the above-ground biomass and its composition, and how these have changed with time.

6.1.1 SITES

Besides owning the Rothamsted Experimental Farm, the Lawes Agricultural Trust has a long-term tenancy on the Woburn Experimental Farm started in 1876 by the Royal Agricultural Society of England (Johnston, 1977). Rothamsted farm is about 330 ha (grid reference TL 130137). The farm lies in a 'semi-rural' location in Western Europe (Figure 6.1), about 42 km in a northerly direction from central London and within 2 km of two major trunk roads and a motorway (M1). The immediately surrounding area is primarily agricultural, although large urban and light industrial areas have been developed within 5-10 km since the 1950s. Woburn Experimental Farm is about 70 ha (grid reference SP 962360). The farm lies in a rural location about 27 km in a north-westerly direction from Rothamsted. The farm is adjacent to the Woburn Abbey Park, and the surrounding area is agricultural.

6.1.2 SOILS

The soils at Rothamsted and Woburn are silty clay loams and sandy loams respectively. In recent years average annual rainfall has been 700 and 640 mm.

Figure 6.1 Location of Rothamsted, Woburn and Saxmundham Experimental Stations

Both soils are free draining; those at Woburn have no free calcium carbonate, and to grow arable crops successfully regular dressings of chalk are needed (7.5 t/ha once every 6 years to maintain pH at about 6.5).

The Rothamsted soil has developed in clay-with flints overlying chalk. The soil is naturally acid, but because the chalk is within 1 to 2 m of the surface in places, it was the practice in the eighteenth and early nineteenth centuries to dig and spread the chalk on the surface soil at rates up to 250 t/ha. However, this was expensive, and the chalk was applied only to fields growing arable crops, where it produced soils with pHs ranging from 7 to 8 and with reserves of free calcium carbonate. Grassland fields were not chalked and so remained acid. Soil pHs given here were measured in a 1:2.5 soil:water suspension.

6.1.3 EXPERIMENTS

The experimental sites at Rothamsted are all on the same soil series and within 1-2 km of each other. The Park Grass experiment, which assessed the effects of manurial treatments on permanent grassland, started in 1856 on soil which was at pH 5.6 to 5.8. (Warren and Johnston, 1964). The Broadbalk experiment, in which winter wheat is grown each year, was started in 1843 on a soil with pH about 8 (Johnston and Garner, 1969). The Geescroft experiment started in 1847. Grain legumes were grown until 1878, the site was then fallowed for four years and clover grown from 1883 to 1885. Like Broadbalk, Geescroft was shown to be in arable cropping on an estate map of 1623 but it probably received less chalk than Broadbalk. Parts of both Broadbalk and Geescroft were fenced off in 1882 and 1886 respectively. Where the land has subsequently remained untended, it has reverted to deciduous woodland except where, on Broadbalk, one area is grazed each year by sheep, and another has the tree and shrub seedlings removed annually (Jenkinson, 1971). These experiments are now known respectively as the Broadbalk and Geescroft Wildernesses.

6.2 SOIL ACIDIFICATION

6.2.1 CHANGES IN SOIL pH DURING 100 YEARS

Johnston *et al.* (1986) recently summarized changes in soil pH for a period of more than 100 years in soil under permanent grassland on Park Grass (Figure 6.2(a)) and deciduous woodland in the Geescroft Wilderness (Figure 6.2(b)), and also assessed the relative contribution of various acidifying inputs. In 1856 the Park Grass soil had a pH of about 5.6 to 5.8. The unmanured soils acidified slightly between 1856 and 1876 and then showed a slight increase in pH from 5.4 (1876) to 5.7 (1923) because small dressings of chalk were applied between 1881 and 1896. From 1923, pH declined steadily and has now reached a predicted equilibrium value of pH 5.1 $\pm$ 0.2. The change in these soils, due to natural acidifying inputs, can be compared to that on plots receiving ammonium sulphate supplying 48, 96, or 144 kg N/ha each year. These soils have also reached a predicted equilibrium, at a pH of about 3.6. The soil given the largest amount of N had reached a pH of 3.6 by 1923 (Figure 6.2(a)); the other two soils have reached this value more slowly. Another soil receives sodium nitrate, supplying 96 kg N/ha each year, and its pH is now less acid than that of the unmanured soils (Figure 6.2(a)). One explanation could be that the added sodium, rather than calcium, was leached accompanying anions in the drainage water. A similar but smaller effect of applying sodium nitrate was observed on the sandy loam soil at Woburn (Johnston and Chater, 1975). Differences in pH between otherwise similar soils in maritime and non-maritime climates could be explained by such an effect.

In 1886, the pH of the Geescroft Wilderness soil was probably just over 7. As deciduous woodland has developed, all three 23-cm horizons down to 69 cm

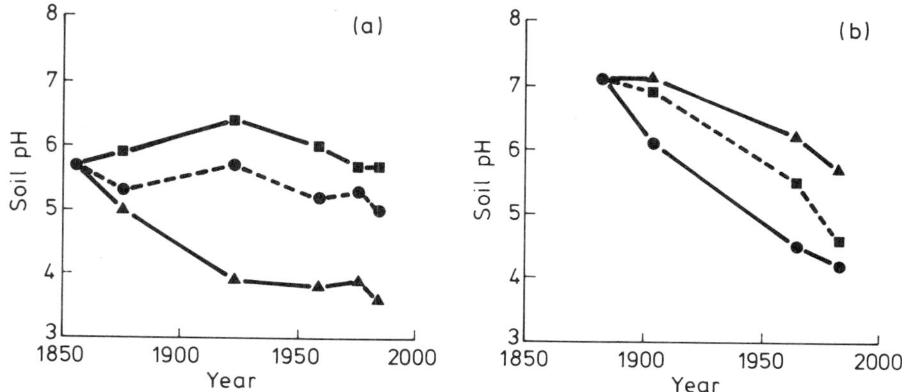

Figure 6.2 (a) The pH values of soil samples taken from the unmanured (●), ammonium N (▲) and nitrate-N (■) treated plots of the Park Grass Experiment (Goulding *et al.*, 1988); (b) the pH values of soil samples taken from the 0–23 cm (●), 23–46 cm (■) and 46–69 cm (▲) horizons of Geescroft Wilderness (Goulding *et al.*, 1988)

have acidified (Figure 6.2(b)). While there was a delay in the onset of acidification in the lower horizons, acidification of subsoils was not delayed until the surface horizon reached its present low value. Both the 0 to 23 and 23 to 46 cm horizons have acidified more than the unmanured permanent grassland soil (compare Figures 6.2(a) and 6.2(b)) and are probably still acidifying. It appears that the woodland soils will have a pH nearer to that of grassland soils given ammonium sulphate (pH 3.6) than to that of the unmanured soil (pH 5.1). Thus, the pH of soil under regenerating deciduous woodland has fallen faster than that under grassland (2.9 units in 100 years compared with 0.7 units in 125 years) and to a lower equilibrium value (pH 3.6 compared with 5.1). The difference in equilibrium pH may be due to the trees being more efficient than grass at capturing wet and dry deposition. The acidifying inputs may now be much greater than previously, and could increase further.

On the sandy loam soil at Woburn under continuous cereal cultivation, the pH of unmanured soils fell from 6.3 in 1876 to 5.3 in 1927 and then remained at this value for the next 27 years (Johnston and Chater, 1975). This equilibrium value of 5.3 is similar to that on the heavier textured soil under grassland at Rothamsted. This suggests that cereals and grassland differ little in their ability to trap aerial deposition, and more important, that the lime potential ($\text{pH} - \frac{1}{2}\text{p(Ca+Mg)}$) of the incoming rainfall had a greater effect on equilibrium soil pH than the texture of the soil (Johnston *et al.*, 1986).

6.2.2 ACIDIFYING INPUTS

Johnston *et al.* (1986) also assessed the contribution to soil acidification of (1) soil-derived natural sources of acidity, (2) atmospheric deposition, (3) nutrient uptake by crops, and (4) fertilizer additions. Nutrient uptake is only important if crops are harvested and removed, although there is some retention in the standing crop under permanent vegetation. Fertilizer inputs need only be considered when applied in agricultural systems.

Soil-derived natural sources were assumed to include dissolution of soil-derived carbonic acid, nitrification of ammonia from mineralized organic matter, and the loss by leaching of base cations with sulphate and nitrate from organic matter. Atmospheric deposition included H^+ deposition in rain, wet and dry deposition of oxides of sulphur and nitrogen, and nitrification of deposited ammonium. Wet deposition of H^+, true acid rain, was calculated to be a negligible source of acidity at Rothamsted, contributing less than 10% of total atmospheric input and less than 2% of total inputs.

Goulding *et al.* (1988) have recently estimated acidifying inputs from both wet and dry sources as keq H^+/ha/yr at Rothamsted (Table 6.1). The potential acidity from the nitrification of NH^+ in precipitation is becoming increasingly important. There are probably two main sources of this ammonia: volatilization from animal excreta, and the combustion of fossil fuels.

Table 6.1 Estimated amounts of wet and dry deposited H^+ at Rothamsted (adapted from Goulding *et al.*, 1988)

	Amounts (keq H^+/ha/yr) for the periods:		
	1860s	1920s	1980s
Input			
Wet deposited			
H^+	0.05	0.07	0.10
NH_4^+	0.4	0.4	2.0
Dry deposited			
SO_2 plus NO_x	0.6	1.2	1.8
Total	1.1	1.7	3.9

Some of the inputs of acidity have changed both in absolute amounts and relative importance during the course of both the Park Grass and Geescroft experiments (Table 6.2). The minimum H^+ input was calculated from Bolton's (1977) equation relating calcium losses to pH. The amount of soil-derived natural sources is the difference between minimum H^+ input and the sum of atmospheric deposition and nutrient uptake. Atmospheric deposition has increased, but the percentage contribution of soil-derived natural sources has decreased, probably because on acid soils the dissociation of carbonic acid is negligible, the nitrification of ammonia is very small, and there is less H^+–Ca^{2+} exchange (the calcium will be held more strongly and H^+ will be buffered by aluminum). Thus, these two Rothamsted soils under woodland and grassland have settled at different equilibrium pH values, but those for the Rothamsted (grassland) and Woburn (arable) soils are similar. This suggests that the amount and intensity of aerial inputs are more important than soil buffering ability and texture, and that vegetative cover is important in trapping

Table 6.2 Estimated amounts and relative contributions of the acidifying inputs to unmanured grassland and woodland (adapted from Goulding *et al.*, 1988). Amounts (keq H^+/ha/yr) and percentage contributions in parenthesis

	Grassland			Woodland		
	1860s	1920s	1980s	1890s	1930s	1980s
Acidifying input						
1. Minimum H^+ input[a]	6.0	7.0	5.5	14.0	9.0	5.5
2. Total deposition	1.1(18)	1.7(24)	3.9(71)	1.1(8)	1.7(19)	3.9(71)
3. Nutrient uptake	0.5(8)	0.5(7)	0.5(9)	0.1(1)	0.2(2)	0.3(5)
4. Soil-derived natural sources[b]	?(74)	?(69)	?(20)	?(91)	?(79)	?(24)

[a]Calculated from Bolton's (1977) equation relating calcium losses to pH

[b]Assuming the sum of 2, 3 and 4 equals 1

these inputs. The results also suggest that once the surface soil has reached its equilibrium value, all the atmospheric acidifying inputs are likely to be leached to acidify subsoils or to be transferred to aquatic ecosystems.

At these different equilibrium pH values, soil physico-chemical properties may well be different, soil microbial properties will be affected, and yield of above-ground biomass will differ.

6.3 EFFECT OF SOIL pH ON SOME SOIL PROPERTIES

Soil physico-chemical properties and microbial activity have considerable effects on nutrient losses from soil. In the 1870s to 1880s drainage from each of the Broadbalk winter wheat plots was collected and analyzed for the nutrients it contained. The data showed that where ammonium sulphate was applied, mainly nitrate was lost. The drainage contained large quantities of calcium but very little potassium even though this nutrient, rather than calcium, was applied as fertilizer. Very little phosphorus was lost.

6.3.1 FIXATION OF PHOSPHORUS

As soil acidity increases, phosphate is usually precipitated as iron and aluminum phosphates which are very insoluble (White, 1979). At low soil pH plants will become much more dependent on mycorrhizal associations and the cycling of phosphorus through soil organic matter. The benefit of inoculation with vesicular arbuscular mycorrhizal (VAM) fungi in the presence of rock phosphate on yields of cassava and sorghum grown on very acid, impoverished soils of Colombia has recently been demonstrated (Table 6.3). In the absence of rock phosphate, there was no yield of either cassava or sorghum. Similar benefits from combined inoculation with VAM fungi and *Rhizobium* have been reported for white clover grown on nutrient-poor, acidic soils in Wales (Table 6.4).

Table 6.3 Effect of rock phosphate and VAM inoculation (+M) on final yields of cassava and sorghum in Colombia (I. Arias, personal communication)

	Dry matter (kg/ha)		
	Cassava	Sorghum	
	Roots	Foliage	Seeds
RP−M	3600	1200	308
RP+M	4500	1450	650

Table 6.4 Effect of inoculation with mycorrhizal fungi on dry matter production from upland swards at pH 4 sod seeded with white clover[a] (adapted from Hayman, 1984)

Treatment	Dry matter, kg/ha
Control	423
Superhosphate 90 kg P_2O_5/ha	829
Mycorrhizae	631
Mycorrhizae plus superphosphate	1800[b]

[a] All clover seed inoculated with *R. trifolii* strain RCR 221

[b] Represents about 30% clover component in the first season

6.3.2 RELEASE OF CATIONS

Potassium, calcium, and magnesium can all be released from soil minerals as soils acidify. This is shown indirectly in Table 6.5, where more of the potassium added to soil remained exchangeable in 1N-ammonium acetate in acid soils than in neutral ones. Goulding and Stevens (1988) have estimated potassium reserves in acid upland soils under forest systems that were either clear-felled or whole-tree harvested. While the latter removed all nutrients, any returned under clear felling (trunks only removed) were poorly absorbed, and rapidly leached from this stagnopodzol at pH 4 with low cation exchange capacity and in a high-rainfall area. However, other data suggest that, at this particular forest at Beddgelert in North Wales, K-bearing minerals will weather quickly enough to supply the K requirements of conifers for tens of cycles of tree growth. The authors considered that other nutrients, like calcium and phosphorus, were more likely to become growth limiting.

Russell (1961) used unpublished results obtained by Schofield at Rothamsted to demonstrate that in acid soils, aluminum becomes an important exchangeable cation. Using a mildly acid (pH 5.6) and a very acid (pH 3.7) soil from Park Grass, Schofield showed that the titration curve relating soil pH and millequivalents of base taken up were identical for the very acid soil and the mildly acid soil after the latter had been pretreated with aluminum chloride. However the reverse process also held—the titration curves for the mildly acid and the very acid soils were similar after the latter had been pretreated first with dilute acid, to remove aluminum, and then with lime-saturated water to replace H^+ by Ca^{2+}.

Table 6.5 Effect of soil pH and past K manuring on the percentage of added K which remained exchangeable in soils which were alternately wet and dry for 12 weeks (adapted from Johnston, 1986)

Past K manuring	Soil pH	Exchangeable K mg/kg in unamended soil	Percentage added K remaining exchangeable
None	5 to 6	80	70
None	7 to 8	140	40
Fertilizer K	5 to 6	200	90
Fertilizer K	7 to 8	360	60

Iron and aluminum, released in soluble form as soils acidify, may be transferred into aquatic ecosystems. Such transfers may well increase if land currently maintained at near neutral pH for arable food crops is set aside into forestry. The data in Figure 6.2(b) show that such land could become very acid over a period of years. The quantities of iron and aluminum released will depend on soil texture and composition of the clay.

6.3.3 EFFECTS OF LIMING

The extent and rapidity of change in pH of field soils depends on the degree of mixing of the neutralizing material. In Britain this is often chalk (calcium carbonate) or lime (calcium hydroxide). In arable farming systems, mixing by ploughing and cultivation will often change pH throughout the plough layer within a year or so. In undisturbed soil, raising the pH to depth may take much longer. Table 6.6 shows amounts of chalk applied to Park Grass soils in attempts to change soil pH to either 5 or 6. In the presence of a mat of partially decomposed plant debris lying on the surface soil, the pH of the underlying mineral soil had not changed after six years. This was because the added calcium was held firmly on the cation exchange sites

Table 6.6 Effects of additions of calcium carbonate on soil pH (adapted from Johnston, 1972)

	pH Nov. 1959 Horizon[a]				Chalk added t/ha			pH Nov. 1971 Horizon			
	'Mat'[b]	1	2	3	1965	1967	1968	'Mat'	1	2	3
Target pH6											
	–	5.8	5.5	5.3	3.3	0	1.6	–	6.4	6.1	5.6
	5.5	4.2	4.1	4.4	12.5	6.2	6.2	6.4	4.9	4.6	4.6
	5.2	4.7	4.5	4.6	7.5	3.8	3.8	–	6.5	6.2	5.6
Target pH5											
	–	4.7	4.6	4.9	2.5	0	1.2	–	5.8	5.2	5.2
	3.8	3.6	3.8	4.0	8.7	4.4	4.4	6.4	4.5	4.1	4.3

[a] 'Mat' partially decomposed dead plant material lying on the soil surface, 1, 2, 3, 0–7.5, 7.5–15, 15–22.5 cm depth of soil

[b] No mat present

on the organic material. However, raising the pH of the mat increased biological activity and eventually the mat was decomposed. Once this had happened the calcium was leached into the mineral soil below and raised its pH.

6.4 EFFECT OF SOIL pH ON FLORA

6.4.1 EFFECT ON SOIL FLORA

Examples of complex interactions between soil nutrient status, pH, and beneficial soil organisms, like VAM fungi and *Rhizobium*, have been discussed earlier in this chapter. However, soil acidity can also affect pathogenic organisms. Mary D. Glynne surveyed fields on Rothamsted and Woburn farms for the incidence of various cereal diseases including take-all, caused by *Gaeumannomyces graminis*, a fungus attacking the stem base. She found very little take-all on soils at pH 5, but considerably more at higher pHs (Glynne, 1935).

6.4.2 EFFECT OF SOIL REACTION ON ABOVE-GROUND BIOMASS AND ITS COMPOSITION

It is well established that plants differ in their tolerance to varying soil reaction. Differences between the Broadbalk and Geescroft Wildernesses and within the Park Grass experiment are good examples because climate and soil type are identical and the effects observed must be due to differences in pH or nutrient status.

The causes and extent of acidification at Geescroft have been discussed above, while mention was made that the Broadbalk soil has been buffered at about pH 8 for nearly 200 years. The effects of the large difference in pH are clearly seen in variations in stand composition and dry weight, which are smaller on Geescroft than on Broadbalk (Table 6.7). Geescroft is almost entirely composed of oak (58%) and ash (26%), with some hawthorn (8%) and elm (5%). Broadbalk has ash (38%), hawthorn (32%), sycamore (18%), and some maple (8%). Both ash and maple are less tolerant of acid soils and the long-term decline of pH on Geescroft could affect their ability to compete and regenerate, making oak even more dominant. However, some recent unpublished evidence suggests that there may be inhibition of regeneration of oak at low pH.

The yield and species composition of the Park Grass plots strikingly demonstrate both the effects of acidity and of ameliorating that acidity with dressings of chalk since 1903. (For details of chalking see Warren and Johnston, 1964, and for species changes Thurston *et al.*, 1976.) Absolute yield levels, but not the effect of treatment, have been affected by a change in the method of harvesting from 1960 (Table 6.8). Before 1960, hay was made on the field in late June with attendant dry matter losses during haymaking. Since then herbage has been cut green, weighed, and dry matter determined immediately; yields were therefore apparently increased. On the unmanured plots, the small differences in yield between limed (pH 7.1) and

Table 6.7 Stand measurements (dry weight in t/ha) of Broadbalk and Geescroft Wilderness sites (adapted from Jenkinson, 1971)

Species	Broadbalk	Geescroft
Ash (*Fraxinus excelsior*)	104.1	46.7
Elder (*Sambucus nigra*)	0.1	1.1
Elm (*Ulmus* spp.)	0	8.4
Hawthorn (*Craetagus monogyna*)	88.6	14.5
Hazel (*Corylus avellana*)	6.1	0.5
Maple (*Acer campestre*)	21.5	1.5
Oak (*Quercus robur*)	3.6	105.1
Sycamore (*Acer pseudoplatanus*)	50.3	1.6
Silver birch (*Betula pendula*)	0	1.4
All species	274.3	180.8

Table 6.8 Effect of acidity on the yield of herbage from the Park Grass Experiment (adapted from Goulding *et al.*, 1988)

Date 1856	Yield of dry matter, t/ha, from the treatments: 2.8 over whole site				
	None	None+lime	NH_4-N	NH_4-N+lime	NO_3-N
1886–1895	2.1	–	4.9	–	5.7
1920–1959	1.5	1.6	4.7	5.6	6.2
1965–1973[a]	3.0	3.3	6.7	9.2	9.2
1974–1982[a]	2.5	3.3	5.3	7.6	7.3

[a] Yields of hay estimated from total herbage yields, i.e., no haymaking losses

Table 6.9 Effect of acidity on the botanical composition of the permanent pasture of Park Grass (adapted from Goulding *et al.*, 1988)

Date	Treatment	Soil pH	Plant species: Number	% by weight: Grasses	Legumes	Herbs
1856	–	5.7	23[a]	76	5	19
1877	None	5.3	52	71	8	21
	NH_4-N	5.0	28	95	0	5
	NO_3-N	5.9	28	88	1	11
1948/9	None	5.3	36	53	7	40
	None+lime	7.1	32	36	16	48
	NH_4-N	3.8	6	99	0	1
	NO_3-N	6.1	16	94	2	4

[a] Not known accurately, certainly an underestimate

Table 6.10 Effect of manuring and soil reaction on grass species, Park Grass, by 1947–49 (adapted from Warren and Johnson, 1964)

Plot	Soil reaction Treatment	pH 3.7-4.1		pH 4.2-6.0		pH 6.0-7.5	
1	N1	*A. tenuis*	79			*D. glomerata*	29
		F. rubra	16			*F. rubra*	24
						H. pubescens	19
17	N_1*			*D. glomerata*	36	*F. rubra*	27
				A. pratensis	20	*D. glomerata*	25
				F. rubra	13	*H. pubescens*	25
				A. odoratum	12		
				H. lanatus	12		
18	N_2 K Na Mg	*A. tenuis*	88			*D. glomerata*	50
		F. rubra	10			*A. elatius*	30
4/2	N_2P	*A. tenuis*	36	*F. rubra*	60		
		F. rubra	35	*A. pratensis*	25		
		H. lanatus	18				
		A. odoratum	10				
10	N_2 P Na Mg	*A. tenuis*	52	*F. rubra*	58		
		H. lanatus	22	*A. pratensis*	30		
		A. odoratum	10				
		F. rubra	10				
9	N_2 PK Na Mg	*H. lanatus*	91	*A. pratensis*	42		
				A. elatius	16		
				D. glomerata	13		
				P. pratensis	10		
11/1	N_3 P K Na Mg	*H. lanatus*	100	*A. pratensis*	84		
11/2	N_3 PK Na Mg Si	*H. lanatus*	93	*A. pratensis*	59		
				A. elatius	18		
				P. pratensis	11		
				D. glomerata	10		
14	N_2* P K Na Mg			*A. elatius*	39	*A.elatius*	48
				A. pratensis	34	*D. glomerata*	14
				C. cristatus	15	*C. cristatus*	14
						A. pratensis	13

D. glomerata	*Dactylis glomerata*	Cocksfoot
C. cristatus	*Cynosurus cristatus*	Crested dogstail
H. pubescens	*Heliotrotrichon pubescens*	Downy oat
A. tenuis	*Agrostis tenuis*	Fine bent
A. pratensis	*Alopecurus pratensis*	Meadow foxtail
F. rubra	*Festuca rubra*	Red fescue
P. pratensis	*Poa pratensis*	Smooth stalked meadow grass
A. odoratum	*Anthoxanthum odoratum*	Sweet vernal
A. elatius	*Arrhenatherum elatius*	Tall oat
H. lanatus	*Holcus lanatus*	Yorkshire fog

N* as sodium nitrate, all other plots receive ammonium sulphate.
N_1, N_2, N_3, 48, 96, 144 kg N/ha/yr

Each species expressed as per cent by weight of the grass fraction; species present in amounts less than 10% are omitted from the table unlimed (pH 5.1) sections suggests that lack of nutrients, mainly nitrogen, rather than acidification was the more important determinant of yield. On plots receiving ammonium sulphate (144 kg N/ha), there was a large benefit from lime; in the absence of lime, pH fell to 3.6, and yield was considerably diminished.

Soils receiving sodium nitrate had a higher pH than unmanured soils (Figure 6.2(a)), and yields were as good as those on soils receiving ammonium sulphate plus lime. The same comparisons for botanical composition (Table 6.9) suggest that there is a complex interaction between pH and nutrition (for further details see Thurston *et al.*, 1976). Supplying N greatly increased yield (Table 6.8) but also enormously decreased species diversity (Table 6.9). Species composition can be expressed as the proportion of each species as a per cent by weight of each group: grasses, legumes, or herbs. For the various grass species this is shown in Table 6.10. Not all combinations of pH and nutrient effects are available because not all exist in the current plot treatments. On soils with pH values 3.7 to 4.1 and where the manuring was deficient in phosphorus, the main grass was *Agrostis tenuis*. When P was given together with ammonium nitrogen, *Agrostis tenuis* was still the dominant species but the grass fraction of the herbage contained about 20% of *Holcus lanatus*. With complete NPK manuring, all or nearly all of the grass was *Holcus lanatus*. With the same complete manuring, but on less acid soils at pH 4.2 to 6.0, this grass was replaced chiefly by *Alopecurus pratensis* and small amounts of *Arrhenatherum elatius*. Omitting K decreased the *Alopecurus pratensis* on these soils (pH 4.2 to 6.0) and *Festuca rubra* became the dominant grass, except where only 48 kg N/ha (as sodium nitrate, Plot 17) was given, and *Dactylis glomerata* then replaced *Alopecurus pratensis*. In the third group of soils (pH 6.0 to 7.5), where P and K were omitted, *Alopecurus pratensis* was replaced by *Dactylis glomerata*, *Festuca rubra*, and *Heliotrotrichon pubescens*. With complete NPK (Plot 14), *Alopecurus pratensis* was present but there was also much *Arrhenatherum elatius* and *Dactylis glomerata*. The results showed that *Alopecurus pratensis* needs P and K in addition to N fertilizer, and that there is a critical limit of soil acidity (pH 4.2) below which it does not grow; but from these data less is known about its growth in neutral and slightly calcareous soils.

While the above results on the effects of changing pH and nutrient status on yield and competition between species are themselves interesting, recent work suggests that we may not always be dealing with the same genetic population. *Anthoxanthum odoratum* has always been one of the most widely distributed but not necessarily dominant species, occurring in both limed and unlimed, fertilized and unfertilized plots. Recently, it has been one of the most abundant species on some of the more acid soils. Its wide distribution could be due to the fact that morphologically and physiologically different populations have evolved on the various plots. Recent information, discussed by Thurston *et al.* (1976), suggests that there have been significant changes within the species. Morphological differences exist between the populations and are apparently adaptive. The differences are usually correlated with environmental conditions on the plots from which they were collected (Snaydon, 1970; Snaydon and Davies, 1972).

Populations also differ in their response to mineral nutrients; again the differences appear to be adaptive (Davies and Snaydon, 1973a,b, 1974; Davies, 1975). The adaptive nature of the differences between populations was confirmed by growing them under uniform conditions for several years and then transplanting them back into either their own or other plots. Each population survived and grew fastest on its own plot (Davies and Snaydon, 1976).

Similar studies of *Anthoxanthum odoratum* populations from subplots which were not limed until 1965 indicate that genetic changes have occurred within seven years, and that populations collected less than 1 m apart from opposite sides of plot boundaries differ both morphologically and physiologically (Snaydon and Davies, 1976). These differences have developed despite appreciable gene flow caused by pollen drift and seed dispersal. Similar differences have also been demonstrated between populations of *Lolium perenne* (Goodman, 1969), *Holcus lanatus*, and *Dactylis glomerata* (Remison, 1976).

6.5 OTHER INPUTS TO SOIL

Many industrial processes release compounds into the atmosphere in particulate or gaseous forms, and these may eventually be deposited on soil. Improvements in analytical techniques now allow us to investigate such occurrences by analyzing crops and soils from the Rothamsted archive which dates back to 1843, an excellent reason for maintaining such an archive.

6.5.1 TEMPORAL CHANGES IN CADMIUM CONTENT

Recently, soils from Park Grass and three of the Classical experiments growing arable crops have been analyzed to investigate time trends in cadmium content due either solely to atmospheric deposition or a combination of atmospheric deposition and various soil treatments (Jones *et al.*, 1987). Since the 1850s, there has been an increase in cadmium concentration in the soil plough layer of between 0.7 and 1.9 μg/kg, equivalent to an increase of 1.9 to 5.4 g Cd/ha/yr, due to atmospheric deposition. The changes in soil cadmium concentrations since 1846 at one site corresponded well to predicted increases in the plough layer cadmium burden based on assumptions about the temporal trends in atmospheric cadmium emissions (Figure 6.3).

In addition, subsamples of a selection of rock phosphates of known origin and superphosphates mainly from one supplier, collected and stored in the archive since 1925, were also analyzed for cadmium. The concentrations ranged from 3.6 to 92 (mean 36) mg Cd/kg for rock phosphates and from 3.3 to 40 (9.7) mg Cd/kg for superphosphates. On the basis of these data and their known application rates, the estimated annual input of cadmium to P-treated plots at Rothamsted was 2 g/ha. However, there was little further increase in soil cadmium due to this addition in

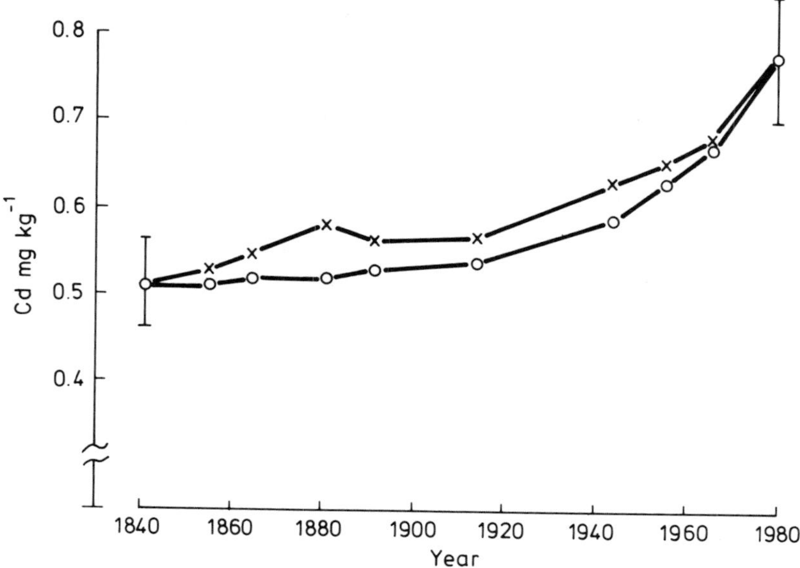

Figure 6.3 Change in soil cadmium concentration with time on the unmanured soil on Broadbalk. Measured data (X—X), predicted soil Cd concentrations (○—○). Error bars relate to analyses of eight separately digested samples for both 1846 and 1980 (Jones *et al.*, 1987)

the three arable experiments where soil pH was >6.5. On these P-treated plots, the mean increase in soil cadmium was 1.2 g/kg/yr which is equivalent to an increase in the plough layer burden of 3.1 g Cd/ha/yr. This suggests there may have been some loss by leaching and highlights the need for comprehensive balance studies. The results also raise another question, namely has cadmium in aerial inputs behaved differently from that in superphosphate? Is there a difference in speciation or are the differences due to only limited adsorption sites at the level of organic matter in these soils? By contrast, P-treated soils under permanent grassland, with a higher organic matter content and a lower pH, have increased their cadmium content by 7.2 g/ha/yr. When permanent grassland soils ranging in pH from 5 to 7 were examined, it was found that organic matter had a larger effect on cadmium concentration than pH, and the effects of pH were not consistent (Jones *et al.*, 1987).

Data in Figure 6.3 show that there has been a very large increase in soil cadmium burden since the 1940s, and other results indicate that there has been some retention of this cadmium in soils with large amounts of organic matter. Jones *et al.* (1987) also observed that farmyard manure, applied to some experimental plots at Rothamsted, appeared to have been a more significant source of cadmium than combined atmospheric and phosphate fertilizer inputs. Larger inputs may have been responsible; for example, McGrath (1984) reported values of 1.8 g Cd/t dry

FYM, but it may be that the increased organic matter content of the soil, as a result of adding farmyard manure (35 t/ha/yr) over a period of more than one hundred years, has strongly held the cadmium applied in FYM or has retained more of the aerial inputs against leaching.

Cereal grain and herbage samples in the archive have been analyzed for their cadmium content because, in part, the significance of the long-term increases in soil cadmium depend on whether increases in plant cadmium concentration are detectable (Jones and Johnston, 1989b). There was little evidence (with one exception) of a long-term increase in grain cadmium concentrations at Rothamsted. Herbage removed from Park Grass has always had a larger concentration of cadmium than cereal grain, approximately 200 and 40 μg/kg dry weight respectively. The herbage was not washed prior to analysis and its larger cadmium concentration may be due to surface contamination. However, the values are probably too large for this to be the sole reason. Wherever the cadmium is located, it will be a source of dietary cadmium to animals.

6.5.2 TEMPORAL CHANGES IN POLYNUCLEAR AROMATIC HYDROCARBONS

The changing levels of polynuclear aromatic hydrocarbons (PAHs), which are mutagenic and carcinogenic, have been investigated. The total PAH burden of the plough layer has increased approximately fivefold since the 1880s to 1890s with some compounds showing substantially greater increases (Jones *et al.*, 1989d). Average rates of increase for individual PAHs over the century since 1880 to 1890 vary between 0.01 to 0.67 (mean 0.21) mg/m^2/yr. These fluxes are similar to contemporary atmospheric deposition rates at other semi-rural locations in Britain. This fall-out of anthropogenically generated PAHs is derived from combustion of organic materials. What is of concern is the observed large increase in soil PAH content in the latter half of this century which may well be representative of other soils in the industrialized countries or regions (Figure 6.4). The similarities between the average annual rate of increase in the soil PAH burden and the likely average deposition flux at Rothamsted suggest that losses via five possible mechanisms (microbial breakdown, photo-oxidation, vaporization, crop offtake, and leaching) effectively remove only relatively small proportions of the total annual input. This implies long residence times for PAHs in soils.

In general, compounds with complex structures have increased more than those with simpler ones, suggesting that microbial breakdown and soil retention of PAHs may well depend on molecular structure. This poses the question of whether some of these more complex compounds may be toxic to the soil microbial population. Clearly there is a need to quantify both inputs to, and outputs from, sites of ecological importance if it can be shown that these materials have adverse effects on soil microbial processes.

There has been no change in the PAH concentration of wheat and barley grains at Rothamsted over many years. This suggests that either plants can exclude PAH

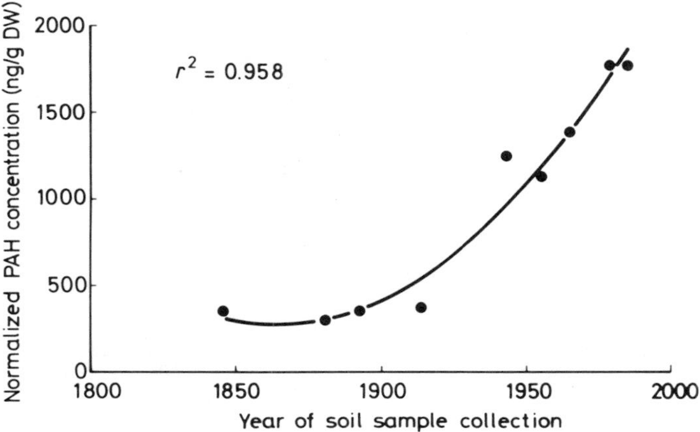

Figure 6.4 Change in soil PAH concentration with time on the unmanured soil on Broadbalk (Jones *et al.*, 1989d)

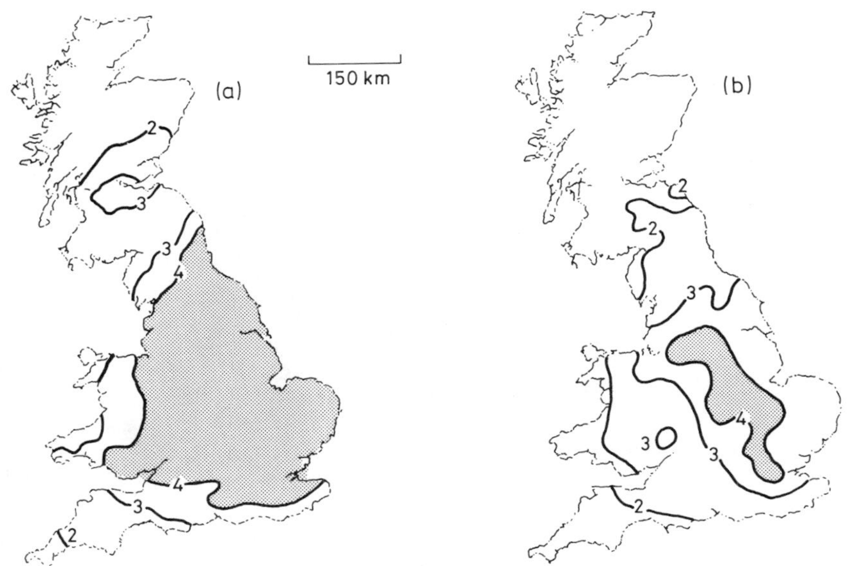

Figure 6.5 Modelled total (wet plus dry) sulphur deposition (a) for 1970 and (b) for 1983 (gSm^{-2} year $^{-1}$) (Warren Spring Laboratory, 1985)

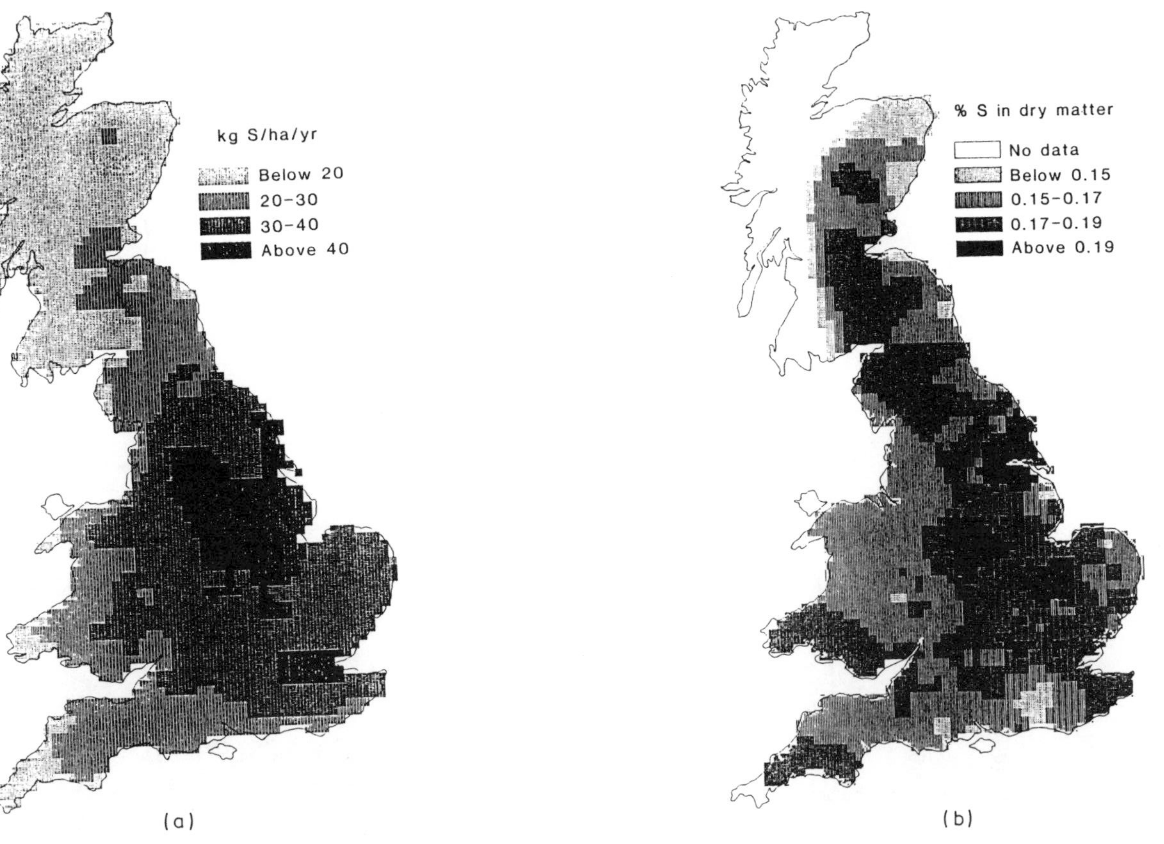

Figure 6.6 (a) Wet and dry sulphur deposition and (b) sulphur in wheat grain in England, Wales and Scotland (McGrath and Johnston, 1986)

uptake at the root surface or they do not translocate them into seeds. Varying amounts of PAHs in herbage samples from Park Grass suggest that this may be surface contamination (Jones *et al.*, 1989a). For herbivores, dietary intake may arise from these surface deposits. Soil profile samples collected from the same plots on Broadbalk in 1893, 1944, and 1987 have been analyzed for their PAH content (Jones *et al.*, 1989c). The total PAH content in the 1893 samples showed little enrichment of surface soil relative to that in the subsoil. The values were similar to those in contemporary isolated rural locations in Britain. By 1987, the surface soil had been enriched in all PAH compounds measured by a factor of between 1.3 (acenaphthalene) and over 20 (benzo[a]pyrene). Increases in the PAH content of the 23- to 46-cm sub-surface layer indicated some migration of PAHs from the plough layer. Net average annual migration rates range from 0.01 to 0.14 mg/m^2 for individual PAHs, and the rate appeared to be primarily a function of the plough layer PAH content rather than the physical/chemical properties of the individual compounds. This suggests that particle-bound translocation is the dominant mechanism for PAH migration. Such movement through soils or by erosion into rivers has implications deserving further study if these compounds affect plants or animals in aquatic ecosystems.

6.5.3 SULPHUR CYCLING

The changing patterns of man's anthropogenic activities is seen also in the cycling of sulphur. There has been outcry that sulphur emissions, from the combustion of fossil fuels, are acidifying inputs which can rapidly acidify some soils. As a result of changing technology, atmospheric deposition of sulphur to many soils in the UK has declined rapidly (Figure 6.5). At the same time, other sulphur inputs to agricultural soils have also diminished (McGrath and Johnston, 1986). Much sulphur used to be added in either single superphosphate or ammonium sulphate, two fertilizers which are now no longer generally available in the UK, and little sulphur is added in other fertilizers. Because both of these sulphur inputs have declined there is now concern about the possible need to apply sulphur to soils growing arable crops with a high sulphur demand, like oilseed rape. Perhaps more important may be the need to supply sulphur to maintain protein quality of cereal grains (Figure 6.6) for bread making and for those people dependent on grain legumes for their protein intake.

6.6 THE EFFECT ON SOILS OF CHANGING LAND USE

One response to the surplus of agricultural products within Western Europe has been the suggestion that land should be taken out of farming to revert to other use. Reference has already been made to how deciduous woodlands have regenerated at Rothamsted when arable soils are totally unmanaged and the effect on soil pH, and, in its turn, that of pH on biomass production and species composition. In addition,

Table 6.11 Mean annual gains (kg/ha/yr) in organic C, N, S and P and in inorganic S in the topsoil (0–23 cm) of Broadbalk and Geescroft Wilderness soils (adapted from Jenkinson, 1971)

		Broadbalk	Geescroft
Organic	C	530	250
	N	45	13
	S	6.9	3.7
	P	5.6	2.3
Inorganic	S	0.2	1.6

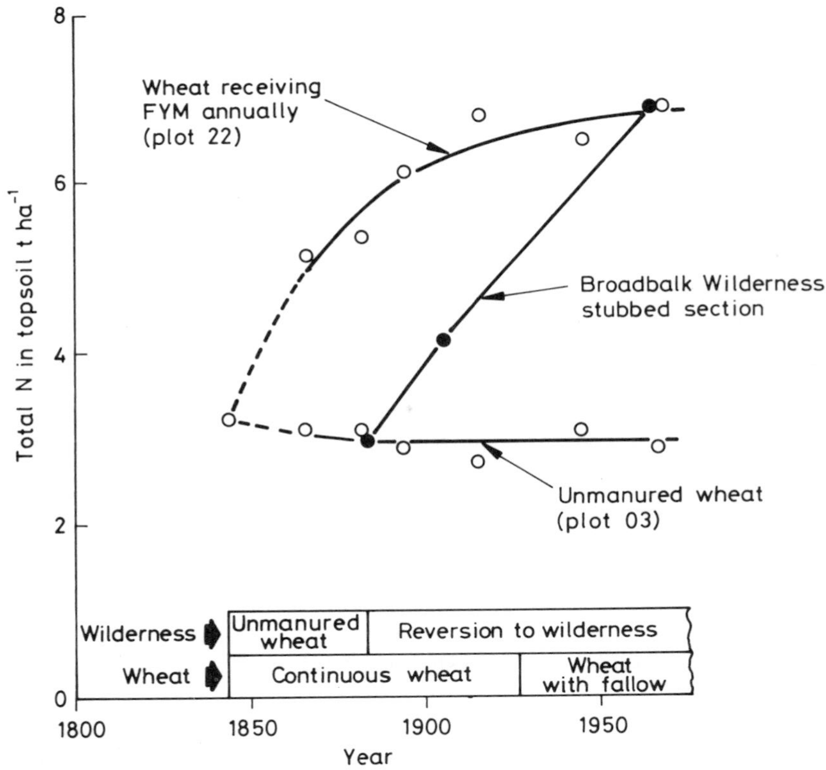

Figure 6.7 Change in the total nitrogen content of the top 23 cm of soil under wheat either unmanured or receiving 35 t/ha each year and in the section of the Broadbalk Wilderness from which tree seedlings are removed each year (Jenkinson, 1971)

the soils have accumulated large, but different, amounts of organic matter, sulphur, and phosphorus. Build-up in the Geescroft Wilderness soil, where the pH has become gradually more acid, is much less than in the Broadbalk soil (Table 6.11). This may be because inputs to the Geescroft site have been smaller than those on Broadbalk and/or decomposition has been slower on Broadbalk. However, the difference between these two sites, which are in close proximity, is so large that there is a need for a greater understanding of the processes which have enhanced accumulation of organic matter, and therefore fertility, in one soil rather than the other. The one factor which is explicable is the greater gain of inorganic sulphur in the Geescroft soil, which is due to the larger number of positive, pH-dependent absorption sites on a soil with low pH.

An interesting comparison can be made between the change in total nitrogen content in the top 23 cm of soil under the Broadbalk Wilderness (pH about 8), where tree seedlings have been removed each year, and that on the rest of the field where wheat is grown year after year (Jenkinson, 1971). Figure 6.7 shows that under continuous wheat the total nitrogen content of the unmanured soil has remained approximately constant over the last hundred years. Where FYM (35 t/ha) has been applied annually the nitrogen content has increased appreciably and is now near equilibrium. However, at the same time much nitrogen has been lost, some by crop uptake and some by leaching. Johnston *et al.* (1989) have recently calculated that about 125 kg N/ha/yr has been lost during the course of the experiment, probably as nitrate by leaching. The amount of total nitrogen which has accumulated in the Broadbalk Wilderness soil during the 100 years since 1886 is about as much as has accumulated in the soil given FYM each year. Whether the soil under woodland will have the same equilibrium organic matter content as that growing wheat and given FYM is not known, but if it is the same then there is a risk that nitrate could leach from soils once equilibrium has been reached.

6.7 CONCLUSIONS

Current funding strategies in biological research tend to favor short-term projects. However, long-term experiments are needed to quantify the effects of the many changes which affect soil fertility over the 25- to 100-year time scale. Funding agencies often disregard or do not appreciate these needs.

A good example of the short- and long-term approach is work on soil microbial biomass and organic matter at Rothamsted. Current results show that the microbial biomass fraction of soil organic matter changes proportionally more rapidly than does total soil organic matter (Powlson and Brookes, 1987). Changes in amounts of soil microbial biomass can be used to predict the direction of change in organic matter as a result of changes in management. However, they cannot be used to predict the equilibrium value at which total soil organic matter will settle. It is not until appreciable differences in total soil organic matter have been established that its effect on soil productivity can be measured (see Johnston, 1990). Another example, one of the inability of short-term observations to predict long-term changes in soil pests, was given by Johnston (1989).

This paper summarizes some further examples illustrating the importance of long-term experiments and others are given by Johnston (1990). All the examples have been taken from data accumulated at Rothamsted since 1843, Woburn since 1876, and Saxmundham since 1899. These experiments were mainly concerned with managed agricultural systems. However, some treatments allow the measurement of effects under less managed conditions. For example, we can estimate the effects of man's anthropogenic activities on soils which receive no agricultural treatment.

The results discussed here can be strictly applied only to similar soils under similar climatic conditions and farming systems. One of the most difficult problems is to assess the general applicability of the results. The danger of assuming similar effects on all soils is well illustrated above, where effects of acidifying inputs on tree growth are discussed. Johnston (1990) also showed very different effects of ammonium sulphate on spring barley yields at the Rothamsted and Woburn sites. The effects were due to differences in soil texture and initial soil pH. Results from similar long-term research would be invaluable, and, indeed, are essential to the production and validation of models to describe on a global scale some of the underlying chemical and biological processes occurring in soil.

The Rothamsted experiments have a number of important features worth careful consideration by research workers planning, and managers funding, such long-term research. The experiments are the responsibility of the Lawes Agricultural Trust Committee, which can continue in perpetuity. There is thus security of tenure of both the Rothamsted and Woburn sites which has been vital not only to the continuation of the long-term experiments but also to the realization by staff that their commitment to a long-term research program would not suddenly stop because the site was no longer available.

The scientific management of the experiments is delegated to a group of scientists rather than an individual. Each group is multidisciplinary and must refer major management changes to the Trust Committee for their approval. The multidisciplinary approach ensures that data are collected for a wide range of factors and that many possible interactions are studied.

A unique feature of the Rothamsted experiments is the archive of crop and soil samples. Samples of harvested cereal crops, grain and straw, and herbage exist for most years and treatments from the start of the experiments. Soil samples have also been taken periodically. Lawes and Gilbert sampled by 9-inch depth horizons, although in the 1840s to 1860s cultivation would not have exceeded 4 to 5 inches. This has proved to be a major benefit because plough depth has since been increased to about 9 inches. The constancy of sampling depth means that comparisons are possible for total soil constituents even though plough depth has changed.

The foresight by Lawes and Gilbert highlights a major need in new experiments where changes in soil properties must be an important factor. At the outset, a soil sampling protocol must be established and then adhered to. The interval between taking samples is less important, and can be related to the length of a crop rotation in agricultural experiments, but it should probably not be longer than every 10 years. Both crop and soil samples should be dried and stored under conditions which prevent deterioration.

Current analytical techniques now allow the opportunity to measure the concentration of a wide range of inorganic and organic constituents. One reason for maintaining archived samples is to allow other constituents to be determined or past analyses updated using more accurate methods. We cannot improve on Lawes and Gilbert's data for total nitrogen in soil, but we can now analyze for zinc, copper, nickel, and cadmium, and polynuclear aromatic hydrocarbons whose significance was not appreciated in the last century.

Such long data sets are essential to estimate small but consistent long-term changes in soil composition, and their effects on crop growth and composition. Without such data sets it is impossible to discuss with conviction the effects of agriculture on the environment, and man's anthropogenic activities on the soil. In the final analysis, soil is one of a country's most important natural assets, on which the ability to feed its population, or to provide them with an acceptable landscape, depends. This chapter has attempted to show by examples that long-term fertility trends in soil can be measured and understood only by using data from long-term experiments.

6.8 REFERENCES

Bolton J. (1977). Changes in soil pH and exchangeable calcium in two liming experiments on contrasting soils over 12 years. *Journal of Agricultural Science, Cambridge*, **89**, 81–86.

Davies, M.S. (1975). Physiological differences among populations of *Anthoxanthum odoratum* L. collected from the Park Grass Experiment, Rothamsted. IV. Response to potassium and magnesium. *J. of Appl. Ecol.*, **12**, 953–964.

Davies, M.S. and Snaydon, R.W. (1973a) Physiological differences among populations of *Anthoxanthum odoratum* L. collected from the Park Grass Experiment, Rothamsted. I. Response to calcium. *J. of Appl. Ecol.*, **10**, 33–45.

Davies, M.S. and Snaydon, R.W. (1973b). Physiological differences among populations of *Anthoxanthum odoratum* L. collected from the Park Grass Experiment, Rothamsted. II. Response to aluminium. *J. of Appl. Ecol.*, **10**, 47–55.

Davies, M.S. and Snaydon, R.W. (1974). Physiological differences among populations of *Anthoxanthum odoratum* L. collected from the Park Grass Experiment, Rothamsted. III. Response to phosphate. *J. of Appl. Ecol.*, **11**, 699–707.

Davies, M.S. and Snaydon, R.W. (1976). Rapid population differentiation in a mosaic environment. III. Measurements of selection pressures. *Heredity*, **36**, 59–66.

Glynne, M.D. (1935). Incidence of take-all on wheat and barley on experimental plots at Woburn. *Ann. Appl. Biol.*, **22**, 225–235.

Goodman, P.J. (1969). Intra-specific variation in mineral nutrition of plants from different habitats. In Rorison, I.H. (Ed.) *Ecological Aspects of the Mineral Nutrition of Plants*. Blackwell Scientific, London, 237–253.

Goulding, K.W.T., Johnston, A.E. and Poulton, P.R. (1988). The effect of atmospheric deposition, especially of nitrogen, on grassland and woodland at Rothamsted Experimental Station, England, measured over more than 100 years. In Mathy, P. (Ed.) *Air Pollution and Ecosystems*. Proceedings of an EEC International Symposium, Grenoble, France, May 1987. D. Reidel, Dordrecht, 841–846.

Goulding, K.W.T. and Stevens, P.A. (1988). Potassium reserves in a forested, acid upland soil and the effect on them of clear-felling versus whole-tree harvesting. *Soil Use and Management*, **4**, 45–51.

Hayman, D.S. (1984). Improved establishment of white clover in hill grasslands by inoculation with mycorrhizal fungi. In Thomson, D.J. (Ed.) *Forage Legumes*. British Grasslands Society Occasional Symposium No. 16. BGS Press, Hurley, Maidenhead, 44–47.

Jenkinson, D.S. (1971). The accumulation of organic matter in soil left uncultivated. *Rothamsted Experimental Station Report for 1970*, Part 2, 113–137.

Johnston, A.E. (1972). Changes in soil properties caused by the new liming scheme on Park Grass. Rothamsted Experimental Station Report for 1971, Part 2, 177–180.

Johnston, A.E. (1977). Woburn Experimental Farm: a hundred years of agricultural research devoted to improving the productivity of light land. Lawes Agricultural Trust, Harpenden.

Johnston, A.E. (1986). Potassium fertilization to maintain a K-balance under various farming systems. In *Nutrient Balances and the Need for Potassium*. 13th Congress of the International Potash Institute, Reims, France, 177–204.

Johnston, A.E. (1989). The value of long-term experiments—a personal view. In Likens, G.E. (Ed.) *Long-Term Studies in Ecology: Approaches and Alternatives*. Springer-Verlag, New York, 175–179.

Johnston, A.E. (1990). The value of long-term experiments in agricultural research. Proceedings of the centennial of the Sanborn Plots, July 1988, University of Missouri-Columbia.

Johnston, A.E. and Chater, M. (1975). Experiments made on Stackyard Field, Woburn, 1876-1974. II. Effects of the treatments on soil pH, P and K in the Continuous Wheat and Barley experiments. *Rothamsted Experimental Station Report for 1974*, Part 2, 45–60.

Johnston, A.E. and Garner, H.V. (1969). The Broadbalk Wheat Experiment: Historical Introduction. *Rothamsted Experimental Station Report for 1968*, Part 2, 12–25.

Johnston, A.E. and Goulding, K.W.T. (1988). Rational K manuring. *Journal of the Science of Food and Agriculture*, **43**, 319–320.

Johnston, A.E., Goulding, K.W.T. and Poulton, P.R. (1986). Soil acidification during more than 100 years under permanent grassland and woodland at Rothamsted. *Soil Use and Management*, **2**, 3–10.

Johnston, A.E., McGrath, S.P., Poulton, P.R. and Lane, P.W. (1989). Accumulation and loss of nitrogen from manure, sludge and compost: long-term experiments at Rothamsted and Woburn. In Hansen, J.A. and Henriksen, K. (Eds) *Nitrogen in Organic Wastes Applied to Soils*. Academic Press, London, 126–139.

Jones, K.C., Grimmer, G., Jacob, J. and Johnston, A.E. (1989a). Changes in the polynuclear aromatic hydrocarbon (PAH) content of wheat grain and pasture grassland over the last century from one site in the U.K. *Science of the Total Environment*, **78**, 117–130.

Jones, K.C. and Johnston, A.E. (1989b). Cadmium in cereal grain and herbage from long-term experimental plots at Rothamsted, U.K. *Environmental Pollution*, **57**, 199–216.

Jones, K.C., Stratford, J.A., Tidridge, P., Waterhouse, K.S. and Johnston, A.E. (1989c). Polynuclear aromatic hydrocarbons in an agricultural soil: long-term changes in profile distribution. *Environmental Pollution*, **56**, 337–351.

Jones, K.C., Stratford, J.A., Waterhouse, K.S., Furlong, E.T., Giger, W., Hites, R.A., Schaffner, C. and Johnston, A.E. (1989d). Increases in the polynuclear aromatic hydrocarbon (PAH) content of an agricultural soil over the last century. *Environmental Science and Technology*, **23**, 95–101.

Jones, K.C., Symon, C.J. and Johnston, A.E. (1987). Retrospective analysis of an archived soil collection. II. Cadmium. *The Science of the Total Environment*, **67**, 75–89.

McGrath, S.P. (1984). Metal concentrations in sludges and soil from a long-term field trial. *Journal of Agricultural Science*, **103**, 25–35.

McGrath, S.P. and Johnston, A.E. (1986). Sulphur—crop nutrient and fungicide. *Span (Progress in Agriculture)*, **29**, 57–59.

Powlson, D.S. and Brookes, P.C. (1987) Measurement of soil microbial biomass provides an early indication of changes in total soil organic matter due to straw incorporation. *Soil Biology and Biochemistry*, **19**, 159–164.

Remison, S.U. (1976). A study of root interactions among grass species. PhD Thesis, University of Reading.

Russell, E.W. (1961). *Soil Conditions and Plant Growth*. 9th edition. Longman, London.

Snaydon, R.W. (1970). Rapid population differentiation in a mosaic environment. I. Response of *Anthoxanthum odoratum L.* populations to soils. *Evolution*, **24**, 257–269.

Snaydon, R.W. and Davies, M.S. (1972). Rapid population differentiation in a mosaic environment. II. Morphological variation in *Anthoxanthum odoratum L. Evolution*, **26**, 390–405.

Snaydon, R.W. and Davies, M.S. (1976). Rapid population differentiation in a mosaic environment. IV. Populations of *Anthoxanthum odoratum L.* at sharp boundaries. *Heredity*, **37**, 9–25.

Thurston, J.M., Williams, E.D. and Johnston, A.E. (1976). Modern developments in an experiment on permanent grassland started in 1856; effects of fertilizers and lime on botanical composition and crop and soil analyses. *Annales Agronomiques*, **27** (5-6), 1043–1082.

Warren, R.G. and Johnston, A.E. (1964). The Park Grass Experiment. *Rothamsted Experimental Station Report for 1963*, 240–262.

Warren Spring Laboratory (1985). Acid Deposition in the United Kingdom. Second Report of the United Kingdom Review Group on Acid Rain. Department of Trade and Industry, London.

White, R.E. (1979). *Introduction to the Principles and Practice of Soil Science*. Blackwell, London.

7 Long-term Ecological Research—A Pedological Perspective

DARWIN W. ANDERSON
Saskatchewan Institute of Pedology, University of Saskatchewan, Saskatoon, Canada S7N 0W0

7.1 INTRODUCTION

Paul B. Sears (1935), in his work *Deserts on the March*, wrote of the just-developing field of ecology and ecologists:

> His [an ecologist's] work involves analysis, of course, but only as a means to a final synthesis and interpretation. When he enters a forest or a meadow he sees not merely what is there, but what is happening there. To him there is afforded a glimpse of continuity, integration and destiny which is indispensable to management and control in any real sense.

Despite its being written more than 50 years ago, amid the spectre of the dust-bowl that had ravaged agriculture and the land resources of the Great Plains, this message is as appropriate today as it was then.

Soil scientists may be criticized in many ways for not fulfilling this vision of the ecologist as they pursued pragmatic and generally successful (at least temporarily) solutions to problems of crop production (Nikiforoff, 1959), or assembled the descriptive data required to characterize the world's soils. Soil

Long-term Ecological Research. Edited by Paul G. Risser

scientists, however, have initiated long-term studies, mainly dealing with crop rotations and the monitoring of soil erosion, that have provided valuable material to those working to synthesize, organize, and predict in today's environment. Earlier practical approaches to soil mapping (Ellis, 1932) that recognized soils at different levels of detail have elements of hierarchical theory which are consistent with more refined treatments of today. Other studies have monitored systems which are highly variable both in time and space, such as saline soils, with some success. The objective of this chapter is to discuss briefly some principles that are appropriate to studying ecosystems, and illustrate their application (or their being ignored) in several soils-related, long-term studies, and to present some current ideas on monitoring, understanding, and modeling mainly agricultural ecosystems in the Great Plains region of the United States and Canada.

7.2 A HIERARCHICAL PARADIGM FOR SOILS

There is little doubt that soil scientists, in common with scientists in general, have the technology to measure accurately a multitude of properties and to evaluate processes on a sample experimental unit or point location. The key question is how to extend or extrapolate the findings to the remainder of the universe for which they are appropriate, and not to impose them where they are not. Soil scientists, through the carrying out of soil surveys, have been attempting to do that for several decades. Interestingly, an early soil-classification system for use in soil survey has many elements of hierarchical theory, despite the system's development as a pragmatic solution to soil survey and land management problems (Ellis, 1932). The basic class dealing with soil individuals was an associate, or a particular combination of soil horizons and accessory features, especially drainage, on a specific kind of geological material. Associates were organized at the next higher level to form associations, basically the various interrelated soil associates within a landscape. An associate, by necessity, could occur in more than one association and was a real and functioning portion of the association; in this manner, this classification differs from taxonomic systems in that higher integrative levels represented real bodies, not constructs in the minds of the classifier. Associations, in turn, were organized into combinations based on major physiographic and landform properties, resulting in an approach consistent with recent ideas on landscape (Swanson *et al.*, 1988). The most general level or soil zone integrated the regional effects of climate and vegetation, but grouped all soils within a region. This kind of hierarchical system or variants thereof have been and are being used as valuable elements in land management.

Some soil maps have been criticized as not realistically describing the soils of particular areas. Much of this criticism is well founded, and can usually be traced to methods of generalization rather than aggregation as scale decreases, coupled with overuse of typical or modal individuals to represent large areas where this individual may or may not occur, and a too-rigid application of taxonomic limits when locating soil boundaries on a map (Witty and Arnold, 1987). Soil maps are most useful when the complexity of the soil is organized with the paradigm of hierarchical systems. These systems have elements of both spatial and functional scale, with lower levels representing units that are relatively small with a quick time response, and high levels representing units that are larger and slower (Urban *et al.*, 1987).

Soil systems represent this well. A particular lowland may have saline or sodic soils because of regional piezometric systems that move soluble components back to the surface (Henry *et al.*, 1985) or limit the leaching of weathering products (Keller and van der Kamp, 1988). Water in large, regional groundwater flow systems may have residence times of years to centuries and millennia (van der Heijde, 1988). Within the lowlands, however, not all soils will be saline. Saline soils occur mainly on slight convexities, where runoff limits the water available for leaching, with non-saline soils in the effectively more moist concave areas that receive the runoff (Sandoval *et al.*, 1964). Certain layers of the soil, particularly near the surface, may be affected by what happened yesterday, with soluble salts leaching after rains and moving upwards during dry periods.

Some researchers propose that objective and numerical models based mainly on geostatistics are the preferred method of extrapolating point data to larger universes (Kachanoski, 1988). These methods, that describe elements of ecosystems, particularly landscape and soil properties, using techniques such as autocorrelation, spectral analyses, and kriging, permit the determination of spatial relation or pattern of single factors, or groups of related factors. The methods probably are most valuable where the factors occur at random and a pattern is not intuitively evident. These statistical methods also have promise as a test of biased and intuitive methods. A recent study of paired cultivated and native soil areas used systematic sampling along transects and statistical analyses such as correspondence analysis, autocorrelation, and non-parametric methods to evaluate relationships between landscape, soil properties, erosion, and crop yield (Moulin, 1988). The sampling method and analyses indicate the considerable importance to crop yield of calcareous soils on planar, lower slopes, organic matter content and composition, degree of erosion, and other factors. These calcareous, lower slope soils account for only a small proportion of the land, and have usually been ignored in biased or stratified sampling methods.

7.3 ECOSYSTEM PROCESSES, TIME, AND THRESHOLD RESPONSE

Several features of ecosystem function and study are related to elements of time. Despite an all-too-frequent view of soils as relatively static bodies that have developed gradually over very long time spans, most soil scientists now view soils as dynamic entities with rapid changes in some properties, and very gradual effects in others. The second concept considers that the present characteristics of a soil represent an integration of a large number of temporally variable processes that may not always move soil formation in the same direction. Many soil properties may be the consequence of episodic events that can initiate a non-linear or threshold response which results in considerable change in properties over short time spans. Soil formation, for example, is in a dynamic balance with soil erosion on uplands, and it is likely that erosion events are episodic, whereas the continued formation of new soil at the weathering front is much more gradual.

Soils, in common with many ecosystems (Holling, 1985), have variables that operate at slow, intermediate, and fast rates, and it is important to recognize the nature of the variable(s) studied. Variables such as soluble salts are highly dynamic, varying over a season and reaching tentative equilibrium in a few years, whereas organic matter levels have a time dimension of decades to centuries, with carbonate and, in particular, clay weathering having a scale of millennia in semi-arid climates (Anderson, 1977). Interestingly, considering organic matter processes to represent a medium time-scale is, in itself, an integration. At another level of detail, it is possible to differentiate fast (mainly microbial) processes, intermediate components where turnover is dampened by physical sorption to clay, and slow or chemically stabilized humus components in soils (Jenkinson and Rayner, 1977; Anderson, 1979).

Related to time is the concept of threshold-controlled responses or mechanisms. Threshold-controlled responses are considered to be important to the understanding of the development of above-ground ecosystems and geomorphic surfaces, but have only recently been discussed in soil science (Stoner and Ugolini, 1988). A threshold-controlled response occurs when a low-frequency, high-intensity impact or combination of impacts impinge on a system, resulting in a new course of development for a previously stable system. Stoner and Ugolini (1988) found that a large and rare rainstorm (740 mm over 4 days) initiated an intense or catastrophic leaching event, resulting in a considerable surge in subsurface translocation in Spodosol soils of Alaska. Formerly stable horizons were disrupted, and normally insoluble organo-metallic bodies were dissolved and moved downwards in the profile in association with suspended particulate matter to form unusually deep B horizons enriched in both sesquioxides and humus. Such episodic pulse mechanisms are important in other soils. Translocation in Luvisol or Alfisol soils occurs mainly when unusually heavy precipitation initiates leaching into relatively dry soil (Howitt and Pawluk, 1985). Solonetzic soils, which combine strongly leached surface horizons with subsoils enriched in soluble salts and sodium, may represent a balance between intense leaching (at snowmelt or following intense storms) through larger pores and fissures, balanced by upward fluxes due to capillary rise in much finer pores (Anderson, 1987).

The highly dynamic and temporal variability of many soil processes indicates the need for continuous and long-term monitoring of soil processes. These studies must recognize the possibility of rare and intense threshold-controlled responses, in that they may be critical to understanding the system.

7.4 LONG-TERM STUDIES OF SOIL SALINITY

7.4.1 BACKGROUND

Saline soils occur generally throughout the sub-arid, semi-arid, and sub-humid regions of the Great Plains of North America. These soils are a natural phenomenon that are the result of high salt content in soil parent materials containing shales of marine origin, the build-up in the soil of the products of present-day weathering (Keller and van der Kamp, 1988), or the redistribution of soluble components due to groundwater flows on the discharge of salt-carrying water from near-surface or glacial aquifers and deep bedrock aquifers (Henry *et al.*, 1985).

Several studies have examined the effect of agriculture on the areal extent and severity of saline soils. Ballantyne (1963), in assessing an apparent increase in saline land during a wet period, the 1950s, in sub-humid south-eastern Saskatchewan, presented evidence for increasing salinity in lower slope soils; the most strongly affected soils occurred on planar, very gentle slopes slightly above and adjacent to depressions that held temporary ponds during wet periods. The saline land had no growth of the intended wheat crop, and salt concentrations that increased towards the soil surface. Ballantyne (1963) considered that the build-up of salts in lower slopes represented a downslope movement of salts within the soil, resulting from the saturation of the upper slope soils. The extra water entering the upper slope soils was considered a result of four years of much-above-average precipitation.

A study in the semi-arid region of southern Alberta reported increasing salinity in agricultural regions, with Solonetzic soils in lower areas most strongly affected in that salts had moved upwards from a normally saline C horizon to re-salinize B and A horizons. Formerly non-saline Dark Brown soils on gentle slopes had also become more saline. The increase in salinity evident in comparing aerial photographs from 1951 and 1962 was attributed to an increased incidence of wet years with greater than the mean precipitation, but implicated the practice of bare fallow and land-use changes on adjacent rangeland.

A more comprehensive assessment involving annual sampling of 64 sites in Saskatchewan (Ballantyne, 1978) indicated the highly variable and dynamic nature of soil salinity. There were always changes in salt concentration for individual soil profiles that were opposite to the average change in the area. Yearly variations were not related to any single factor such as cropping practice, topography, or type of soil profile. The largest annual changes in salinity (30% increase in soluble salts) occurred in soils under low knolls in bare fallow fields.

Despite these studies that indicate the complexity and dynamic nature of soil salinity, the most often quoted estimates of increases in salinity due to agriculture were based on comparisons of aerial photographs of the same areas taken ten years apart (van der Pluym *et al.*, 1981). Estimates of soil salinity were based on visible evidence (light-colored areas) or restricted vegetative growth, and indicated that salinity was increasing at a rate of 10% annually. The total area of land that is affected is not known accurately, with estimates for the prairie region of Canada ranging from about 2 million to 5.4 million ha. Combining the 10% rate of increase with 5.4 million ha gives an annual rate of increase which is probably at least an order of magnitude too high, and indicates the need for long-term studies to monitor soil salinity and the factors that control its occurrence. The temporally variable nature of soil salinity, and assessments that are based mainly on proxy data such as the severity of crop growth restriction, make accurate assessments of areal extent difficult and generally unreliable.

7.4.2 CORONACH SALINITY MONITORING

This study involved the monitoring of saline soils in an area where predicted and actual increases in artesian pressure or piezometric level of the groundwater had occurred. The study was situated in a local basin, with a semi-arid climate and Brown (*Aridic Haploboroll*) soils occurring on mixed glacial till, glaciofluvial, and alluvial deposits. The study was initiated because the filling of a reservoir constructed to provide cooling water for a thermal power station had resulted in a consistent rise of 2–4 m in the piezometric head within land just below the reservoir (Figure 7.1). The response was most rapid on the groundwater directly connected to the reservoir through the Empress gravel, but gradually affected other stratigraphic units, including the surficial glacial till deposits (Figure 7.2).

Soil salinity was monitored by sampling in early August, at marked sites at seven different locations, with 26 individual sites in total. There were variable changes in soil salinity in the 1979 to 1983 period. There was no change or a slight decrease in salinity in the 0- to 30-cm depth at 12 sites, a variable response at six, and a gradual increase in salinity at eight sites. Results for a silty alluvial soil with a piezometric head at ground level and a water table at less than 2 m depth indicate a marked increase in salinity as shown by increases in the electrical conductivity (EC) of saturation extracts of the soil (Figure 7.3(a)) and the soluble sodium concentration (Figure 7.3(b)). EC and soluble sodium values were highest in surface layers, and generally decreased with depth, indicating a pronounced upward flux of soluble constituents that is driven by artesian pressure and evaporation of soil water at the surface.

Other sites, usually Solonetzic soils, that had little or no salts in surface layers but saline subsoils had salt move upwards, particularly in 1980 and 1981 when summer precipitation was well below average. These salts were flushed out of the soil in 1982 by several heavy rains (Figure 7.4). Sampling in 1983 showed that the salts were moving back towards surface layers. These kinds of dynamic fluctuations were evident at several sites, particularly those with low to moderate concentrations of salts, and indicate the dynamic nature of salt movement in soil. Upward fluxes of salts were greatest in silty soils with high water contents, consistent with earlier observations of upward movement of sodium by convection (carried along by

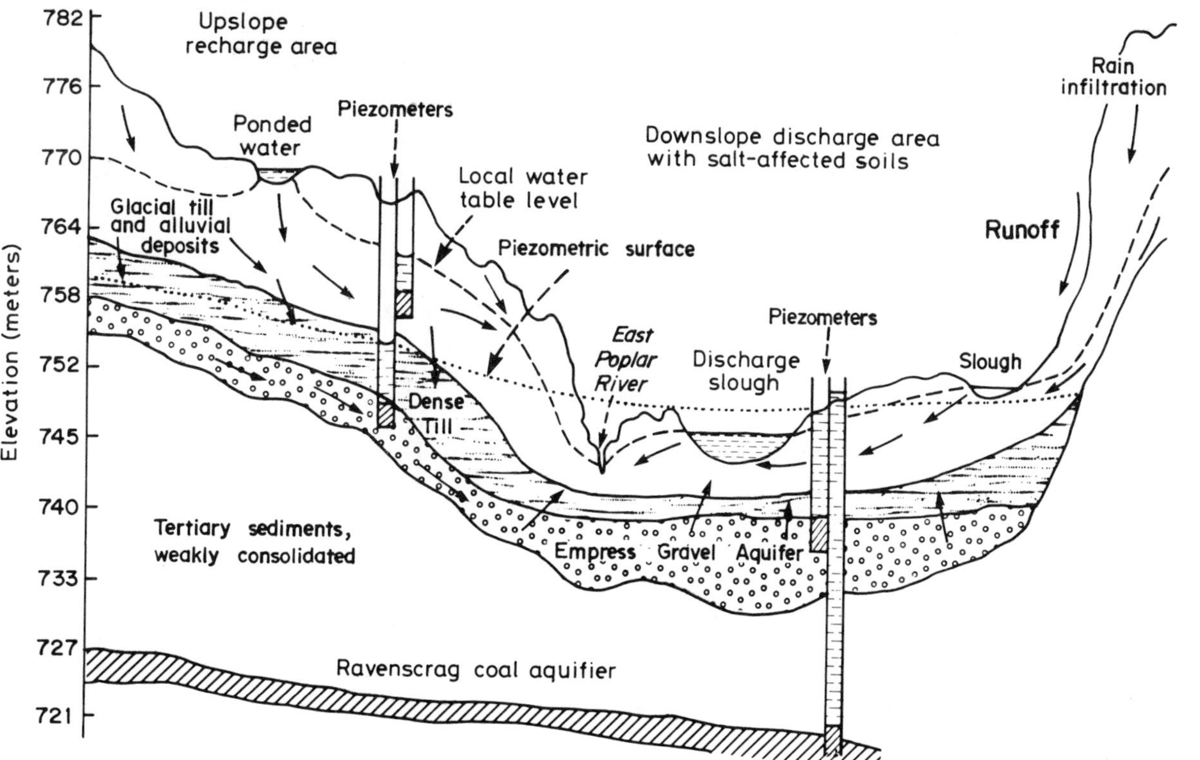

Figure 7.1 Generalized hydrogeology and salinity at the Coronach soil salinity study area

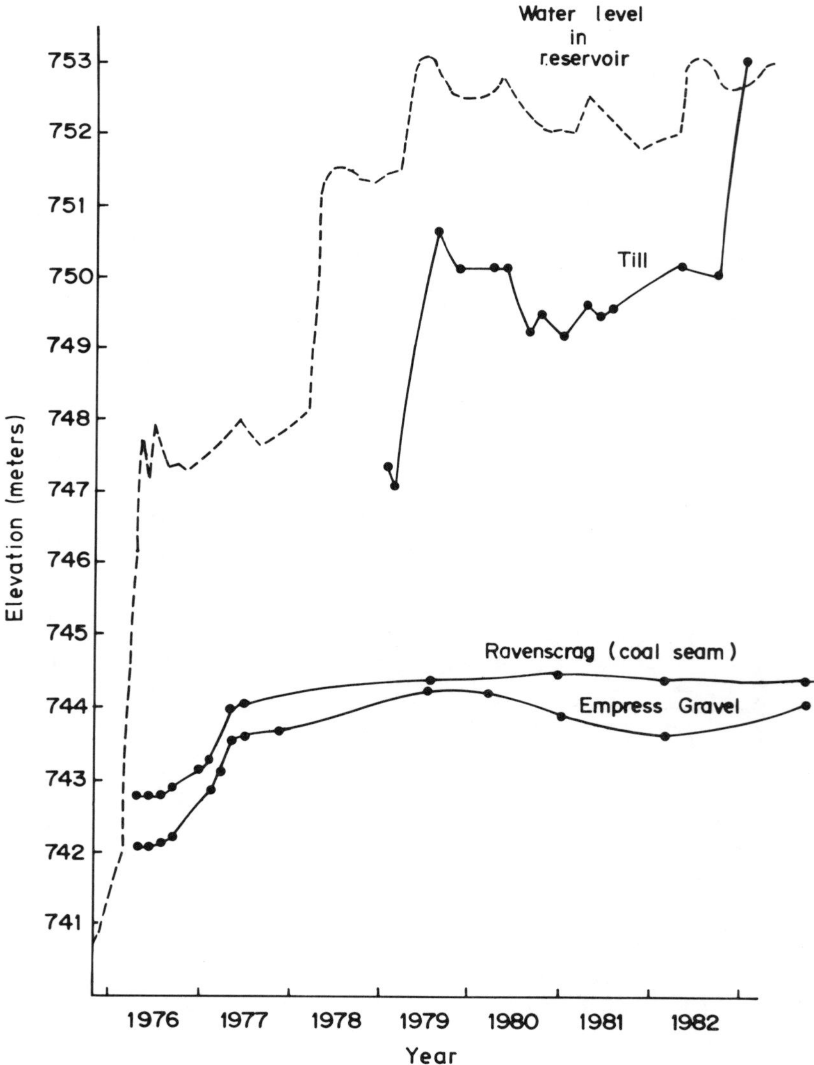

Figure 7.2 Changes in the piezometric level of main aquifers following the filling of a reservoir upstream

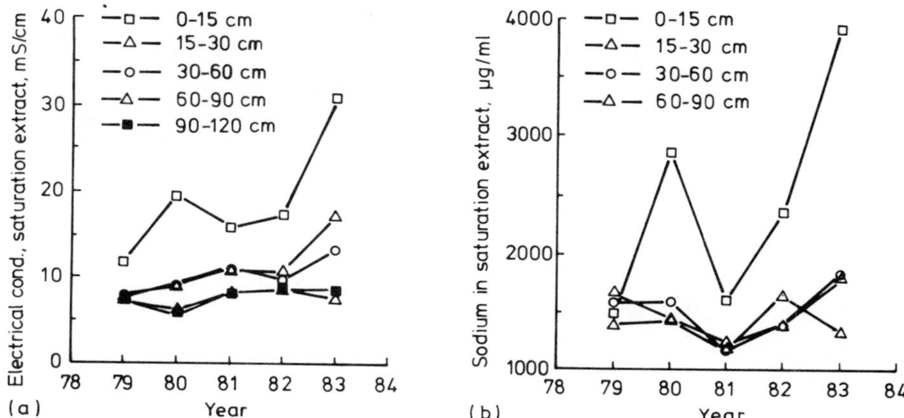

Figure 7.3 Changes in (a) soil salinity and (b) soluble sodium in a saline Brown soil with time

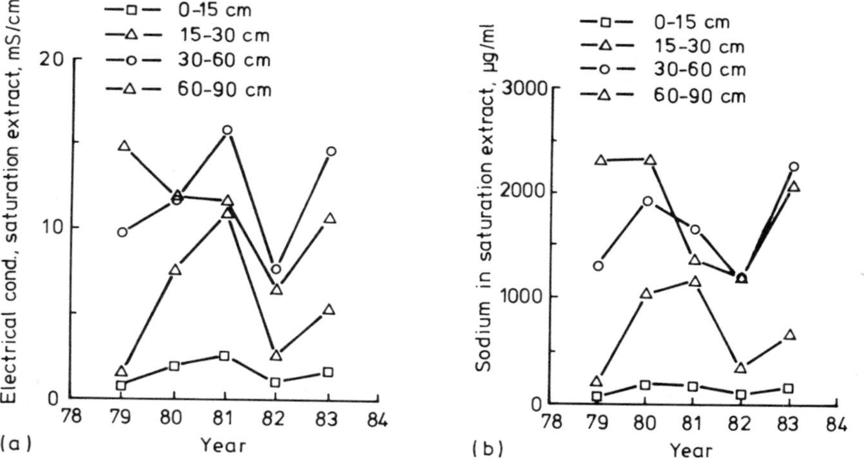

Figure 7.4 Changes in (a) soil salinity and (b) soluble sodium in a saline Brown soil with time

moving water) and diffusion or moving within the soil solution in response to a concentration gradient (Merrill *et al.*, 1983).

Aerial photographs taken each year within a few days of 10 July indicate the gradual but highly variable increase in saline soils, particularly in silty and sandy materials at low elevations within the basin (Figure 7.5). Those soils that were most severely affected include Solonetzic soils, which are soils that normally have saline

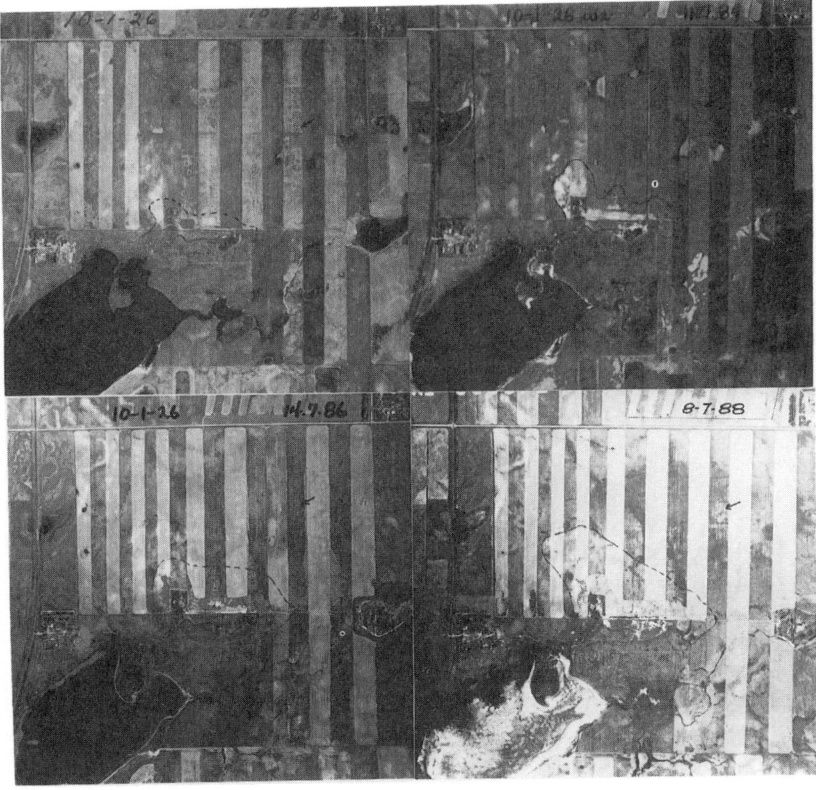

Figure 7.5 Color infra-red aerial photographs of the Coronach salinity study area for 1982, 1984, 1986 and 1988. The black line indicates the extent of saline soils as determined with the EM salinity meter

subsoils because of upward fluxes of soluble constituents due to artesian pressure. Rego or A/C soil profiles, where strong accumulations of calcium carbonate and, occasionally, soluble salts are considered to be indicative of artesian influences in the past, were also strongly salinized. Field checking, mainly in areas of limited crop growth, using a hand-portable, non-contacting conductivity meter (EM38, Geonics Ltd) indicated that previously non-saline Orthic Brown soils that had developed under moderate leaching regimes were becoming saline by about 1983 to 1988. The soils that were most strongly affected had sandy to silty clay loam textures, and occurred on slight convexities or very gentle slopes at low elevation, similar to those observed by Ballantyne (1978). Soils in concave areas, despite their occurring at slightly lower elevations and with supposedly stronger artesian fluxes, often were much less saline. These differences, similar to those observed by Sandoval *et al.* (1964), were attributed to runoff from sloping or convex lands that results in a reduced leaching potential, coupled with enhanced leaching in runoff-collecting concave areas, that counterbalanced upward fluxes due to artesian pressure.

The results of this study are not directly applicable to other areas as most land has not had the increase in artesian pressure that occurred in the study area. There are, however, principles that can be applied to long-term studies of soil salinity. At the start of this study, the investigators had a more mechanistic view of salinity which considered that salinity in the area was due to artesian pressure, that it would uniformly worsen as pressure increased, and that the process would be unidirectional and gradual. This work, despite its shortcomings, has indicated the highly complex and dynamic nature of salinity and the factors that control it.

The dynamic nature of soil salinity indicates that monitoring salinity at different depths of the soil requires more than annual sampling and careful records of precipitation at the monitoring sites. Sampling the soil by removing cores will alter the permeability of the soil, and eventually the salt concentration. The best technique may be to insert electrodes to measure electrical conductivity *in situ*,

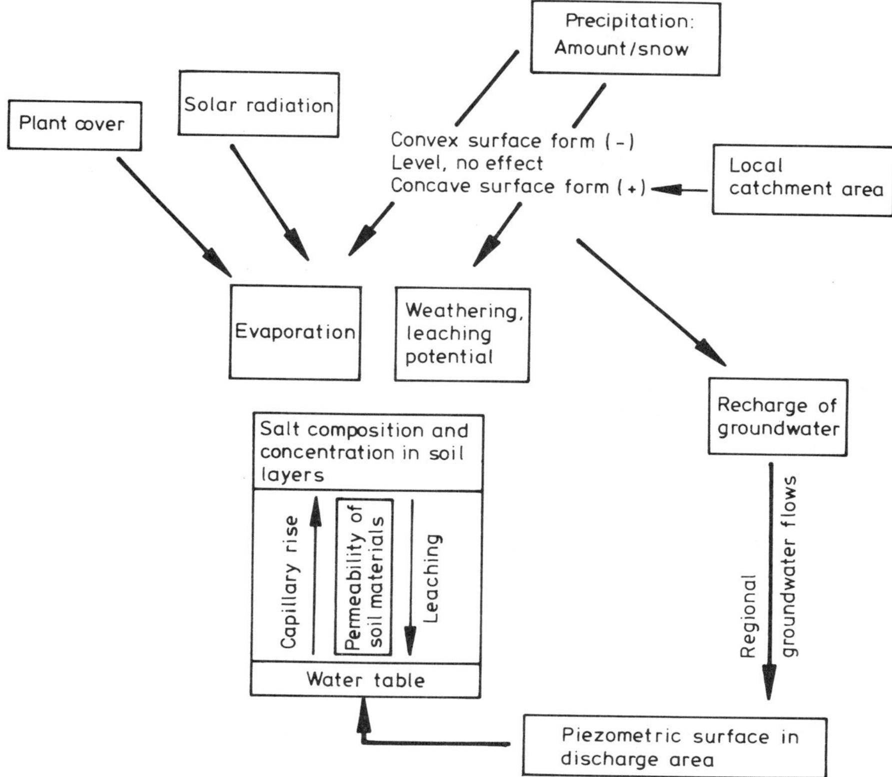

Figure 7.6 A conceptual model describing the factors and processes that affect soil salinity

rather than removing soil cores. The non-contacting, portable salinity meters are of limited value, but could be used if soil moisture was also monitored.

A conceptual model that describes the factors and indicates processes (Figure 7.6) is based partly on our experience in the area. This model, that considers landscape and regional hydrological processes that are relatively slow, and changes in the soil that may be rapid, is presented as an initial step to modeling soil salinity. Submodels will be required to describe processes at greater detail, such as models for salt flow (Childs and Hanks, 1975), salt-transport and chemical equilibria (Robbins *et al.*, 1980), and upward fluxes by convection or diffusion (Merrill *et al.*, 1983). The basic approach, however, is to key on the processes critical to change, namely the relative balance between downward flux (leaching) and upward flux (capillary rise or artesian pressure), in order to predict the impacts of agricultural practices, including irrigation, and long-term effects due to change in climate. Changes in climate due to increased greenhouse gases that result in more moist winters and drier, warmer summers may result in marked changes in salinity. Recharge to the groundwater is mainly depression-focused and could be enhanced by the accumulation of snowmelt in depressions, resulting in higher artesian pressure after some years or decades. Higher rates of evapo-transpiration and less rainfall in summer would decrease downward fluxes and slightly increase upward fluxes near the surface, resulting in an increase in saline land over time. Evaluations of the effects of changes in artesian pressure on soil salinity will be dependent on long-term records of piezometric levels in major aquifers. This kind of record is available in Saskatchewan, a result of far-thinking hydrologists who began keeping systematic records more than 25 years ago (Meneley *et al.*, 1979).

7.5 LONG-TERM CROP ROTATION STUDIES

The soil science research that has been ongoing for long periods of time includes crop rotation studies. The studies at Rothamsted were begun more than a century ago, and records, soil analyses, and soil samples have been used in recent times to evaluate processes such as the turnover of organic matter (Jenkinson and Rayner, 1977; Johnston, Chapter 6, this volume). The designation of rotation studies as ecological research may be questionable in that the studies included monocultures under highly managed conditions in which response rather than process was measured. However, several aspects of these studies indicate their value to discerning gradual and long-term changes to soils.

Of several long-term rotations in Western Canada, three will be discussed in this chapter. One study, which was made to compare the effects of green manures and farmyard manures on yield and soil properties, was begun at the University of Manitoba in 1919 on clayey Black soils (Poyser *et al.*, 1957), and has more detailed soil sampling and yield records available from 1930 until the late 1960s. The records show organic carbon and nitrogen in the soil decreased steadily from 1930 to 1955, with smaller losses on rotations where manure was added or legumes grown. Crops grown following manures generally yielded higher than crops given other treatments, with interactions between enhanced nitrogen supply and the moisture-depleting effects of green manures showing up as reduced yields after green manures during dry periods. A later evaluation showed the organic matter-depleting effects of more frequent bare fallow, and indicated that soil fertility and yields could be sustained or increased over time in continuous cereal rotations, provided that adequate manure or chemical fertilizer was applied (Ridley and Hedlin, 1968). These rotations were discontinued in about 1970, partly because the rotations being studied were no longer considered pertinent to Manitoba agriculture, and the land was required for athletic fields. Since many of the long-term effects on soils were not evaluated, the value of the studies was diminished.

Continuous wheat and fallow–wheat rotations were begun in 1912 on Dark Brown, loamy soils at the Agriculture Canada Research Station at Lethbridge, Alberta. Recent soil analyses have provided data that furthers our understanding of the effects of bare fallow, fertilizer, and crop rotation on soil organic matter (Janzen, 1987). The application of inorganic nitrogen fertilizer in continuous wheat or wheat–wheat–fallow rotation significantly increased organic carbon and nitrogen in the soil, with an even more marked increase in labile or mineralizable forms of carbon and nitrogen. This effect was attributed to increased production where nitrogen was applied, resulting in greater return of residues to the soil, greater microbial activity, and more efficient organic cycling of nitrogen. Applying ammonium nitrate fertilizer at 45 kg N ha^{-1}, however, decreased the pH of surface soils from 7.2 to 6.9, indicating the potential for acidification problems on less buffered and more acidic soils.

A five-year rotation that included forage and cereal and a two-year wheat–fallow rotation were set up at Breton, Alberta, in 1929. Soils are Gray Luvisols on loamy glacial till, with a pH of about 6.0 and an organic carbon content of 1.2% (McGill *et al.*, 1986). Organic carbon and nitrogen contents of the Ap horizon have increased in rotations containing forages, and yields of cereal grains are higher following forages. These soils have surface horizons that form compact crusts upon drying after heavy rains and are difficult to till; soil tilth was improved as indicated by larger and more stable soil aggregates in the forage rotations (Toogood and Lynch, 1959). Chemical fertilizers, mainly ammonium sulphate, ammonium nitrate, and ammonium phosphate compounds, were applied at rates to supply 11 kg N ha^{-1} and 9 kg S ha^{-1} annually. Despite a natural gradient in pH across the systematically arranged plots, detailed evaluations indicated that the chemical fertilizers had increased acidity and easily extractable forms of aluminum and manganese, and decreased exchangeable calcium (McCoy and Webster, 1977). Manures had no effect on soil reaction. The decrease in pH from approximately 6.0 to about 5.2 resulted in reduced proportions of legumes in forage stands and yields, and was remedied by applying lime to the soil.

A later study examined both the short-term dynamics of microbial biomass and the longer-term processes affecting organic matter (McGill *et al.*, 1986). The five-year cereal-forage rotation contained 38% more total nitrogen, but 117% more microbial nitrogen, than the wheat–fallow soils. Manured plots had twice the microbial nitrogen of chemically fertilized or control plots. The average quantity of biomass appears to be controlled by long-term additions of carbon, whereas seasonal fluctuation in environmental conditions, particularly moisture, appears to control short-term biomass dynamics. Average turnover rates of the biomass were 0.2 to 3.9 years, being 1.5 to 2 times faster in the two-year wheat–fallow rotation. Average carbon additions are insufficient to account for annual turnover in the two-year rotation, indicating that native soil organic matter may still be decreasing. These studies use modern-day methods but are made possible by the legacy of experimental material initiated more than five decades ago.

The rotation studies can be criticized because they were systematically rather than randomly arranged, thereby limiting statistical analysis (Ridley and Hedlin 1968), because soil samples representative of initial conditions often were not taken, and because response rather than process was measured. In addition, it is difficult to allocate research funds to continue rotations that are no longer pertinent to the region. Despite these limitations, these rotation studies have provided much of the data that evaluate long-term trends, and have provided stored soil samples that can be analyzed by modern methods (Jenkinson and Rayner, 1977; Anderson and Paul, 1984). Many of the findings are currently being used to develop and validate simulation models that describe long-term processes such as organic matter turnover. In other instances, the measurement of surprises, such as the acidification of soil by even low additions of nitrogen fertilizer to poorly buffered soils, has provided findings the originators had not envisaged.

7.6 LONG-TERM ECOLOGICAL RESEARCH IN WESTERN CANADA

The Great Plains region of Canada includes the former grasslands that are now mostly converted to farmland, and the boreal forest and wetlands of the northern portion. Both areas require long-term ecological study to address problems of both local and global perspectives.

The sustainability of agriculture is the question of greatest concern in the prairie region, with a perception of increasing soil deterioration due to erosion, loss of organic matter and fertility, and increasing salinization. Questions related to the severity and extent of soil deterioration and current economic costs to farmers and society have not been adequately resolved and require more study (van Kooten and Furtan, 1987). There is, however, a consensus that soil deterioration is a serious problem, and concerted action is required to address the problem. Knowledge of the effect on soil quality of recent, conservation-oriented tillage systems such as stubble-mulch tillage is lacking, although studies of farmers' fields in North Dakota (Bauer and Black, 1981) and the aforementioned rotation studies (Janzen, 1987) suggested that soil quality is not deteriorating under good crop management. Climatic change due to the greenhouse effect may make agriculture impossible in the prairie region and extend the agricultural frontier in the north, and may have complex effects on conditions such as soil salinity, as discussed earlier in this chapter.

Determining long-term changes in soils will require a number of carefully selected and adequately sampled sites, with annual monitoring of climate and land management. These monitoring sites should be established as soon as possible, to provide some data in the medium term on processes such as organic matter dynamics, salinization, and erosion. The real value, however, of these sites may be decades in the future, when they will provide the baseline data required to evaluate the changes and surprises that undoubtedly will occur.

The area of concern is approximately 500 000 km^2, and it certainly will not be possible to locate long-term study sites on all combinations of land, climate, and agricultural use. Simulation models are envisaged as the means to address change and potential change in these areas. The models will build on the findings gathered by long-term rotation studies that were discussed earlier, utilize research that has examined soil development and response using an environmental gradient approach, and be extrapolated to the region within a hierarchical paradigm based mainly on soil and landscape maps. Models may be tested by comparing simulations to present conditions in soils that have been managed under known rotations by farmers. An example of such a model is the Century model, that simulates the effects of climate and texture on production and organic matter quality and quantity (Parton *et al.*, 1987). This model integrates three submodels: soil and decomposition, nitrogen, and plant production. The variables required include climate (temperature and moisture), soil texture, and nature of plant residues. This model was developed based on the findings of soils along environmental gradients (Anderson 1979), and relies on the results of a wide variety of studies, both basic and more applied.

Other models will be required. Current models of soil erosion, such as the Universal Soil Loss Equation, for example, do not work well in Western Canada because of differential effects of both wind and water erosion on soils within a landscape (Kiss *et al.*, 1986) and the complex nature of landscapes with both erosion and deposition occurring in the same field (Johnson, 1988). Current efforts are to relate erosion losses to landform and the composition of soils. Erosion is estimated by determining the current amounts of ^{137}Cs in soils and relating that amount to the background levels. The ^{137}Cs in the soil resulted mainly from atmospheric testing of nuclear weapons about 1960. These studies have identified the kinds of landscapes most susceptible to erosion, the landscape regimes where erosion is most severe, and areas of deposition (Martz and de Jong, 1987; Pennock and de Jong, 1987). The objectives of this research are empirically based models that can be used to rate various combinations of landform, soils, and climate for susceptibility to erosion under various management alternatives, and to predict the effects of conservation practices on erosion. Soil salinity models generally are not available, but may be possible by integrating hydrological, weathering, leaching, and landscape sub-models as depicted earlier (Figure 7.6).

The Boreal Forest is of considerable importance, not only because of possible

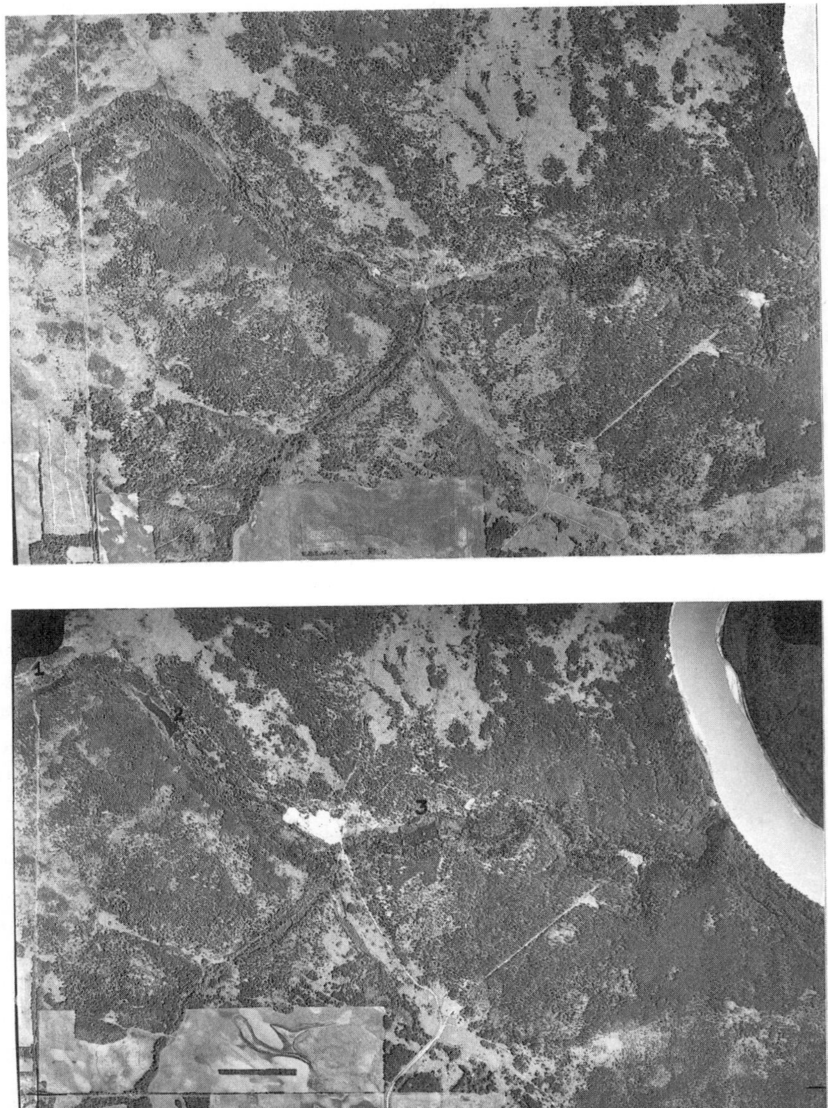

Figure 7.7 Aerial photograph showing a stream with no evident beaver wetlands in 1950 (upper photo) and three wetlands in 1960 (lower photo)

impacts due to acidic precipitation or climatic change, but because of its significance to the global atmosphere, particularly methane additions (Matthews and Fung, 1987). Northern wetlands appear to contribute significantly to methane generation, and the strong recovery of beaver (*Castor canadensis*) population in recent decades may be an important factor in the increase (Nisbet, 1988). Wetlands altered by beaver activity produce orders of magnitude more methane than beaver-free wetlands. The marked increase in beaver population is indicated (Figures 7.7 and 7.8) by the construction of 11 beaver ponds along a stream in the Boreal Forest where no ponds existed in 1950 and three existed in 1960. Each beaver pond forms a wetland of at least one hectare, and influences land upstream by reducing drainage.

A significant drying of the climate in Boreal regions could result in a lowering of the watertable in wetlands and substantially increased rates of aerobic decomposition to release CO_2, coupled with reduced methane evolution (Moore and Knowles, 1989). The significance of these kinds of wetlands, plus the bogs and fens of this region, to methane and CO_2 cycles is important. The annual flux of methane was estimated at 0.1 to 0.6 g C m^{-2} which, although low compared to other wetlands, becomes substantial because of the large area of subarctic peatlands (Moore and Knowles, 1987).

Other questions of concern are the rates of sulphur emission from anaerobic wetlands and from plants under saline or high sulphur stress. The sulphur cycle of Boreal regions is incompletely understood. Most sulphur is retained by Luvisol soils in organic form, mainly in organic surface layers but also as organic sulphates

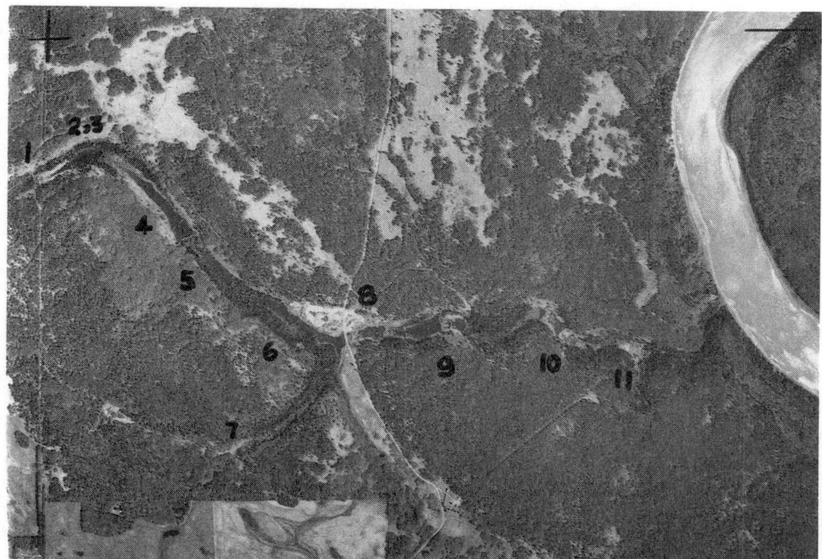

Figure 7.8 An aerial photograph taken in 1985, showing the same stream as in Figure 7.7, with eleven beaver wetlands evident

that appear to have leached into subsoils (Schoenau and Bettany, 1987). The acidic but only moderately weathered Luvisol soils contain low total amounts of sulphur, indicating somewhat inefficient storage mechanisms for sulphur received in precipitation. Some well-humified peats within concave areas in lands dominated by sandy soils have very high sulphur contents, and much sulphur in relation to carbon. This anomalously high concentration of sulphur indicates that these bogs may be a sink for sulphur within those landscapes. Similarly, estimates of losses of phosphorus during the formation of Luvisolic soils are generally larger than the amount of phosphorus in surface waters (St Arnaud *et al.*, 1988). This suggests that phosphorus lost from mineral soils may be retained in the organic soils of wetlands.

7.7 CONCLUSIONS

This chapter has dealt mainly with long-term studies of the nature of soil, with no review of the several ecological research efforts in which soil scientists have participated with other scientists. The long-term crop rotation studies have provided key data beyond the original and practical objectives of the scientists who established them, and still remain a valuable resource for those requiring records, analyses, or samples that were taken decades ago. Improved data to determine the direction, severity, and extent of soil-deteriorating processes such as salinization, erosion, and decline in soil organic matter are required so that research efforts can be appropriately directed to solve the problems of most serious long-term concern. Some long-term monitoring sites are required, but the main impetus should be to build on those data already available from detailed studies, to construct simulation models that key on the variables most critical to long-term processes within ecosystems, and to extrapolate these models to larger universes within a hierarchical paradigm, based mainly on soil, landscape, and climatic maps.

7.8 ACKNOWLEDGMENTS

The author gratefully acknowledges the Saskatchewan Power Corporation for allowing use of data from the Coronach study, and D.R. Cameron for his contribution to the hydrogeology at Coronach. Careful reviews of the manuscript by D.F. Acton and J.R. Bettany are appreciated.

7.9 REFERENCES

Anderson, D.W. (1977). Early stages of soil formation on glacial till mine spoils in a semi-arid climate. *Geoderma*, **19**, 11–19.

Anderson, D.W. (1979). Processes of humus formation and transformation in soils of the Canadian Great Plains. *J. Soil Sci.*, **30**, 77–84.

Anderson, D.W. (1987). Pedogenesis in the grassland and adjacent forests of the Great Plains. *Adv. Soil Sci.*, **7**, 53–93.

Anderson, D.W. and Paul, E.A. (1984). Organo-mineral complexes and their study by radiocarbon dating. *Soil Sci. Soc. Am. J.*, **48**, 298–301.

Ballantyne, A.K. (1963). Recent accumulation of salts in the soils of southeastern Saskatchewan. *Can. J. Soil Sci.*, **43**, 52–58.

Ballantyne, A.K. (1978). Movement of salts in agricultural soils of Saskatchewan 1964 to 1975. *Can. J. Soil Sci.*, **58**, 501–509.

Bauer, A. and Black, A.L. (1981). Soil carbon, nitrogen and bulk density comparision in two cropland tillage systems after 25 years, and in virgin grassland. *Soil Sci. Soc. Am. J.*, **45**, 1166–1170.

Childs, S.W. and Hanks, R.J. (1975). Model of salinity effects on crop growth. *Soil Sci. Soc. Am. Proc.*, **39**, 617–622.

Ellis, J.H. (1932). A field classification of soils for use in the soil survey. *Sci. Agric.*, **12**, 338–345.

Henry, J.L., Bullock, P.R., Hogg, T.J. and Luba, L.J. (1985). Groundwater discharge from glacial and bedrock aquifers as a soil salinization factor in Saskatchewan. *Can. J. Soil Sci.*, **65**, 749–768.

Holling, C.S. (1985). Simplifying the complex: The paradigms of ecological function and structure. *European J. Oper. Res.*, **30**, 139–146.

Howitt, R.W. and Pawluk, S. (1985). The genesis of a Gray Luvisol within the Boreal Forest region. II. Dynamic pedology. *Can. J. Soil Sci.*, **65**, 9–19.

Janzen, H.H. (1987). Effect of fertilizer on soil productivity in long-term wheat rotations. *Can. J. Soil Sci.*, **67**, 165–174.

Jenkinson, D.S. and Rayner, J.H. (1977). The turnover of soil organic matter in some of the Rothamsted classical experiments. *Soil Sci.*, **123**, 298–305.

Johnson, R.R. (1988). Putting soil movement into perspective. *J. Prod. Agric.*, **1**, 5–12.

Kachanoski, R.G. (1988). Processes in soils—from pedon to landscape. In Rosswall, T., Woodmansee, R.G. and Risser, P.G. (Eds) *Scales and Global Change*, SCOPE 35. John Wiley, Chichester, 153–178.

Keller, C.K. and van der Kamp, G. (1988). Hydrogeology of two Saskatchewan tills, II. Occurrence of sulfate and implications for soil salinity. *J. Hydrol.*, **101**, 123–144.

Kiss, J.J., de Jong, E. and Rostad, H.P.W. (1986). An assessment of soil erosion in west-central Saskatchewan using cesium-137. *Can. J. Soil Sci.*, **66**, 581–590.

Martz, L.W. and de Jong, E. (1987). Using cesium-137 to assess the variability of net soil erosion and its association with topography in a Canadian prairie landscape. *Catena*, **14**, 439–451.

Matthews, E. and Fung, I. (1987). Methane emission from natural wetlands: Global distribution, area, and environmental characteristics of sources. *Global Biochemical Cycles*, **1**, 61–86.

McCoy, D.A. and Webster, G.R. (1977). Acidification of a Luvisolic soil caused by low-rate, long-term applications of fertilizers and its effects on growth of alfalfa. *Can. J. Soil Sci.*, **57**, 119–127.

McGill, W. B., Cameron, K.R., Robertson, J.A. and Cook, F.D. (1986). Dynamics of soil microbial biomass and water soluble organic C in Breton L after 50 years of cropping to two rotations. *Can. J. Soil Sci.*, **66**, 1–20.

Meneley, W.A., Maathuis, H., Jaworski, E.J. and Allen, V.F. (1979). SRC Observation wells in Saskatchewan, Canada. *Saskatchewan Research Council Report No. 19*, Saskatchewan Research Council, Saskatoon, Canada.

Merrill, S.D., Doering, E.J., Power, J.F. and Sandoval, F.M. (1983). Sodium movement in soil-minespoil profiles: Difussion and convection. *Soil Sci.*, **136**, 308–331.

Moore, T.R. and Knowles, R. (1987). Methane and carbon dioxide evolution from subarctic fens. *Can. J. Soil Sci.*, **67**, 77–82.

Moore, T.R. and Knowles, R. (1989). The influence of water table levels on methane and carbon dioxide emissions from peatlands. *Can. J. Soil Sci.*, **69**. In press.

Moulin, A.P. (1988). Spatial variability of soil properties in hummocky terrain as affected by cultivation. PhD Thesis, University of Saskatchewan, Saskatoon, Canada.

Nikiforoff, C.C. (1959). A re-appraisal of the soil. *Science*, **129**, 186–196.

Nisbet, E.G. (1989). Some northern sources of natural atmospheric methane: production, history and future implications. *Can. J. Earth Sci.*, **26**, 1603–1611.

Parton, W.J., Schimel, D.S., Cole, C.V. and Ojima, D.S. (1987). Analysis of factors controlling soil organic matter levels in Great Plains Grasslands. *Soil Sci. Soc. Am. J.*, **51**, 1173–1179.

Pennock, D.J. and de Jong, E. (1987). The influence of slope curvature on soil erosion and deposition in hummocky terrain. *Soil Sci.*, **144**, 209–217.

Poyser, E.A., Hedlin, R.A. and Ridley A.O. (1957). The effect of farm and green manures on the fertility of blackearth-meadow clay soils. *Can. J. Soil Sci.*, **37**, 48–56.

Ridley, A.O. and Hedlin, R.A. (1968). Soil organic matter and crop yields as influenced by the frequency of summerfallowing. *Can. J. Soil Sci.*, **48**, 315–322.

Robbins, C.W., Wagenet, R.J. and Jurinak, J.J. (1980). A combined salt transport-chemical equilibrium model for calcareous and gypsiferous soils. *Soil Sci. Soc. Am. J.*, **44**, 1191–1194.

Sandoval, F.M., Benz, L.C., George, E.J. and Mickelson, R.H. (1964). Microrelief influences in a saline area of Glacial Lake Agassiz: I. On salinity and tree growth. *Proc. Soil Sci. Soc. Am.*, **28**, 276–280.

Schoenau, J.J. and Bettany, J.R. (1987). Organic matter leaching as a component of carbon, nitrogen, phosphorus, and sulfur cycles in a forest, grassland, and gleyed soil. *Soil Sci. Soc. Am. J.*, **51**, 646–651.

Sears, P.B. (1935). *Deserts on the March*. University of Oklahoma Press, Norman, OK.

St Arnaud, R.J., Stewart, J.W.B. and Frossard, E. (1988). Application of the 'pedogenic index' to soil fertility studies, Saskatchewan. *Geoderma*, **43**, 21–32.

Stoner, M.G. and Ugolini, F.C. (1988). Arctic pedogenesis: 2. Threshold-controlled subsurface leaching episodes. *Soil Sci.*, **145**, 46–51.

Swanson, F.J., Kratz, T.K., Caine, N. and Woodmansee. R.G. (1988). Landform effects on ecosystem patterns and processes. *BioScience*, **38**, 92–98.

Toogood, J.A. and Lynch, D.L. (1959). Effect of cropping systems and fertilizers on mean weight diameter of aggregates of Breton Plot soils. *Can. J. Soil Sci.*, **39**, 151–156.

Urban, D.L., O'Neill, R.V. and Shugart, H.H., Jr (1987). Landscape ecology. *BioScience*, **37**, 119–127.

van der Heijde, P.K.M. (1988). Spatial and temporal scales in groundwater modelling. In Rosswall, T., Woodmansee, R.G. and Risser, P.G. (Eds) *Scales and Global Change*, Scope 35. John Wiley, Chichester, 195–224.

van der Pluym, H.S.A., Paterson, B. and Holm, H.M. (1981). Degradation by salinization. In *Agricultural Land: Our Disappearing Heritage*. A Symposium, Proc. 18th Alberta Soil Science Workshop.

van Kooten, G.C. and Furtan, W.H. (1987). A review of issues pertaining to soil degradation in Canada. *Can. J. Ag. Econ.*, **35**, 33–54.

Witty, J.E. and Arnold, R.W. (1987). Soil taxonomy: An overview. *Outlook on Agriculture*, **16**, 8–13.

8 Long-term Ecological Research and Fluvial Landscapes

HENRI DÉCAMPS and MADELEINE FORTUNÉ
Centre d'Ecologie des Ressources Renouvelables, CNRS, 29 rue Jeanne Marvig, 31055 Toulouse Cedex, France

8.1 INTRODUCTION

Few ecological systems are as variable as fluvial landscapes. The hydrological characteristics which rule these landscapes fluctuate in an as yet often unpredictable manner at different time scales. The settlement of human societies along alluvial plains, which has produced numerous changes in the past, is a constant cause of continuous, and sometimes acute, change. Clearly, the dynamics of a fluvial landscape, with its constant fluctuation and change, can be understood only through long-term research.

This chapter focuses on the ideas that long-term ecological research must take into account to provide an historical perspective of fluvial landscape dynamics, and at the same time to be included in a global and comparative perspective. The first

Long-term Ecological Research. Edited by Paul G. Risser

part of this chapter gives the results obtained on the River Garonne, France, as an example of the importance of an historical perspective. The second part discusses the necessity of global and comparative perspectives. The third and fourth parts of the chapter discuss four questions to be addressed by long-term ecological research on fluvial landscapes, and the possibility of monitoring by remote sensing.

8.2 AN HISTORICAL PERSPECTIVE

An historical perspective is particularly necessary in river ecology. Rivers and their floodplains have been utilized by man for a long time, and many human societies have settled and developed along large rivers. Fluvial landscapes were early utilized for transport, water and fish availability, and fertility of the soils in the floodplain. In many parts of the world it is impossible to answer the question 'Why are the rivers like they are now?' without a knowledge of their past utilization by man. The length of time involved in man's utilization of rivers is quite different on different continents. Clearly, Europe has been the most significantly modified for the longest period of time.

8.2.1 THE EXAMPLE OF THE RIVER GARONNE

The River Garonne is 580 km long and drains a basin of 57 000 km^2. Figure 8.1 shows the diversity of the hydrological network. According to Strahler (1957), the Garonne is a seventh-order river after it receives the water from its last Pyrenean tributary (River Ariège) and an eighth-order river after its first tributary from the Central Massif (River Tarn). The hydrology is characterized by large variations of flow between sudden and violent floods in spring and very low waters in summer (Table 8.1).

The choice of the River Garonne as an example is justified by the fact that it is one of the largest West European alluvial rivers in a non-industrialized region. Agriculture is responsible for most of the modifications of the fluvial landscape in

Table 8.1 Physical characteristics of the River Garonne

	Downstream to the Ariège tributary	Downstream to the Tarn tributary	Downstream to the Lot tributary
Distance from source (km)	190	300	380
Drainage basin (km^2)	9980	32 350	52 000
Mean annual flow (m^3/s)	200	460	630
Exceptional mean daily flow			
1875	8000		8000
1952	4300	6200	5300
Mean flow in August (every 10 years)	51	70	92

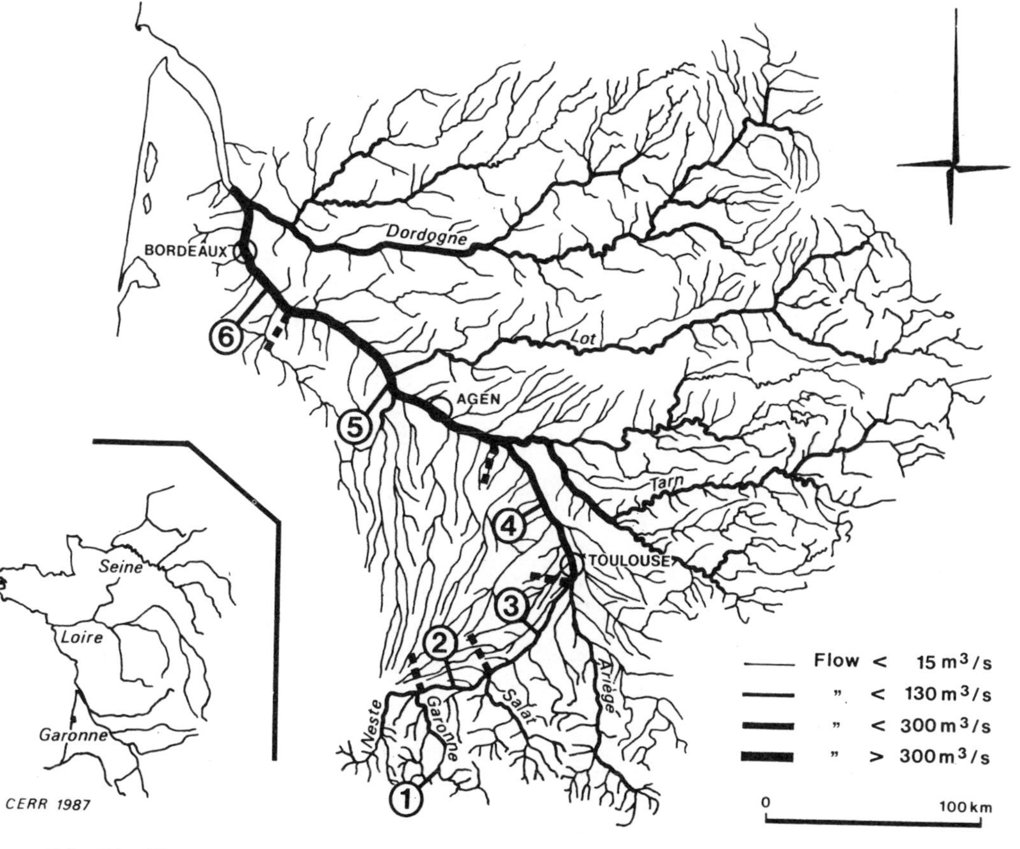

Figure 8.1 The River Garonne and its tributaries in south-west France

the floodplain, and it is, therefore, possible to consider the floodplain as an echo from the past (Heal, 1991). This echo comes from devices for flood and erosion control in the seventeenth century, construction of a navigation channel in the eighteenth and nineteenth centuries, and agricultural and urban developments of the nineteenth and twentieth centuries. The River Garonne, then, gives an example of a fluvial landscape essentially modified by agriculture and, more recently, by urbanization. The main purpose of the research program on the river is to explain the causes and consequences of man's modifications of the fluvial landscape.

The modifications of the fluvial landscape have been of two kinds. The first modification lies in an ever-increasing control of the water dynamics. The control has led to a reduction of the flooded surfaces and so to a transformation of the physio-chemical conditions of soils and waters. Moreover, the isolation of the river from its floodplain has greatly reduced the transport of organic matter to and from the river. The disappearance of numerous secondary channels and backwaters has equally contributed to this reduction. As a consequence, the exchange dynamics within the fluvial landscape in the floodplain has been modified. A second kind of modification, linked to the previous one, lies in the intensification of agriculture and urbanization. This intensification has resulted in a fragmentation of the riparian woods in the floodplain. The consequences of this are many, and among them are a lesser retention capacity of the organic matter in the floodplain forest, an alteration of the succession of plant communities in the riparian area, and a change in the structure of the animal communities, the birds in particular (Décamps *et al.*, 1987).

8.2.2 RIVER CHANGES IN AN URBAN AREA

The morphology of the River Garonne in Toulouse has undergone drastic changes since the end of the seventeenth century. These changes are the result of the interaction between the river flow and the activities of man designed to harness the river through the city of Toulouse (Fortuné, 1988). For example, it has long been impossible to maintain bridges across the River Garonne (Chalande, 1912). The oldest known bridge in Toulouse, constructed in the twelfth century of wood on a Roman foundation, is reported to have been destroyed and reconstructed six times after floods, and finally destroyed in 1523 never to be rebuilt. Such destructions of bridges during floods have been frequent. The first bridge able to resist floods has required long efforts. It was begun in 1543, and inaugurated in 1660, after more than a hundred years in construction (Lotte, 1982).

As in the case of bridges, dams constructed for mill activities were an obstacle to water flow. Mills were built in 1182 and 1190 (Sicard, 1953). Since the Middle Ages, these mills have had the greatest long-term effects on the river bed. In exchange for supplying the region with flour, millers were given the right to manipulate flows and to utilize the banks. Two stone dams were built in Toulouse at the end of the twelfth century to direct the river flow towards the mills. A consequence of the dams was an increase in the sedimentation of alluvial deposits and, therefore, an elevation of the level of the river bed. These modifications rendered the left bank liable to floods, provoking catastrophic destruction in the city during the eighteenth and nineteenth centuries. New embankments were built to prevent erosion and flood damage in the city. This construction and the consequent channel changes resulted again in an increase of flood levels, which required further work for flood and erosion control.

The analysis of ancient maps reveals that the number of islands was reduced from 17 to four over a 2.2-km distance during a 300-year period. At the same time, the shorelength was reduced by 4 km. The degree of these reductions lessened between 1777 and 1874, after the better flood control of the eighteenth century and before the connection of the islets to the banks during the twentieth century (Table 8.2). The historical facts show that man has modified the river bed in the city of Toulouse since at least the twelfth century. Ravaging floods which followed the construction of dams inundated portions of the city and, during the same time, regularly destroyed bridges and dams. Not until 1660, after a hundred-year period of construction, did a secure bridge join the two banks. This success marks the beginning of a period during which man achieved a better control of the river flow. Since this control resulted in increased erosion and sedimentation and an elevation of the level of the water during floods, more channelization was required, leading to a reduction in the number of islands and of the length of the banks in the urban area of Toulouse (Table 8.3).

Table 8.2 Modification of the fluvial landscape

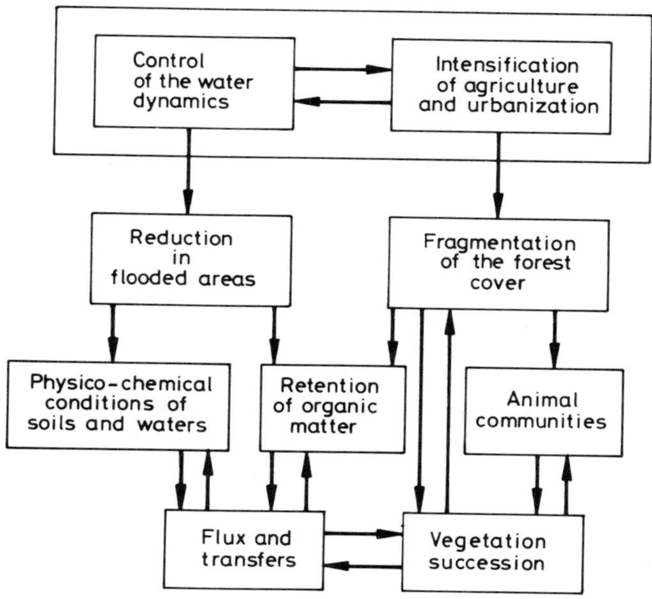

Table 8.3 Changes of morphological characteristics of the river bed as shown by the maps

	1680	1774	1847	1982
Number of islands	17	9	9	4
Area of islands in the river bed (%)	25	24	25	17
Length of shoreline (km)	11.4	9.5	9.6	7.2

The ecological consequences of river manipulations appeared gradually during the last three centuries. The number of habitats available to fish and other aquatic fauna diminished because of the reduction in the length of the banks and the replacement of a stony river bed by a uniform rocky bottom. Moreover, the organic matter contribution from the riparian vegetation became practically nil. In the long run, the utilization of the river resulted in economic prosperity, but also in disturbances in the aquatic environment.

8.2.3 CHANGES OF THE FLUVIAL LANDSCAPE

Three successive influences characterize the history of the entire Garonne fluvial landscape. These are navigation, agriculture, and urbanization and industrialization. A flourishing activity developed around navigation in the eighteenth and first part of the nineteenth centuries. This activity reached a maximum in 1840 to 1850, and took several forms as regards to the river and its floodplain. First, the consolidation of the banks and the construction of a towing-path resulted in a transformation of the riparian woods. Elms which had existed since the Middle Ages, for example, were replaced by more flexible cultivated willows, which were regularly cut down. Second, the channel itself was simplified by removal of secondary arms, and by a straightening of some sections. Third, the river bed itself was transformed to allow the passage of boats through rapids. Finally, the consequences of human concentrations in ports appeared along the river, one of which was the acceleration of deforestation along the floodplain.

During the eighteenth and nineteenth centuries, agriculture became a priority in the floodplain. Poplar plantations replaced natural riparian woods in many places, and cultivated areas approached the river. As a consequence, riparian woods became more fragmented along the River Garonne. Another consequence of the development of agriculture was an increasing need of water for irrigation of the alluvial plain, particularly during the twentieth century. Figure 8.2 shows the considerable increase of irrigation during the last few decades.

Finally, during the twentieth century, the river contributed to the development of urbanization in the alluvial plain. Solid material taken from the river bed more than doubled in some parts of the valley from 1966 to 1979 (Table 8.4). A spectacular effect of this removal has been the deepening of the water table (Figure 8.3) and, as a consequence, a further decline of the riparian woods.

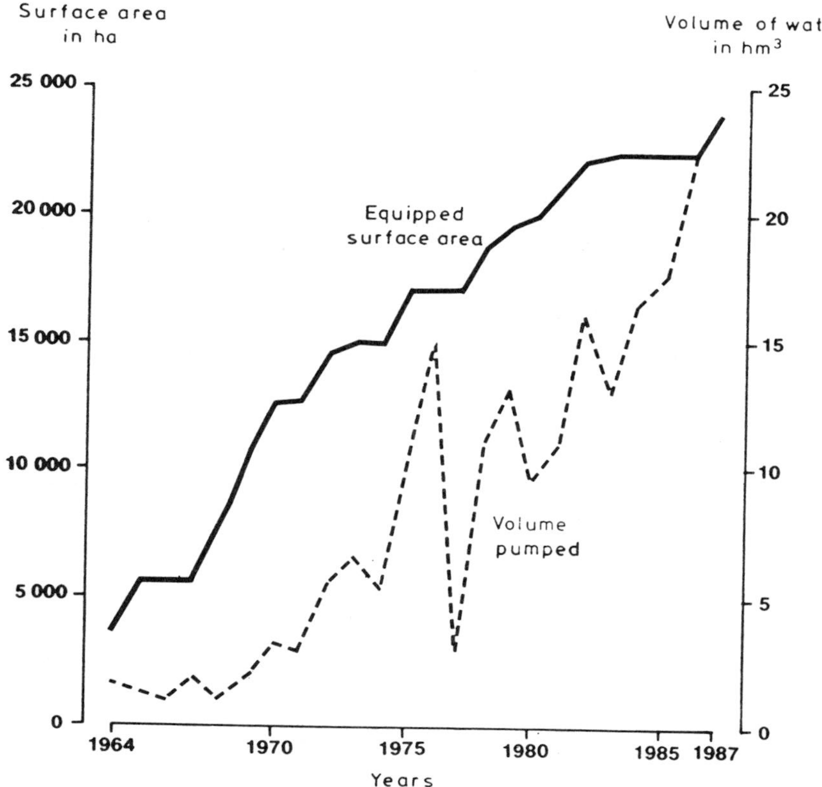

Figure 8.2 Increase of equipped areas for irrigation and pumped volume of water in the middle part of the Garonne floodplain

Table 8.4 Extraction of sand and gravel from the River Garonne and its environment within the limits of the administrative department of the Haute-Garonne from 1966 to 1979

Year	Extraction (10^6 tons)	Year	Extraction (10^6 tons)
1966	3.50	1973	4.76
1967	3.54	1974	6.17
1968	3.77	1975	6.32
1969	3.83	1976	8.17
1970	4.79	1977	7.77
1971	5.68	1978	8.03
1972	6.02	1979	8.87

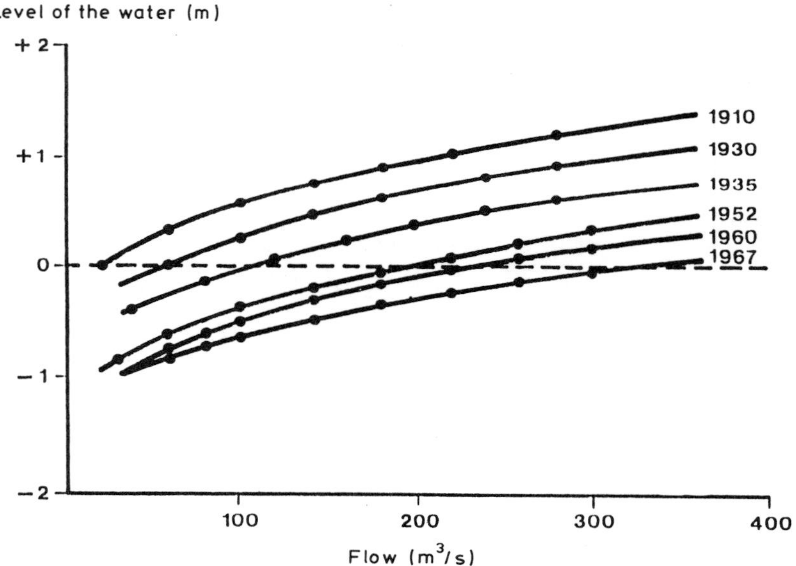

Figure 8.3 Diminution of the water level of the River Garonne from 1910 to 1967 at various flows (from Beaudelin, 1987)

8.2.4 CONTRIBUTION OF THE HISTORICAL PERSPECTIVE

The example of the River Garonne illustrates the contribution of a historical perspective to the understanding of the present tendencies in fluvial landscapes. The socio-economic development along the floodplain has led to a channelization of the river. This phenomenon was accompanied by a disappearance of the secondary arms, by a straightening of the river channel and by a stabilization of the banks and of the river bed. A first consequence of this was the increase of the hydrological contrasts between catastrophic spates and more severe low waters. A second consequence was the reduction in length of the boundary between land and water and, therefore, a reduction in the exchanges between the river and the floodplain. It is important to realize that this complex evolution, essentially due to human influences, began, at least significantly, at the end of the seventeenth century. The effects of these disturbances affect the dynamics of the fluvial landscape even to the present.

8.3 GLOBAL AND COMPARATIVE PERSPECTIVES

Global and comparative perspectives are also necessary in floodplain ecology to acquire an understanding of at least four phenomena: socio-economic influences, consequences of changes in the land use of drainage basins, and both fluvial and ecological dynamics.

8.3.1 SOCIO-ECONOMIC DEVELOPMENT

A fluvial valley is a natural as well as a social space, and the result of this is a great complexity of the systems involved. Clearly, a knowledge of the regional socio-economic development is essential to explain the present state of fluvial landscapes, but understanding of this socio-economic development requires a broader context (Richards, 1986). The main difficulty is to place local events in a more general context, at the level of the valley for example. A river like the Garonne is the sum of its geomorphology, hydrology, climate, and man's utilization, from the springs to the mouth. Each sector gives only partial answers to the understanding of the evolution of the whole. Also, each sector maintains with the other sectors complex and diverse relationships. It is necessary to take into account the upstream–downstream sequence of these systems as shown by the river continuum concept (Vannote *et al.*, 1980). It is also necessary to consider these systems in their nesting in larger systems. As an example, it is difficult to understand the evolution of the River Garonne in Toulouse without mentioning the successive constructions of weirs across the river beginning in the twelfth century. In turn, these local events are only understood in terms of economics. The supplying of the entire region depended on the functioning of the mills in the city of Toulouse. The maintenance of these mills and their associated dams, in addition to the flood damage, has deeply marked the fluvial environment at the level of the city.

More generally, from local events repeated over the decades, it is possible to reconstruct economic trends. The history of the river and the economic development have interacted, and together provide an explanation of the observed changes. However, it is necessary to go beyond such studies and make comparisons between diverse fluvial landscapes in order to interpret global and international patterns.

8.3.2 LAND-USE CHANGES

At the level of a drainage basin, changes in land use have proven to be of paramount importance for river ecology. As an example, the River Ain, a tributary of the River Rhône, was studied by Bravard (1987). The lower part of the River Ain is sinking into its sediments, particularly since the beginning of the century. This sinking has been about 1 m in the last 10 years. Several factors explain this phenomenon, among which are the replacement of cultivated areas by forest in the upper drainage basin, the construction of a hydroelectric dam in the middle course of the river 20 years ago, and a change in the river discharge regime. A drop in the base level of the River Rhône due to embankment construction since the nineteenth century has accelerated the tendency of the River Ain to incise its downstream course. Moreover, because of man's activities in the upper part of the drainage basin, the load of solid material has practically disappeared in the River Ain. Therefore, the energy of this river is utilized to incise its own sediments in its lower part. As a consequence, the river does not meander in its alluvial plain and is not able to rejuvenate the riparian communities which characterized the system in early times (Pautou and Girel, 1986). Only an approach that considers the entire basin, taking into account past utilizations of the upper part of the drainage basin, may help to understand the present dynamics of the River Ain system in its lower part.

8.3.3 FLUVIAL DYNAMICS

The fluvial dynamics of most European rivers has been modified by man's activities. Braga and Gervasoni (1983) have reconstituted the natural variations of the River Pô since the sixteenth century. They demonstrated that the meanders of the River Pô moved largely within the floodplain at the level of Plaisance, Italy. Such natural dynamics are now impeded by the riverbanks built to stabilize the river. Similarly, the Rhine was transformed for navigation from a meandering river before 1830 to a straightened channel after 1860, and then to a channelized river after 1950. As a result of these modifications a unique and linear course has replaced a braided, anastomosed and meandering section between Bâle and Mainz (Carbiener *et al.*, 1986). This tendency of channelization and linearization of the water courses under the influence of man is general in developed countries, as illustrated by works on various rivers: the Danube (Bacalbasa-Dobrovici, 1989), the Rhône (Bravard *et al.*, 1986), the Garonne (Fortuné, 1988), the Rio Grande (Boggs, 1940), the Williamette (Sedell and Froggatt, 1984), and various British rivers (Petts, 1988). A comparative approach is needed to understand the various ecological consequences of this general trend.

Clearly, the best theoretical basis for such a comparative approach is given by the river continuum concept proposed by Vannote *et al.* (1980), and refined by various authors (Ward and Stanford, 1983; Minshall *et al.*, 1983, 1985; Statzner and Higler, 1985; Naiman *et al.*, 1987).

8.3.4 ECOLOGICAL DYNAMICS

Before human influence, riparian systems were characterized by periodic successions dependent upon flood cycles and their severity. Human activities have modified, and in some cases interrupted, the previous succession patterns. Along many large rivers, embankments have isolated the channels from their terrestrial surroundings, and therefore modified the natural dynamics of the riparian communities. To understand the ecological consequences of these modifications, we need to improve our theoretical knowledge of lotic ecotones, in particular as regards to their function in a river system. First, the riparian vegetation has its own dynamics, with successions in time and space as studied along many rivers in the world (Pautou and Décamps, 1985; Bravard *et al.*, 1986; Carbiener *et al.*, 1986;

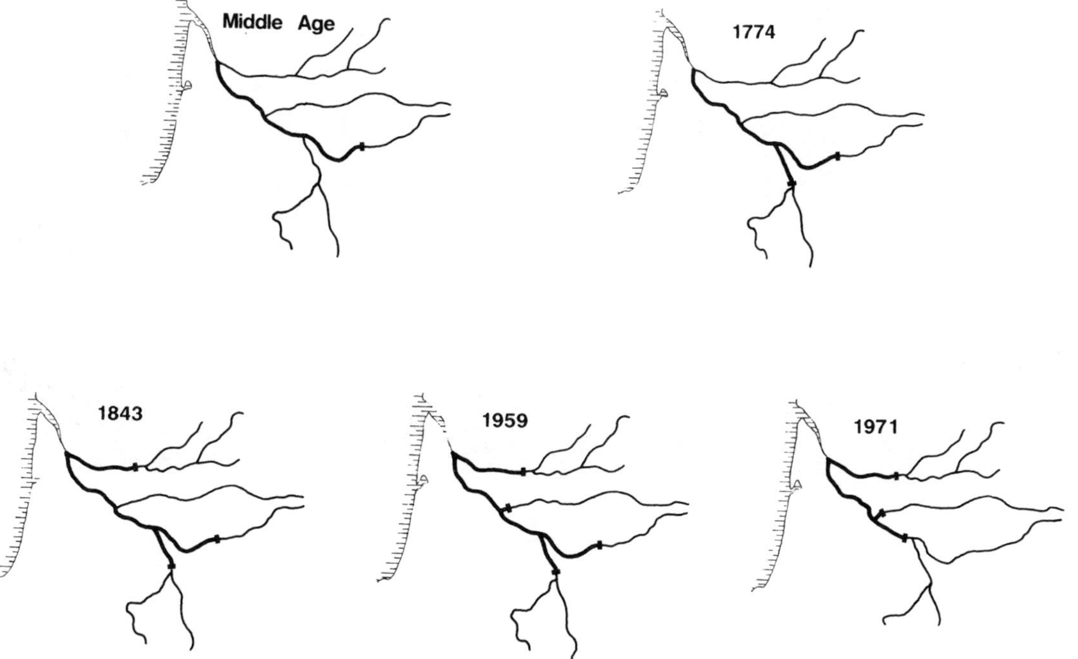

Figure 8.4 Dam construction and reduction of spawning areas for migratory fish in the Garonne network

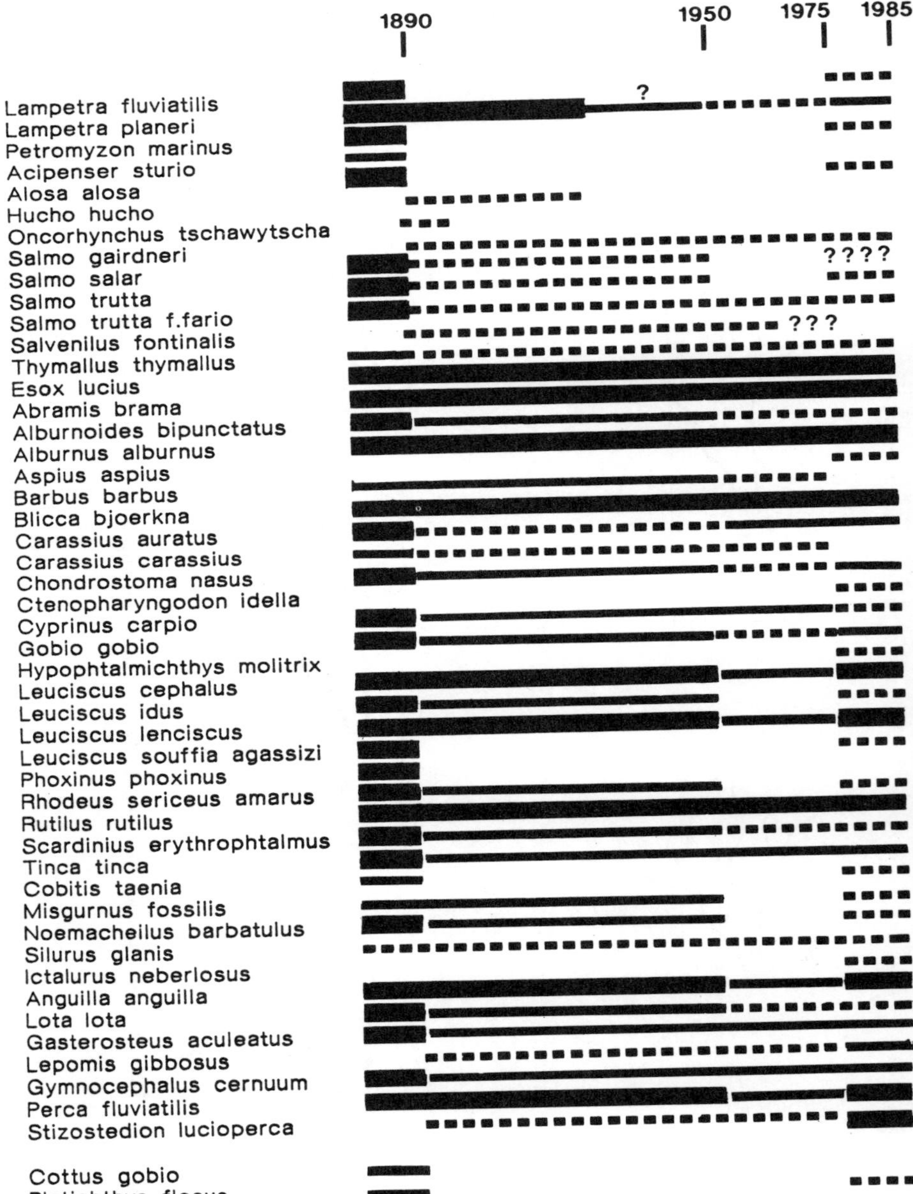

Figure 8.5 Changes in the specific richness of the fish community in the River Rhine. The species are classified as abundant, rare, and sporadic according to Lelek (1989)

Walker *et al.*, 1986; Salo *et al.*, 1986; Décamps *et al.*, 1988). Second, the riparian vegetation plays a role in the retention of nutrients transported downstream (Ward and Stanford, 1988; Pinay *et al.*, 1989). This retention function of the riparian vegetation is to be considered at the different stages of the succession from recent alluvial deposits to pioneer vegetation and to willow and alder communities. The capacity of retention of these different stages must be characterized as well as the turn-over length of nutrients on the floodplain. For this, as discussed by S.T.A. Pickett in Chapter 5 of this volume, space substitutes for chronosequences are not sufficient, i.e. succession cannot be reconstructed by comparing different sites at different stages at the same time. Comparative long-term studies within a network of sites are necessary to get a general understanding of the role of the different stages of the ecological successions on river banks.

The specific richness of fish communities in rivers has been subject to variations due to human activities. A first example, for the River Garonne, is given by the reduction of spawning sites for migratory fish from the Atlantic Ocean (Figure 8.4). The successive constructions of dams have considerably reduced the ability of salmon and aloses to attain the upper parts of the hydrographic network of this river and its tributaries since the Middle Ages. A second example, for the River Rhine, is given by the changes in the specific richness of the fish community due to pollution during the last decades (Figure 8.5). Lelek (1989) classified the fish species according to their richness as abundant, rare, and sporadic. He was able to show that for more than 50 species recensed, 27 changed to an inferior category of abundance between 1890 and 1950, and 17 between 1951 and 1975, whereas 18 species changed to a superior category of abundance between 1976 and 1985, and only two species diminishing during this last period. This recovery since 1976 is confirmed by other observations made in 1985, even though some species practically disappeared during the 1950s.

8.4 QUESTIONS FOR A LONG-TERM ECOLOGICAL RESEARCH PERSPECTIVE ON FLUVIAL LANDSCAPES

Four questions are to be considered in order to predict the dynamics of fluvial landscapes for the next 10 to 15 years. These relate to complex, rare, subtle, and slow phenomena which, according to Pickett (1991), are relevant to long-term ecological research.

Interactions between the resources linked to a river produce cascade effects, so that the utilization of a resource affects the use of other potential resources in the future. The interactions are complex and are not completely understood. As shown for the dynamics of the River Garonne in Toulouse, the interactions are due to links between socio-economic development and the ecology of the river. Conflicts of use have been frequent along rivers; for example, between navigation and agriculture, between industries and fishing, or between the needs of riverine societies in the upper and in the lower parts of the hydrological networks. Because of the complexity of these relationships, the consequences of land-use changes in the ecology of rivers need long-term research. At the same time, various time scales are necessary (Urban *et al.*, 1987), depending upon the subject under study. For a microbial population long-term can be weeks, for invertebrates it can be years, for some fish it can be decades, and for some riparian trees it is centuries.

Due to the interactions, fluvial landscapes appear as systems characterized by strong inertias. Therefore, important environmental tendencies in rivers and their floodplains are practically irreversible over the next decade or two, as stated by Munn (1987) for environmental systems in general. This is the case in the depression of the water table along many alluvial rivers in Europe. For most of these rivers, as yet unrevealed crisis conditions may already be in their initial stages. Because of the surprising character of such crises, a network of research programs is necessary to follow the pattern and cycles of changes in order to predict potential crises, and try to bring them under control.

It must be realized that a number of changes concerning fluvial landscapes appear to be unpredictable because they are rare and sometimes unique. How to adapt to this unpredictability is a very important question in river management. The difficulty is particularly clear when scenarios combining the effects of socio-economic development with environmental effects are to be elaborated at the scale of a fluvial landscape.

The importance of river degradations during the last decades has resulted in a recognition of the problems of river conservation. The public is becoming increasingly aware of these problems in many countries, even though some reactions are purely emotional and not based on a clear scientific knowledge of the dynamics of the fluvial systems. Two points are to be considered in relation to this. The first important point is to explain to the public the scientific basis of the patterns of change in rivers and fluvial landscapes; the second is to ensure full interaction between the public and the scientists in developing coherent decisions for conservation and management.

These four major problems—understanding the interactions between resources, inertia to reversal of geomorphic changes, unpredictability, and changes in public opinion—point to the need for a network of long-term ecological research on rivers, and, in addition, for a linked monitoring of the changes in fluvial landscapes.

8.5 MONITORING OF FLUVIAL LANDSCAPES

Remote sensing offers a particularly promising tool for monitoring the ecology of fluvial landscapes. Sudden, as well as longer-term, modifications of these landscapes often signal changes affecting drainage basins on a much broader scale. As an example, the canopies of riparian forests 'can give clues to ecosystem properties that might be precursors for a major change in structure and species composition' (Waring *et al.*, 1986). Remote sensing can be used to map, monitor, and measure such changes or damage along fluvial landscapes. The ecologists need to improve their ability to interpret remote sensing information, in order to define new questions of research compatible with the capabilities of existing satellites, and also to design sensors for detecting environmental problems concerning rivers and their landscapes. As shown by Cummins *et al.* (1989), a combination of remote sensing and selected on-the-ground measurements may be used to establish patterns of association between riparian vegetation communities and stream shredder population. In this connection, long-term monitoring may make it possible to ask some research questions which cannot otherwise be addressed.

8.6 CONCLUSIONS

The ecology of fluvial landscapes requires long-term research. Several decades are necessary to understand the changes that are taking place, and to predict the results of present tendencies. These predictions will be more precise if they are supported by a good knowledge of the past utilization of the river landscape. Knowledge of the historical effects of human societies is increasingly being taken into account in ecosystems research in North America and in Europe (Sedell and Froggatt, 1984; Bravard, 1987; Fortuné, 1988). Moreover, these predictions have a chance to be successful only if local or regional knowledge is included in a comparative and broader knowledge of the tendencies towards change along various fluvial landscapes in the world (Petts *et al.*, 1989). Only through a network of long-term research can the dynamics of fluvial landscapes be fully understood, and predictive scenarios, particularly of the ecological consequences of socio-economic development, be made.

8.7 ACKNOWLEDGMENTS

This work was supported by the Programme Interdisciplinaire de Recherche sur l'Environnement (PIREN) of the Centre National de la Recherche Scientifique. We thank K.W. Cummins for helpful comments on our manuscript, and A. Lelek for information on the River Rhine.

8.8 REFERENCES

Bacalbasa-Dobrovici, N. (1989). The Danube River and its fisheries. *Canadian Special Publication of Fisheries and Aquatic Sciences*, **106**, 455–468.

Beaudelin, P. (1987). Les méfaits des extractions des galets de la Garonne. *Adour Garonne*, **34**, 10–13.

Boggs, S.W. (1940). *International boundaries*. Columbia University Press, New York.

Braga, G. and Gervasoni, S. (1983). Evoluzione storica dell'alveo del fiume Pô nel territorio Lodigiano-Piacentin: rischi idrogeologici connessi. *Atti Convegno Nazionale 'Il suolo come risorsa'*, **60**, 59–69.

Bravard, J.P. (1987). *Le Rhône du Léman à Lyon*. La Manufacture, Lyon.

Bravard, J.P., Amoros, C. and Pautou, G. (1986). Impact of civil engineering works on the successions of communities in a fluvial system: a methodological and predictive approach applied to a section of the Upper Rhône river, France. *Oikos*, **47**, 92–111.

Carbiener, R., Dillman, E., Dister, E. and Schnitzler, A. (1986). Variations de comportement et vicariances écologiques d'espèces ligneuses en zone inondable: l'exemple de la plaine du Rhin. *Journées d'Hydrologie de Strasbourg Crues et Inondations*, 237–259.

Chalande, J. (1912). Les inondations et les formations alluviales dans le bassin de la Garonne à Toulouse depuis le douzième siècle. *Mem. Acad. Sci. Inscript. Belles Lettres Toulouse*, 10ème série, **12**, 1–16.

Cummins, K.W., Wilzbach, M.A., Gatges, D.M., Perry, J.B. and Taliaferro, W.B. (1989). Shredders and riparian vegetation. *BioScience*, **39**, 24–30.

Décamps, H., Fortuné, M., Gazelle, F. and Pautou, G. (1988). Historical influence of man in the riparian dynamics of a fluvial landscape. *Landscape Ecology*, **1**, 163–173.

Décamps, H., Joachim, J. and Lauga, J. (1987). The importance for birds of the riparian woodlands within the alluvial corridor of the River Garonne, SW France. *Regulated Rivers*, **1**, 301–316.

Décamps, H. and Naiman, R.J. (1989). L'ecologie des fleuves. *La Recherche*, **208**, 310–319.

Di Castri, F. and Hadley, M. (1988). Enhancing the credibility of ecology: interacting along and across hierarchical scales. *Geojournal*, **17**, 5–35.

Fortuné, M. (1988). Historical changes in a large river in an urban area: the Garonne river, Toulouse, France. *Regulated Rivers*, **2**, 179–186.

Heal, O.W. (1991). The role of study sites in long-term ecological research: A UK experience. Chapter 3 in this volume.

Lelek, A. (1989). The Rhine River and some of its tributaries under human impact in the last two centuries. *Canadian Special Publication of Fisheries and Aquatic Sciences*, **106**, 469–487.

Lotte, R. (1982). *Construction d'un Pont Sous la Renaissance, le Pont Neuf de Toulouse*. Ecole Nationale des Ponts et Chaussées, Paris.

Minshall, G.W., Petersen, R.C., Cummins, K.W., Bott, T.L., Sedell, J.R., Cushing, C.E. and Vannote, R.L. (1983). Interbiome comparison of stream ecosystem dynamics. *Ecological Monographs*, **53**, 1–25.

Minshall, G.W., Cummins, K.W., Petersen, R.C., Cushing, C.E., Bruns, D.A., Sedell, J.R. and Vannote, R.L. (1985). Developments in stream ecosystem theory. *Canadian Journal of Fisheries and Aquatic Sciences*, **42**, 1045–1055.

Munn, R.E. (1987). Environmental prospects for the next century: implications for long-term policy and research strategies. *Research Report 15*, International Institute for Applied Systems Analysis.

Naiman, R.J., Décamps, H., Pastor, J. and Johnston, C.A. (1988). The potential importance of boundaries to fluvial ecosystems. *Journal of the North American Benthological Society*, **7**, 289–306.

Naiman, R.J., Melillo, J.M., Lock, M.A., Ford, T.E. and Rice, S.R. (1987). Longitudinal patterns of ecosystem processes and community structure in a subarctic river continuum. *Ecology*, **68**, 1138–1156.

Pautou, G. and Décamps, H. (1985). Ecological interactions between the alluvial forests and hydrology of the Upper Rhône. *Archiv. f. Hydrobiol*, **104**, 13–37.

Pautou, G. and Girel. J. (1986). La végétation de la basse plaine de l'Ain: organisation spatiale et évolution. *Documents de Cartographie Ecologique*, **29**, 75–96.

Petts, G.E. (1988). Ecological management of regulated rivers: a European perspective. *Regulated Rivers*, **1**, 363–369.

Petts, G.E., Möller, H. and Roux, A.L. (Eds) (1989). *Historical Change of Large Alluvial Rivers. Western Europe*. John Wiley, Chichester.

Pickett, S.T.A. (1989). Long-term studies: past experience and recommendations for the future. Chapter 5 in this volume.

Pinay, G. and Décamps, H. (1988). The role of riparian woods in regulating nitrogen fluxes between the alluvial aquifer and surface water: a conceptual model. *Regulated Rivers*, **2**, 507–516.

Pinay G., Décamps, H., Chauvet, E. and Fustec, E. (1989). Lotic ecotones. In Naiman, R.J. and Décamps, H. (Eds) *Land/Inland-Water Ecotones: Strategies for Research and Management*. Parthenon Press, Carnforth.

Richards, J.F. (1986). World environmental history and economic development. In Clark, W.C. and Munn, R.E. (Eds) *Sustainable Development of the Biosphere*. Cambridge University Press, Cambridge, 53–74.

Salo, J., Kalliola, R., Hakkinen, I., Makinen, Y., Niemela, P., Puhakka, M. and Coley, A.D. (1986). River dynamics and the diversity of Amazon lowland forest. *Nature*, **322**, 254–258.

Sedell, J.R. and Froggatt, J.L. (1984). Importance of streamside vegetation to large rivers: the isolation of the Williamette River, Oregon, USA, from its floodplain. *Verh. Internat. Verein. Limnol.*, **22**, 1828–1834.

Sicard, G. (1953). *Aux Origines des Sociétés Anonymes les Moulins de Toulouse au Moyen Age*. Armand Colin, Paris.

Statzner, B. and Higler, B. (1985). Questions and comments on the River Continuum Concept. *Canadian Journal of Fisheries and Aquatic Sciences*, **42**, 1038–1044.

Strahler, A.N. (1957). Quantitative analysis of watershed geomorphology. *American Geophys. Union Trans.*, **38**, 913–920.

Urban, D.L., O'Neill, N.V. and Shugart, H.H., Jr. (1987). Landscape ecology. *BioScience*, **37**, 119–127.

Vannote, R.L., Minshall, G.W., Cummins, K.W., Sedell J.R. and Cushing, C.E. (1980). The river continuum concept. *Canadian Journal of Fisheries and Aquatic Sciences*, **37**, 130–137.

Walker, L.R., Zasada J.C. and Chapin III, F.S. (1986). The role of life history processes in primary succession on an Alaskan floodplain. *Ecology*, **67**, 1243–1253.

Ward, J.V. and Stanford, J.A. (1983). The serial discontinuity concept of lotic ecosystems. In Fontaine, T.D. and Bartell, S.M. (Eds) *Dynamics of Lotic Ecosystems*. Ann Arbor Science Publishers, Ann Arbor, Michigan, 29–42

Ward, K.V. and Stanford, J.A. (1989). Riverine ecosystems: the influence of man on catchment dynamics and fish ecology. *Canadian Special Publication of Fisheries and Aquatic Sciences*, **106**, 56–64.

Waring, R.H., Aber, J.D., Melillo, J. M. and Moore III, B. (1986). Precursors of change in terrestrial ecosystems. *BioScience*, **36**, 433–438.

9 Long-term Ecological Questions and Considerations for Taking Long-term Measurements: Lessons from the LTER and FIFE Programs on Tallgrass Prairie

TIM R. SEASTEDT and J.M. BRIGGS
Division of Biology, Kansas State University, Manhattan, Kansas 66506, USA

We have just enough time left in this century to achieve a major new synthesis and understanding of the Earth System... (NASA, 1988)

9.1 INTRODUCTION

The earth, with its global problems of overpopulation, over-use and abuse of fossil fuel and nuclear energy, and production of toxic wastes, has often been compared to a sick patient. Illness is recognized as a significant deviation from known, long-term trends. Long-term monitoring represents a minimal activity for responsible individuals and agencies interested in placing current environmental problems

Long-term Ecological Research. Edited by Paul G. Risser

into perspective. Long-term measurements are directed at questions involving phenomena not interpretable or perhaps not useful when viewed over short (annual or less) time scales, but are related to the long-term 'health' or functioning of the system. At a minimum, the Long-term Ecological Research (LTER) data therefore provide the context in which short-term observational or experimental results can be interpreted (Magnuson, 1990). A much more interesting, albeit potentially less relevant, use of LTER data involves the study of a set of complex questions that cannot be resolved with short-term studies (Franklin, 1989; Tilman, 1989). The juxtaposition of basic and applied science within the context of a single research effort is a strength of the LTER program.

This chapter attempts to identify a set of long-term ecological questions that are useful to a national or international network of research sites. While there exists a nearly infinite list of interesting questions that could be addressed with long-term studies, a realistic and goal-oriented list of measurements is presented. The criteria for selecting these questions involved identifying variables that (1) are useful for intersite comparisons, (2) are not strongly biased by spatial scaling factors, and (3) can provide the necessary linkages between atmospheric/climatological variables and biological measurements. 'Focused studies of the interactions between the atmosphere and the biosphere that regulate trace gases can improve both our understanding of terrestrial ecosystems and our ability to predict regional- and global-scale changes in atmospheric chemistry' (Mooney *et al.*, 1987). The list of proposed variables for study was developed from the 'core LTER measurements', a guideline used since the inception of the LTER effort (Callahan, 1984) from recommendations suggested in Earth System Science (NASA, 1988), and from practical experience with the recent NASA-ISLSCP (International Surface Land Climatology Project) conducted on the Konza Prairie LTER site (Sellers *et al.*, 1988). While appropriate examples are taken from many systems, particular emphasis has been given to questions that have interested researchers studying grasslands. We build on the work of Strayer *et al.* (1986). Their extensive overview of long-term studies provided useful definitions of research productivity, of what constitutes 'long-term research', and reasons for the 'successes' of previous and existing long-term research efforts. Their findings emphasized that individual scientists and not specific research protocols or experimental designs were largely responsible for successful long-term research efforts. Here, however, we suggest that certain constraints on research designs are important if a goal of the research is to benefit directly a regional or global network.

9.2 APPROPRIATE OBJECTS FOR LONG-TERM NETWORK MEASUREMENTS

The five core areas of the LTER include studies of the following topics (Callahan, 1984):

(1) Spatial and temporal distributions of populations;
(2) Patterns and frequency of disturbance;
(3) Pattern and control of primary production;
(4) Pattern and control of organic matter accumulation; and
(5) Patterns of inorganic input and movements through soils.

While excellent research has been done on some or all of these topics at one or more of the LTER sites, current efforts of linking sites in regional or global networks suggest that certain measurements are likely to be more useful than others.

Many of the most interesting and useful empirical studies of individual species have been long term in nature (Iker, 1983; Strayer *et al.*, 1986). For example, Weaver (1954) documented the response of the North American prairie species to climate and grazing intensities observed over a 40-year interval in the first half of the twentieth century. The rainfall and temperature conditions under which these studies were made are exemplified by data obtained for the Manhattan, Kansas, area (Figure 9.1). These data show that the great drought in North America during the 1930s was accompanied by (and contributed to) relatively high ambient temperatures. During the drought, Weaver documented the eastward advance and expansion of xeric, shortgrass species at the expense of mesic, tallgrass vegetation. The return of relatively wet years in the 1940s reversed this trend. The annual values of temperature and precipitation shown in Figure 9.1 indicate that 'average' conditions for the prairie cannot be expected without an approximately 20-year record. Even then, the factors that govern patterns of species composition and abundance may be misinterpreted or overlooked. Weaver (1954), for example, did not appreciate or acknowledge the role of fire in suppressing the invasion of woody species onto the tallgrass prairie, nor did he notice that the productivity of the dominant species was often enhanced by frequent fires.

Studies of within- and between-habitat species diversity remain of keen interest to many ecologists. Nonetheless, we suggest that individual species, species lists, or indices derived from species lists make poor primary intersite comparison measurements. Many species are not found across large environmental gradients. Those species that do cover regional areas are not physiologically identical across these regions. The relevant units to address intersite comparisons should confer equivalency across sites, and these units should aggregate into meaningful values at different spatial scales. Energy and mass (including elements, trace gases, etc.) are obvious candidates for study. Biologists must still focus on the biota as cause and effect participants in energy and mass transformations, but both the forcing functions and the response variables must employ units common to all sites. Eventually, life history characteristics and physiological responses of the individual species will provide a mechanistic interpretation of site-specific responses. Even then, however, these responses will be governed by spatial patterns not often measured in population studies (Huston *et al.*, 1988).

All LTER sites have been charged with studying 'disturbance' as a core measurement. Our own experience with this topic has suggested serious problems

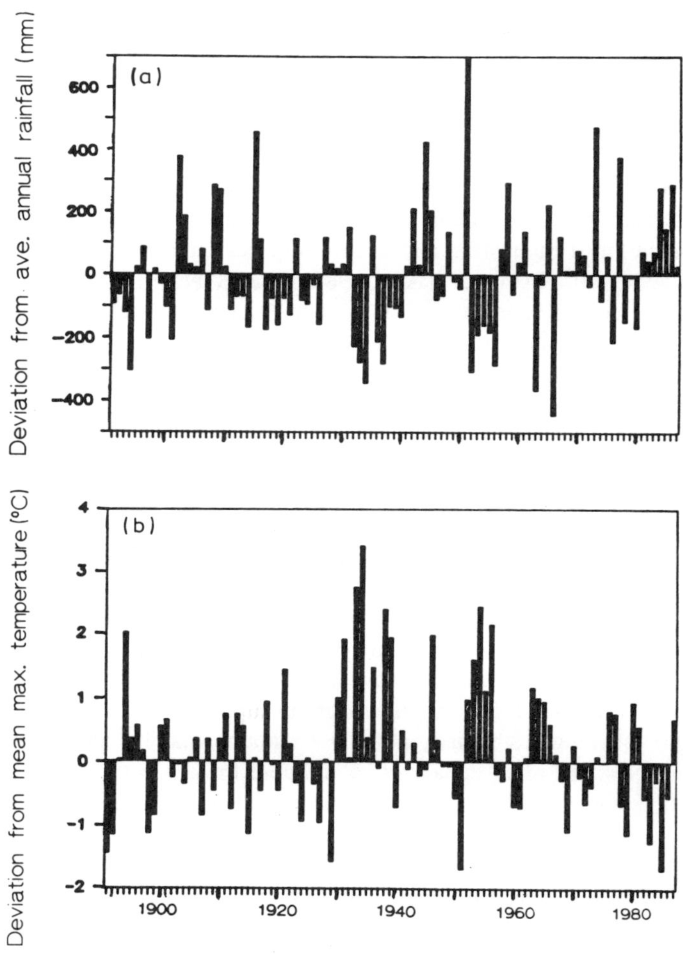

Figure 9.1 Deviations from average annual precipitation (a) and mean maximum temperatures (b) for an area near Manhattan, Kansas and the Konza Prairie LTER site. The data are for the period 1891–1987. Note that the temperature maxima corresponded with low precipitation values during the 1930s

associated with the concept that may prevent 'disturbance ecology' from becoming a major tenet of ecological theory (Evans *et al.*, 1989). One problem has been the popularity of the topic, and the inevitable misuse of the term that comes with popularity. 'Disturbance' is used simultaneously to describe a system input (e.g. a storm) and system output (e.g. species die-off) (Rykiel, 1985). Obviously, the latter is the interaction of the system with an input, and is therefore very much a characteristic of the system while the former is uncontrolled by the state of the system. A second problem with disturbance theory is that identical inputs can produce very different outputs depending upon the initial state of the system and the scales at which the output is measured. For example, fire adversely affects a number of populations of plants and animals in the tallgrass prairie. Nonetheless, certain species are benefited and periodic fires are required for the perpetuation of the system. Is fire or the absence of fire the disturbance in this system? Can systems lacking stable equilibria be disturbed? System-level properties of resistance and resilience to disturbances can be viewed more logically and mechanistically as consequences of structural and life-history characteristics of biological systems. Different disturbances (fire, drought, grazing, etc.) produce very different species and system responses. Konza Prairie researchers found the discussions about species and ecosystem responses to 'disturbances' to be largely an exercise in after-the-fact descriptive ecology and a topic not conducive to the development of predictive models. A much more productive approach to generic disturbance-type questions involves explicit identification of forcing functions and the responses of the system at specific levels of resolution. In other words, we believe that the LTER core area involving disturbance can be adequately addressed within the context of studies focused on the other core areas. This is certainly true in grasslands and agroecosystems where studies use fire, grazing, or tillage practices as experimental manipulations.

The remaining three core areas of the LTER program (net primary production, organic matter, and nutrient dynamics) provide a logical, unified focus for regional and global networks. These core areas employ units that are constants and provide the direct links between biotic and atmospheric processes. A combination of relatively new, spatially explicit measurements, in conjunction with traditional methodologies, will allow ecologists to study biota-climate interactions while concurrently focusing on questions of local interest.

9.2.1 PRIMARY PRODUCTIVITY

Forested sites have considerable potential to demonstrate the linkages among net primary productivity, trace gases, and climatic changes. Dendrochronology studies have used annual woody growth increments to reconstruct recent past climates. Other studies have combined paleobotany, records of lake ash deposition, and dendrochonology to reconstruct forest species composition, fire frequency, and growth relationships. Clark (1988) demonstrated the relationship between climate and fire frequency which, together, shaped the species composition and productivity of the north temperate forests. Of particular interest has been the work of LaMarche *et al.* (1984) which suggests that subalpine forests in western North America began to alter their growth patterns with respect to climatic variables some time in the 1960s. Those authors suggested CO_2 enrichment as a possible factor. Anthropogenic sources of nutrients in bulk precipitation could, perhaps, be an alternative hypothesis. Regardless, the measurement of woody growth and, therefore, a record of the past productivity is possible at many sites, and is a reasonable, partial index of above ground net primary productivity. Such data are particularly desirable since (1) sampling can be accomplished on a very infrequent, year-to-decades basis, (2) large sample sizes can be obtained and potentially interacting variables (soils, species, etc.) can be evaluated, and (3) the samples can be easily archived so that future analyses or reanalysis of the same, original data set are possible. To complete the story of above ground productivity, foliage production should be measured. Litterfall or needle production measurements and procedures are common, but should be supplemented, if possible, with satellite-derived digital images. These images can provide a spatial perspective not possible with microplot measurements, and the types and uses of currently available satellite images are discussed below.

Retrospective analyses of grassland productivity cannot be as easily accomplished as forest studies. Sedimentation rates of glacial lakes, in conjunction with pollen analyses, may provide some useful historical data. Also, carbon isotope studies of sediments, soils (including paleosols), and groundwaters in conjunction with these or other research may also provide an interesting story, particularly with respect to changes in the composition of C_3 and C_4 plants (O'Leary, 1988).

Table 9.1 Characteristics of current satellite imagery[a]

Satellite	Pixel size	Return time	Comments/example of users
NOAA AVHRR	1.1 km	Daily	Five channels with three thermal bands/Tucker *et al.* (1985)
Landsat MSS	80 m	16 days	Four channels with no thermal bands/White and MacKenzie (1986)
Landsat TM	30 m	16 days	Seven channels with one thermal band/Ustin *et al.* (1986)
SPOT	10 m, 20 m	Programable, 26-day nadir	10 m pixel panchromatic, 20 m pixel 3 channels, no thermal measurements/Hardisky *et al.* (1989)

[a] Summarized from Greegor (1986). Future instrumentation descriptions are in NASA (1988)

More recent retrospective analyses of indices of grassland productivity can also be conducted using the satellite imagery. Researchers and sites should move quickly to secure these images lest useful information be lost by agencies not funded as data archives. A listing of potential data sources (Table 9.1) indicates the resolution and information available from each type of satellite. Investigators need to be aware of the various trade-offs involved in using these various types of data, and some important considerations are outlined in Sellers *et al.* (1988). In general, we believe that the high spatial resolution (small pixel size) of the Landsat TM or SPOT satellites is extremely useful in evaluating within-site topoedaphic or experimental (fire or grazing) effects. However, a seasonal time-series of these types of data is expensive or simply unlikely to be obtained due to relatively infrequent overflights in conjunction with moderate to high probabilities of cloud cover. In contrast, the NOAA-AVHRR satellite provides relatively low spatial resolution (large pixel size) but high temporal resolution, making cross-site, cross-year, and seasonal comparisons possible.

The potential for using these images as analogs of regional productivity and for estimating trace gas interactions and energy exchange is just beginning to be developed. Recent improvements of algorithms, particularly those employing the vegetation index (Tucker *et al.*, 1985; Goward *et al.*, 1986) or some combination of the vegetation index in conjunction with thermal measurements (Forrest Hall *et al.*, unpublished results), can demonstrate both seasonal and long-term trends in plant biomass and plant vigor. We expect that the more sophisticated, high-resolution imaging spectrometers scheduled for space orbit in the near future will provide more useful data for measuring both biomass and plant productivity at moderate scales. This enhancement begins with the anticipated launch of Landsat 6 with the Enhanced Thematic Mapper (ETM) on board. Eight bands of spectral information are planned, four in the visible (one being a 15 m pixel panchromatic), two in the near-infrared, and two in thermal portions of the spectrum. This system is reported to be very sensitive to surface temperature changes and should, therefore, be very useful in relating vegetation dynamics to energy dynamics. Subsequent satellite equipment scheduled for the Earth Observing System (EOS) program will make considerable advancements in the spectral resolution of these digital images. These standard products could also be supplemented with aerial photography, including standard panchromatic, color, and color IR images. Photographic records are proving useful for a variety of retrospective analyses.

9.2.2 THE INTERACTION BETWEEN PRODUCTIVITY AND SURFACE CLIMATE

A conceptual model developed by Shugart (1986) (Figure 9.2) suggests how we might think about the relationship of LTER measurements to studies involved in trace gas fluxes. The latter measurements are, by necessity, made on a scale that detects strong diurnal and seasonal fluctuations. In contrast, LTER measurements of NPP, organic matter, or elements have a much coarser temporal scale. However, as suggested by the model, these long-term ecological processes function as constraints on short-term physiological processes, and therefore mediate the response of vegetation to climate. We present an example of this phenomenon to emphasize the need to recognize that changes in ecological constraints such as fire frequency, herbivory, or nutrient availability may temporarily overshadow direct changes in temperature or rainfall.

Our data on temperate grassland plant productivity demonstrate a strong relationship between the type of management treatment and productivity (Figure 9.3). The tallgrass prairie requires periodic fires to maintain its species composition

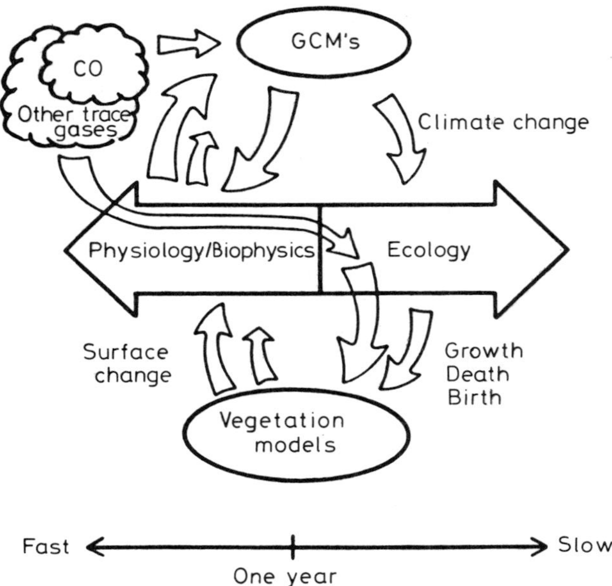

Figure 9.2 Conceptual model by H.H. Shugart suggesting the relationships between LTER-type measurements (right side of figure) and those variables strongly influenced by diurnal variations (left side of figure)

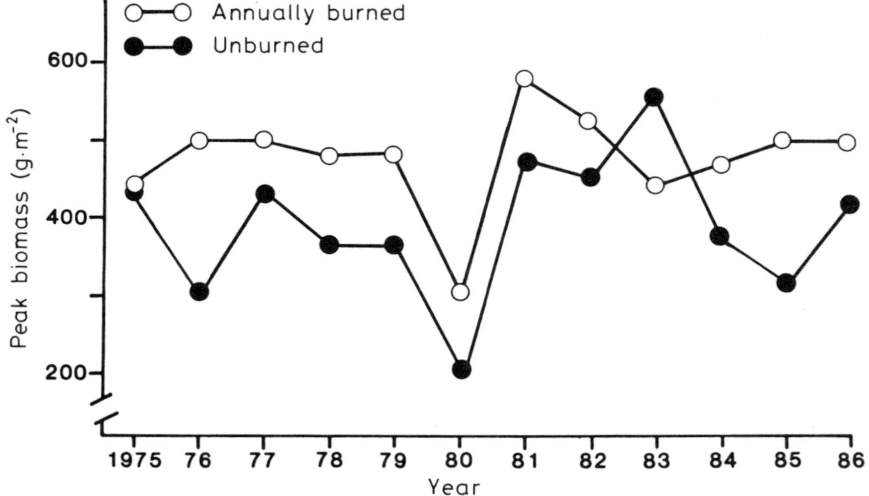

Figure 9.3 Time series of maximum foliage production on annually burned and unburned prairie. Year-to-year climatic fluctuations affect the vegetation response to treatment

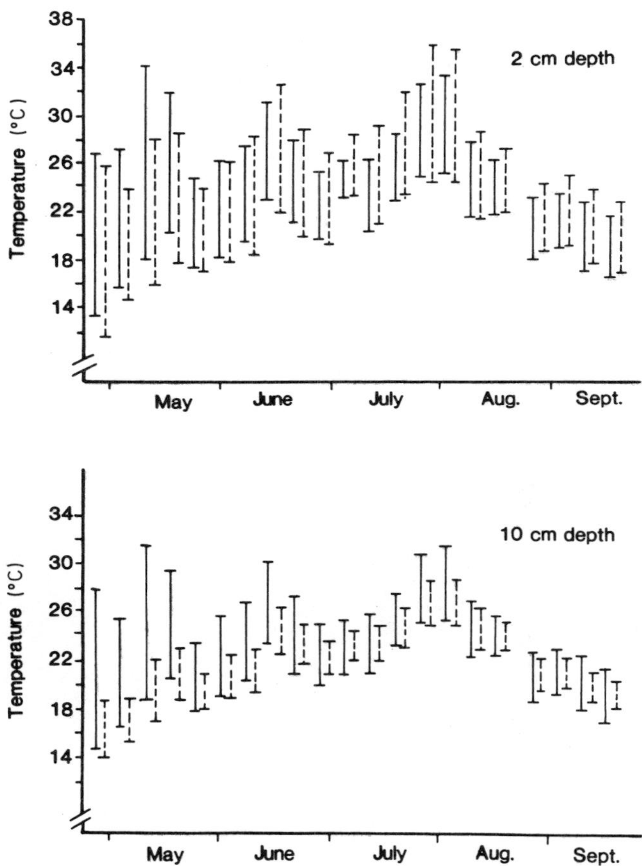

Figure 9.4 Weekly mean minimum–maximum soil temperatures in summer 1987 for burned (solid lines) and unburned (dashed lines) tallgrass prairie at 2 cm and 10 cm soil depths. Note that temperatures are relatively cooler on the burned site at 2 cm but are relatively warmer at 10 cm

and productivity (Knapp and Seastedt, 1986). In average or wet years, annual burning in late spring benefits the C_4 grasses. However, some but not all drier than average years result in more productivity by the combination of C_4 and C_3 grasses, forbs, and woody species found in the unburned prairie. Following a fire, the blackened soil surface of burned prairie is exposed to direct solar radiation and converts much of this energy into sensible heat absorbed by the soil (Figure 9.4). However, by midsummer, the re-establishment of the canopy, in conjunction with greater rates of evapotranspiration, results in a cooler soil surface. This pattern is reversed at the 10 cm depth, where the drier soils on the burned sites lack the thermal inertia of generally moistened, litter-covered soils of the unburned sites. The greater rates of evaporation coupled, perhaps, with higher rates of reflected infrared radiation keep burned areas cooler in midsummer than unburned areas (Figure 9.5). The thermal (channel 6) Landsat TM image in Figure 9.5 shows that the radiometric brightness on burned watersheds is, on average, less than that measured for adjacent unburned areas (Figure 9.6) (Asrar *et al.*, 1988). These data demonstrate that the ecological constraints operating on the vegetation (here, a spring fire) influence both the hydrologic and energy budget. These changes are detectable at both a micro- and macro-scale level. Obviously, a change in the fire frequency of relatively large tracts of grassland could have an impact on the regional climate.

Grazing by cattle also had a measurable affect on canopy temperatures as measured by the radiometric brightness of the TM image (Figures 9.5 and 9.6). Grazed areas were cooler in August, presumably because the grazed vegetation was

Figure 9.5 A Landsat TM thermal (channel 6) photo of Konza Prairie. Watershed boundaries have been superimposed over the image. Burned watersheds or grazed pastures are distinguishable by the darker pixel values

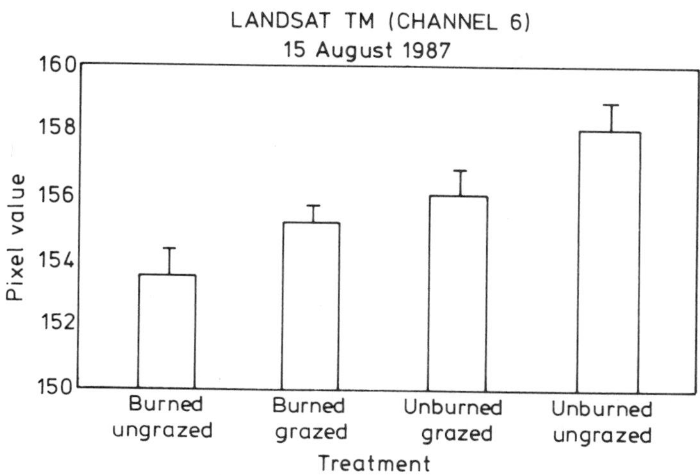

Figure 9.6 Means and standard deviations of pixel values of radiometric canopy brightness obtained from Figure 9.5. All treatments exhibit statistically different values

physiologically more active than a similar amount of ungrazed vegetation and was transpiring relatively greater volumes of water. Consumers affect both the amounts and physiology of the vegetation and thereby can greatly alter vegetation–climate interactions, particularly in grasslands. Investigators should also be aware that interactions between energy and nutrients may affect consumers, so that consumers become important transient controlling factors on net primary productivity (White, 1984). These controls can operate directly via consumption of plant parts or indirectly by controlling plant species composition (Schowalter, 1981). Thus, knowledge of consumer populations may contribute to an understanding of vegetation–climate interactions. This observation also has particular relevance in agroecosystems, where biotic mechanisms of consumer regulation have been severely altered.

9.2.3 NUTRIENTS

Virtually all LTER sites measure nutrient inputs, standing crops, and outputs. The input data may be restricted to analyses of wetfall, often associated with the National Atmospheric Deposition Program (NADP). This measurement is often inadequate because dryfall deposition or deposition associated with dew can be considerable (Lindberg *et al.*, 1986). Most sites obtain pH measurements in conjunction with the inputs of nitrate, ammonia, sulfate, and the major cations. Measurements of the standing crops of the major elements in vegetation were initiated at many sites during the International Biological Program. It is to be hoped that such data have been archived for future analyses or as baselines for future comparisons. Our site archives plant and soil samples along with the numerical data, and this procedure has received endorsement by other research groups (Pace and Cole, 1989). To our knowledge, no LTER site has engaged in long-term monitoring of net inputs or outputs of trace gases (CO_2, NO_x, NH_3, H_2S, or SO_2). However, with the advent of large path-length infrared spectroscopy (Gosz *et al.*, 1988), and procedures to estimate fluxes, this deficiency should be resolved, at least at a few sites. Moreover, as mentioned above, the trace gas fluxes are diurnal phenomena operating under the ecological constraints being studied by the LTER. Empirical results and modeling efforts currently underway as part of FIFE (First ISLSCP Field Experiment) at Konza Prairie should be able to tell us the relationships and sensitivity of measurements such as productivity to short-term and seasonal estimates of gas flux.

Nutrients become constraints on plant growth during periods when energy and water are not limiting, i.e. under conditions otherwise favorable for plant growth. An obvious question of interest to those involved with climatic change studies is the extent that nutrient limitations may mediate vegetation responses to enhanced CO_2 (Tissue and Oechel, 1987). If plant growth is nutrient as opposed to energy limited, then carbon dioxide enrichment and/or increased temperatures should not immediately affect productivity. In tallgrass prairie, an improved energy environment (created by fire) results in a higher nitrogen use efficiency (NUE) of the vegetation (Ojima, 1987). With this greater production, however, comes increased detritus build-up and nutrient immobilization. In several biomes, including the taiga (van Cleve *et al.*, 1983) and tallgrass prairie (Knapp and Seastedt, 1986), plant litter has a direct negative physical effect on energy availability to plants. Detritus production could, therefore, affect productivity both by affecting usable energy inputs and by influencing nutrient availability. Seasonal shifts in energy, nutrient, and water limitations, in conjunction with negative feedbacks resulting from biomass production, prevent these ecosystems from maximizing their production responses.

The need for long-term nutrient measurements relates to the fact that climate can influence both amounts and availability of nutrients. For example, the rainfall data shown in Figure 9.1 indicate that precipitation at the Konza Prairie LTER site was above normal for the period 1981 to 1987. Accordingly, plant productivity was above the long-term average during this interval. Since inorganic nitrogen availability in soils is inversely related to the amount of 'new' fixed carbon present, the organic matter build-up during this interval undoubtedly adversely affected inorganic nitrogen availability to plants. Generalizations about nitrogen availability and cycling and its importance to vegetation made during this wet interval are therefore biased and potentially incorrect, in spite of a seven-year data base.

Agroecosystems have additional nutrient inputs and outputs not found or not important in natural systems. Nutrient supplements from fertilizers and outputs in the form of harvested plant parts tend to create an artificially dynamic system. Areas employing irrigation also have potential additional exports of trace gases or leaching losses, and certain agricultural practices, such as conversion of largely aerobic, vegetated sites to largely anaerobic rice fields, are probably having a large effect on trace gas dynamics (Mooney *et al.*, 1987). A detailed accounting of these nutrients is warranted given the progressive enrichment of groundwaters with undesirable organics and nitrates.

Moreover, the tillage of the soil, the artificial, excessive harvesting of plant nutrients, in conjunction with applications of supplemental water and fertilizers, have created unique situations of nutrient limitation, soil acidification and aluminum toxicity problems for agricultural systems (Adams, 1984). Indeed, many sites have been so totally altered by intensive agricultural practices that moderate changes in temperature, rainfall, rainfall chemistry, or rainfall pH would appear of secondary consideration relative to the direct human manipulations. The relevant emphasis from a network standpoint is, therefore, not how these systems are affected by climate change scenarios, but rather how the systems are affecting regional energy and trace gas dynamics. The ecological constraints of agroecosystems are the crop and tillage manipulations. These, like fire and grazing in the prairie, control the system interactions and responses to climatic inputs.

Measurement of nutrient outputs from ecosystems has proved to be an extremely relevant and useful long-term index of integrated system behavior. Likens *et al.* (1977), Likens (1983) and Driscoll *et al.* (1989) have provided ample examples of these measurements. Their 25+ year effort on the relationships between nitric and sulfuric acid rain inputs and stream pH and stream nutrient responses comprises some of the most relevant and important ecological research of this century. Ironically, this work began with some focused, short-term experimental studies, but the utility of these measurements for questions requiring a longer study period became obvious shortly after the initiation of the experiments. Stream chemical analyses have provided a measurement of the integrated ecosystem response to changes in atmospheric inputs or to those induced by within-system manipulations. In similar fashion, the new generation of remote sensing equipment scheduled for earth orbit within the next 10 years should provide equivalent information for terrestrial systems. Multispectral scanner, high-resolution sensors will provide a spatially explicit measurement of the integrated landscape response to changes in atmospheric inputs and landscape manipulations. Certain chemical properties of vegetation such as water status and nitrogen content can already be measured to some extent with current satellite data (Rock *et al.*, 1986; Waring *et al.*, 1986).

9.2.4 ORGANIC MATTER

Plant detritus and soil organic matter provide the major reservoir of nutrients in most terrestrial ecosystems. This storage component provides the 'resistance' of the system to changes caused by the destruction of the vegetation. The tropics-to-taiga gradient in organic matter is an example of the interaction between net primary productivity, decomposers, and climate (Swift *et al.*, 1979). Any brief interpretation of this pattern is an oversimplification. Nonetheless, plants appear to have dealt better with climatic restraints than have the decomposers. In the United States the east-to-west gradient in soil organic matter observed across the prairie is largely controlled by moisture (Jenny, 1930). Prediction of changes in the organic reservoir, therefore, potentially depends upon the interaction of temperature and moisture, and the net effects that these variables have on production and decomposition (Hunt *et al.*, 1988). Soil organic matter measurements tend to be rather insensitive to short-term manipulations of productivity and decomposition, but should be useful monitors of long-term changes (Jenny, 1980; Ojima, 1987; Ojima *et al.*, 1990). Moreover, such data are generally available on a regional basis, and have been modeled very successfully using climate and management constraints as forcing functions (Parton *et al.* 1987a,b).

Investigators need to recognize that edaphic factors and climatic variables produce interaction effects that add to the complexity of regional patterns. A

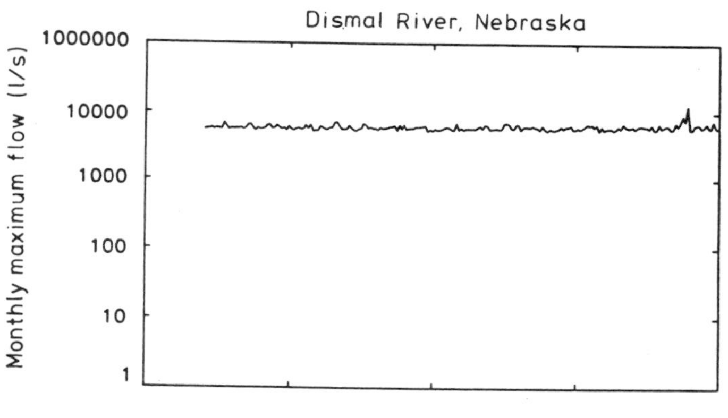

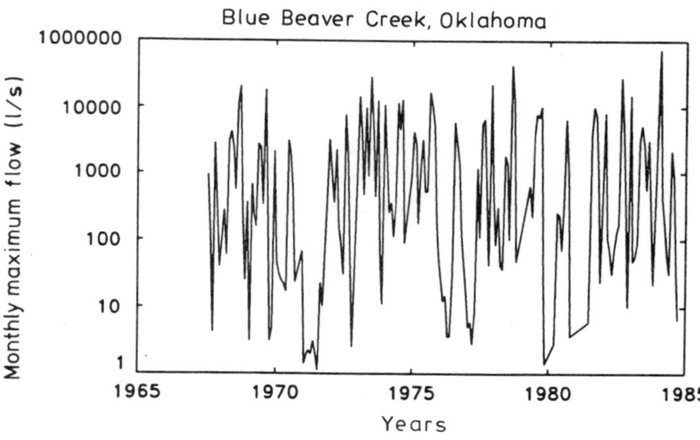

Figure 9.7 An 18-year record of maximum monthly streamflow from two US Geological Survey Benchmark watersheds (data courtesy USGS)

recent example is from Sala *et al.* (1988). That study found that sites with coarse, sandy soils were relatively more productive than fine, clay soils under below-average rainfall, while clay soils were relatively more productive under average or above-average rainfall. The coarse soils lacked the fertility of fine soils, but tended to allow water to penetrate below the zone of evaporation. Hence, the variability in productivity in coarse soils was reduced (i.e. productivity in wet years was diminished while the consequences of drought years were less severe). This phenomenon, when linked to a climatic gradient, produces a complex pattern (Figure 9.7). The relevance of these findings to models linking ecosystems to global climatic models should be particularly obvious.

9.2.5 SCALING AND SAMPLING CONSIDERATIONS

The problem of how to integrate point measurements so that these data can be useful and accurate estimates of regional dynamics remains unresolved. A large literature on scaling is developing (Allen *et al.*, 1984; Urban *et al.*, 1987). By far the most productive approach we have seen involves the use of explicit spatial models to aggregate ecosystem processes (Huston *et al.*, 1988). Successful large-scale regional models of net primary production, nutrient cycling, and organic matter dynamics have to date employed a coarser approach based on the ecological constraints of climate and soils (Parton *et al.*, 1987a,b). However, plans are underway to interface the fine-scale, spatially explicit models as inputs to the larger scaled models (Shugart *et al.*, Chapter 12, this volume). We believe that a minimum of a two-step approach (organismic to ecosystem process level phenomena and ecosystem to global climatic models) will be required. Successful models will include those that adequately portray the operation of temporal and spatial ecological constraints on biotic processes.

Inputs required for global scale models may require large spatial resolution but fine temporal resolution. As discussed above, such data will probably use satellite data and algorithms developed from FIFE-type projects (Sellers *et al.*, 1988). Our own work with that project has convinced us that certain characteristics measured at small plot scales can be directly related to larger scale measurements (Figures 9.3 to 9.5). These measurements can be scaled up to function in input–output relationships with large-scale climate models. However, large errors will be introduced if the ecological constraints (i.e. land management) contributions are not included. In our region, changes in the ecological constraints to net primary productivity, i.e. changes in fire frequency or grazing intensity in prairie or in cropping and tillage practices in agroecosystems, will alter these algorithms.

Rules for minimum sample frequency and minimum sample size must be developed, based on knowledge of the underlying population variance (Wiegert, 1962; Kimmons, 1973; Greig-Smith, 1983). Investigators must decide what constitute significant shifts in their systems given the intrinsic level of variability. This analysis is relatively straight-forward for time-series measurements that are assumed to be measured without error or that have no within-site variance component. For example, given our current precipitation data base (Figure 9.1), annual rainfall at our site must deviate by about 50% of the mean before we have an 'unusual' wet or dry year (i.e. assuming a normal distribution, outside of the 95% confidence). For each measurement, investigators can calculate the necessary sample size needed to detect statistically significant differences given hypothetical differences in population means with known or estimated population variances.

Consider the problems associated with measuring export of nutrients from two watersheds (Figure 9.7). In the Dismal River system, a few baseflow samples accompanied by a storm-event sampler should produce a very accurate measure of export. The lack of inherent variability and high degree of predictability in flow from this sandhills prairie makes detection of slight changes in mean values or in the magnitude of annual variations a potentially easy task. In contrast, the Blue Beaver Creek system in Oklahoma exhibits extreme variability and little predictability. Stream discharge appears to be largely controlled by surface runoff in this mixed-grass drainage. The ability to show some statistically significant change in export in this system as a result of changes in land management or climatological inputs would be difficult if not impossible for studies shorter than a decade in duration.

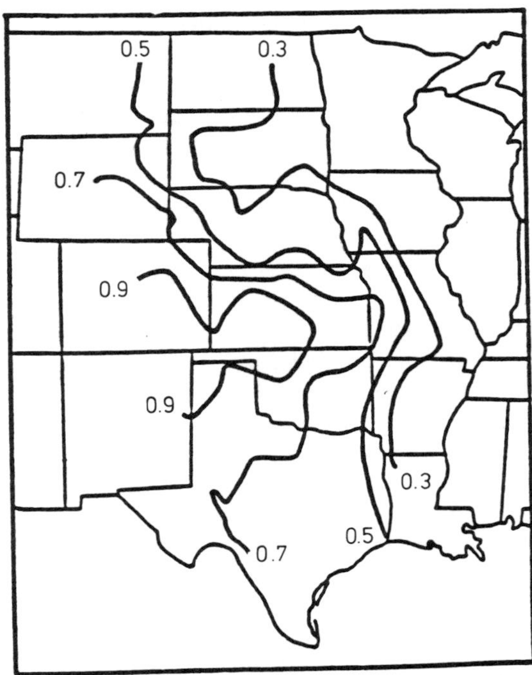

Figure 9.8 Variation in foliage production in the central United States (mean values of good years minus mean values in bad years, divided by the overall mean). Reproduced with permission from *Ecology* (Sala *et al.*, 1988)

Likens (1983) discussed similar problems for measuring element and ion export for northeastern United States streams.

On a regional basis, the variability in stream flow appears to be correlated with the variability in annual foliage productivity. Figure 9.8 illustrates that the source of the Dismal River, the sandhills of Nebraska, is a relatively more stable environment for plant production than the drainage area of Blue Beaver Creek in the western plains of Oklahoma. This variability, itself, can become a long-term measurement, and analyses such as those conducted by Sala *et al.* (1988) should identify sites that have intrinsically lower variability. For studies interested in evaluating directional changes, low variability and high predictability appear to be a desirable characteristic. The Sala *et al.* data also demonstrate the need for a large number of sites to characterize the regional response to year-to-year climatic variability.

9.3 DOCUMENTATION AND DATA BASE MANAGEMENT

The value of creating permanent plots, adequately documenting procedures, and developing a user-friendly data base cannot be overemphasized. With few exceptions, data bases have not outlived the investigators who collected them (Strayer *et al.*, 1986). Those that have survived have become ecological treasures. Tilman (1989) noted that about 90% of all field studies are three years or shorter in duration. Even these short-term studies, if adequately documented and site referenced, could be subsequently resampled for similar or other ecological questions. We believe that we have lost many thousands of dollars' worth of valuable data because a number of ecological studies with a 'short-term focus' were not well documented on our site. Since those projects were terminated, we have come up with a number of questions that could have been addressed using the original data if we could only locate the site where the original measurements were obtained. A similar argument can be made for user-friendly documentation of the data. We have found a variety of new questions for old data sets. These data can be quickly retrieved and reanalyzed, even in the absence of the individual(s) responsible for the original data set. One cannot be serious about measuring decade-to-century-level phenomena without making a serious time and financial commitment to documentation. Researchers are referred to Gurtz (1986) and other references in Michener (1986) for excellent guidelines in this area.

9.4 CONCLUSIONS

The recent Earth System Science Report (NASA 1988), in its recommendations and review of ongoing and proposed research for the IGBP, concluded, 'The overwhelming importance of sustained, long-term measurements of global variables emerges clearly from these studies' (page 137). We contend that a subset of the LTER core measurements, NPP, nutrients, and organic matter dynamics are particularly appropriate for relating vegetation dynamics to surface climatological measurements at a regional or larger scale. Biophysical measurements obtained from small plots, measured under known ecological constraints, will scale up in a fashion conducive to the modeling approaches suggested by Urban *et al.* (1987) and Huston *et al.* (1988). In our region, these ecological constraints include the fire and grazing regimes of the grasslands, or the particular management practice imposed on agroecosystems. We do not mean to ignore biodiversity efforts, and emphasize that individual organisms are driving the spatially explicit site responses (Huston *et al.*, 1988). Nonetheless, these effects must be translated into biophysical rather than simply biological units to be useful at the intersite level.

9.5 ACKNOWLEDGMENTS

We thank M.E. Gurtz, US Geological Survey, for the stream examples. Ideas about ecological constraints developed from discussions with D.S. Schimel. Climatic data were provided by Dr Dean Bark. We appreciate a review of an earlier draft of this manuscript by A.K. Knapp and C.L. Turner. Support for our research and the preparation of this effort was provided by the National Science Foundation (BSR-8514327) and NASA (NAG-5-897) grants to Kansas State University.

9.6 REFERENCES

Adams, F. (1984). *Soil Acidity and Liming*, 2nd Edition, Amer. Soc. of Agronomy, Number 12, Soil Sci. Soc. America, Madison, WI.

Allen, T.F.H., O'Neill, R.V. and Hoekstra, T.W. (1984). Interlevel relations in ecological research and management. Some working principles from Hierarchy theory. USDA Tech. Rep. RM-110, Rocky Mountain Forest and Range Experiment Station, Fort Collins, CO.

Asrar, G., Harris, T.R. and Lapitan, R.L. (1988). Radiative surface temperatures of the burned and unburned areas in a tallgrass prairie. *Remote Sensing of Environment*, **24**, 447–457.

Callahan, J.T. (1984). Long-term ecological research. *BioScience*, **34**, 363–367.

Clark, J.S. (1988). Effect of climate change on fire regimes in northwestern Minnesota. *Nature*, **334**, 233–234.

Driscoll, C.T., Likens, G.E., Hedin, L.O., Eaton, J.S. and Bormann, F.H. (1989). Changes in the chemistry of surface waters. *Environmental Science and Technology*, **23**, 137–143.

Evans, E.W., Finck, E.J., Briggs, J.M., Gibson, D.J., James, S.W., Kaufman D.W. and Seastedt, T.R. (1989). Is fire a disturbance in grasslands? In Bragg, T.J. and Stubbendiech, J. (Eds) *Proceedings of the Eleventh North American Prairie Conference*. University of Nebraska Press, 159–161.

Franklin, J.F. (1989). Importance and justification of long-term studies in ecology. In Likens, G.E. (Ed.) *Long-Term Studies in Ecology*. Springer-Verlag, New York, 3–19.

Gosz, J.R., Dahm, C.N. and Risser, P.G. (1988). Long-path technology for large scale ecological studies. *Ecology*, **69**, 1326–1330.

Goward, S.N., Tucker, C.J. and Dye, D.G. (1986). North American vegetation patterns observed with meteorological satellite data. In Dyer, M.I. and Crossley, D.A., Jr (Eds)

Coupling of Ecological Studies With Remote Sensing. US Department of State Publication 9504, Bureau of Oceans and International Environmental and Scientific Affairs, Washington, DC, 96–115.

Greegor, D.H. (1986). Ecology from space. *BioScience*, **36**, 429–432.

Greig-Smith, P. (1983). Quantitative Plant Ecology. *Studies in Ecology*, Vol. 9. University of California Press, Berkeley.

Gurtz, M.E. (1986). Development of a research data management system. In Michener, W.K. (Ed.) *Research Data Management in the Ecological Sciences*. University of South Carolina Press. Columbia, SC, 33–38.

Hardisky, M.A., Gross, M.F. and Klemas, V. (1986). Remote sensing of coastal wetlands. *BioScience*, **36**, 453–460.

Hunt, H.W., Ingham, E.R., Coleman, D.C., Elliott, E.T. and Reid, C.P.P. (1988). Nitrogen limitation of production and decomposition in prairie, mountain meadow and pine forest. *Ecology*, **69**, 1009–1016.

Huston, M., DeAngelis, D. and Post, W. (1988). New computer models unify ecosystem theory. *BioScience*, **38**, 682–691.

Iker, S. (1983). A lifetime of listening. *Mosaic*, **14**, 8–13.

Jenny, H. (1930). A study on the influence of climate upon the nitrogen and organic matter content of the soil. *Res. Bull 152*, Missouri Agr. Exp. Station.

Jenny, H. (1980). The soil resource. *Ecological Studies* 37. Springer-Verlag, New York.

Kimmins, J.P. (1973). Some statistical aspects of sampling throughfall precipitation in nutrient cycling studies in British Columbia coastal forests. *Ecology*, **54**, 1008–1019.

Knapp, A.K. and Seastedt, T.R. (1986). Detritus accumulation limits productivity of tallgrass prairie. *BioScience*, **36**, 662–668.

LaMarche, V.C., Jr, Graybill, D.A., Fritts, H.C. and Rose, M.R. (1984). Increasing atmospheric carbon dioxide: Tree ring evidence for growth enhancement in natural vegetation. *Science*, **225**, 1019–1021.

Likens, G.E. (1983). A priority for ecological research. *Bull. Ecol. Soc. Amer.*, **64**, 234–243.

Likens, G.E., Bormann, F.H., Pierce, R.S., Eaton, J.S. and Johnson, N.M. (1977). *Biogeochemistry of a Forested Ecosystem*. Springer-Verlag, New York.

Lindberg, S.E., Lovett, G.M., Richter, D.D. and Johnson, D.W. (1986). Atmospheric deposition and canopy interactions of major ions in a forest. *Science*, **231**, 141–145.

Magnuson, J.J. (1990). Long-term ecological research and the invisible present. *BioScience*, **40**, 495–501.

Michener, W.K. (Ed.) (1986). Research data management in the ecological sciences. *The Belle W. Baruch Library in Marine Sciences* Number 16, University of South Carolina Press, Columbia, SC.

Mooney, H.A., Vitousek, P.M. and Matson, P.A. (1987). Exchange of materials between terrestrial ecosystems and the atmosphere. *Science*, **238**, 926–932.

NASA (Anonymous) (1988). Earth system science: a closer view. *Report of Earth System Sciences Committee, NASA Advisory Council*. University Corp. for Atmospheric Research, Boulder, CO.

Ojima, D.S. (1987). The short-term and long-term effects of burning on tallgrass ecosystem properties and dynamics. PhD Dissertation, Colorado State University, Fort Collins, CO.

Ojima, D.S., Parton, W.J., Schimel, D.S. and Owensby, C.E. (1990). Simulating impacts of annual burning on prairie ecosystems. In Collins, S.L. and Wallace, L.L. (Eds) *Fire in North American Tallgrass Prairies*. University of Oklahoma Press, Norman, 118–132.

O'Leary, M. (1988). Carbon isotopes in photosynthesis. *BioScience*, **38**, 328–336.

Pace, M.L. and Cole, J.J. (1989). What questions, systems, or phenomena warrant long-term ecological study? In Likens, G.E. (Ed.) *Long-Term Studies in Ecology*. Springer-Verlag, New York, 183–185.

Parton, W.J., Schimel, D.S., Cole, C.V. and Ojima, D.S. (1987a). Analysis of factors controlling soil organic matter levels in Great Plains grasslands. *Soil Sci. Soc. Am. J.*, **51**, 1173–1179.

Parton, W.J., Steward, J.W.B. and Cole, C.V. (1987b). Dynamics of C, N and P in grassland soils: a model. *Biogeochemistry*, **5**, 109–131.

Rock, B.N, Vogelmann, J.E., Williams, D.L., Vogelmann, A.F. and Hoshizaki, T. (1986). Remote sensing of forest damage. *BioScience*, **36**, 439–445.

Rosenzweig, C. and Dickinson, R. (1986). *Climate-Vegetation Interactions*. Rep. OIES-2, University Corp. for Atmospheric Research, Boulder, CO.

Rykiel, Jr, E.J. (1985). Toward a definition of ecological disturbance. *Aust. J. Ecology*, **10**, 361–365.

Sala, O.E., Parton, W.J., Joyce, L.A. and Lauenroth, W.K. (1988). Primary production in the central grassland region of the United States. *Ecology*, **69**, 40–45.

Schowalter, T.D. (1981). Insect–herbivore relationship to the state of the host plant: biotic regulation of ecosystem nutrient cycling through succession. *Oikos*, **37**, 126–130.

Sellers, P.J., Hall, F.G., Asrar, G., Strebel, D.E. and Murphy, R.E. (1988). The first ISLSCP field experiment (FIFE). *Bull. Amer. Meteor. Soc.*, **69**, 22–27.

Shugart, H.H. (1986). Conclusions. In Rosenzweig, C. and Dickinson, R. (Eds) *Climate–Vegetation Interactions*. Report OIES-2, University Corp. Atmospheric Research, Boulder, CO, 152–154.

Strayer, D., Glitzenstein, J.S., Jones, C.G., Kolasa, J., Likens, G.E., McDonnell, M.J., Parker, G.G. and Pickett, S.T.A. (1986). *Long-Term Ecological Studies: An Illustrated Account of Their Design, Operation, and Importance to Ecology*. Occasional Publ. Inst., Ecosystem Studies Number 2. Institute of Ecosystem Studies, New York Botanical Garden, Millbrook, NY.

Swift, M.J., Heal, O.W. and Anderson, J.M. (1979). *Decomposition in Terrestrial Ecosystems*. Blackwell Scientific, Oxford.

Tilman, D. (1989). Ecological experimentation: strengths and conceptual problems. In Likens, G.E. (Ed.) *Long-Term Studies in Ecology*. Springer-Verlag, New York, 136–157.

Tissue, D.T. and Oechel, W.C. (1987). Response of *Eriophorum vaginatum* to elevated CO_2 and temperature in the Alaskan tussock tundra. *Ecology*, **68**, 401–410.

Tucker, C.J., Townshend, J.R.G. and Goff, T.E. (1985). Continental land cover classification using satellite data. *Science*, **227**, 369–374.

Urban, D.L., O'Neill, R.V. and Shugart, H.H., Jr (1987). Landscape ecology. *BioScience*, **37**, 119–127.

Ustin, S.L., Adams, J.B., Elvidge, C.D., Rejmanek, M., Rock, B.N., Smith, M.O., Thomas, R.W. and Woodward, R.A. (1986). Thematic mapper studies of semiarid shrub communities. *BioScience*, **36**, 446–452.

Van Cleve, K., Dryness, C.T., Viereck, L.A., Fox, J., Chapin, III, F.S. and Oechel, W. (1983). Taiga ecosystems in interior Alaska. *BioScience*, **33**, 39–44.

Waring, R.H., Aber, J.D., Melillo, J.M. and Moore, III, B. (1986). Precursors of change in terrestrial ecosystems. *BioScience*, **36**, 433–438.

Weaver, J.E. (1954). *North American Prairie*. Johnsen Publishing Co., Lincoln, Nebraska.

White, P.S. and MacKenzie, M.D. (1986). Remote sensing and landscape pattern in Great Smoky Mountains Biosphere Reserve, North Carolina and Tennessee. In Dyer, M.I. and Crossley, Jr, D. A. (Eds) *Coupling of Ecological Studies With Remote Sensing*. United States Department of State Publication 9504, Bureau of Oceans and International Environmental and Scientific Affairs, Washington, DC, 52–70.

White, T.C.H. (1984). The abundance of invertebrate herbivores in relation to the availability of nitrogen in stressed food plants. *Oecologia*, **63**, 90–105.

Wiegert, R.G. (1962). The selection of an optimum quadrat size for sampling the standing crop of grasses and forbs. *Ecology*, **43**, 125–129.

10 Long-term Ecological Research in African Ecosystems

SAM J. McNAUGHTON
Biological Research Laboratories, Syracuse University, Syracuse, NY 13244-1220, USA

and

K.L.I. CAMPBELL
Serengeti Ecological Monitoring Programme, PO Box 3134, Arusha, Tanzania

10.1 INTRODUCTION

Long-term ecological research in Africa has had a substantial range of foci and approaches. The International Livestock Centre for Africa, headquartered in Addis Abbaba, has emphasized management, veterinary, and other studies related to development. Various United Nations organizations, including UNESCO, FAO, and UNEP, have also supported ecological research in a wide variety of countries

Long-term Ecological Research. Edited by Paul G. Risser

and with varied objectives. Our focus in this chapter, for reasons of coherence and conceptual integration, emphasizes research on natural ecosystems, largely national parks and game reserves, highlighting research in Tanzania's Serengeti National Park as a case history approach.

Long-term research in natural African ecosystems has been centered on the continent's abundant large mammal populations, but those populations have also been integrated from the outset into a more comprehensive ecosystem perspective (McNaughton and Georgiadis, 1986). Systematic, quantitative research began in the 1950s and early 1960s, most of it initially undertaken by visiting scholars from the United States, Great Britain, and Europe. That research tended to focus on defining the ranges of migratory herds as an aid to drawing boundaries of the newly developing game reserve and national park systems, evaluating reductions in tree cover associated with concentrations of elephants (*Loxodonta africana*) in the protected areas, and taking a census of the game populations to ascertain their numbers.

Most long-term ecological research in Africa, in fact, has an explicit management responsibility. It is performed in national parks and other reserves, is financially supported, at least in part, by African governments, and is explicitly charged with advising management personnel and formulating management plans. Among the principal sites of the programs are Kenya's Amboseli National Park, South Africa's Kruger National Park, Uganda's Queen Elizabeth National Park, Ivory Coast's Lamto Research Station, and Tanzania's Serengeti National Park.

Kruger National Park supports one of the most sophisticated long-term ecological research programs, with extensive computerized records of animal populations, vegetation, weather, fire, and other ecological components important in African ecosystems. Regular aerial censuses of animal populations and tree cover are features of the data bases. Among the longest-term ecological experiments known to us is the factorial burning treatments initiated in the Park in 1954 (Gertenbach and Potgieter, 1979). Kruger also is an intensively and extensively managed park, and it is not accidental that intense management is coupled to sophisticated long-term data records. The Park has an entirely fenced perimeter, is divided into firebreaked landscape blocks subject to a regular burning program, and the animal populations are intensively harvested to maintain their bounds within explicit management goals. Annual reports of the Kruger research staff, however, are published in Afrikaans, and much of the information from the program is inaccessible to most potential readers.

10.2 ECOLOGICAL MONITORING

A core component of most African long-term research projects has been ecological monitoring, the process of maintaining regularly updated records of key parameters in a defined geographic area. Management goals have been explicit elements in monitoring programs. The term 'monitoring' has the implicit implied function in research of warning management personnel of both desirable and undesirable trends, as they are defined by the goals of the regional management. The monitoring process involves maintaining and continually updating records of such key components as rainfall, flora, fauna, human activities, and land use. In most African programs, monitoring is designed to establish comprehensive baseline data, and to quantify basic patterns of geographical distributions and abundances of major large mammal species. Observations are designed to encompass the key phenomena believed to reflect trends and parameters of broad environmental and ecosystem conditions. The monitoring records eventually become the bases for determining trends in ecosystem state and dynamics over extensive temporal sequences and spatial scales.

Ecological monitoring relies heavily upon co-operation between monitoring personnel, managers, and project-oriented research scientists, who together evaluate and modify the monitoring program to pin-point important parameters desirable or essential for long-term evaluation. The ecological monitoring strategy is designed to improve its own cost-effectiveness. Measurements implemented on the ground, common in project research and some management activities, provide detailed information but are impractical and expensive to implement over large areas. For example, they are impractical for such key goals as animal counting, and expensive for monitoring the balance of grass, bush, and trees over large areas. Low-level aerial techniques supply less detailed information, but are cheaper and quicker to apply for such goals as animal censuses. Satellite images can provide low-resolution, but spatially extensive, information of certain types on a regular and cost-effective basis for whole regions, or even continents (Tucker *et al.*, 1985). A comprehensive monitoring program combines data from each of these three levels, ground, light aircraft, and satellite.

Ecological monitoring is an essential, integral element of long-term research. It provides the data base necessary for implementing goal-focused project research and for computer modeling. Because the authors of this chapter are directors of the Serengeti Ecological Monitoring Programme and of a project research program of over 15 years' duration in the Serengeti, the chapter emphasizes the research programs in the Serengeti ecosystem, probably Africa's most intensively studied and comparatively unmanaged ecosystem that preserves reasonable components of the continent's abundant large mammal community (Sinclair and Norton-Griffiths, 1979). The chapter emphasizes three aspects of the Serengeti research:

(1) The history of long-term research there, as an instructive model for the hazards of long-term concepts in the absence of reliable funding;
(2) The accomplishments, for what they reveal about the implementation of concepts; and
(3) Future projections, contrasting the ideal with the financially and logistically feasible in less-developed countries.

10.3 THE SERENGETI

10.3.1 THE ECOSYSTEM

The Serengeti ecosystem (Figure 10.1) is defined by the annual movement of the earth's largest population of migratory large ungulates, over one million (Sinclair, 1987) wildebeest (*Connochaetes taurinus*). This herd moves over approximately 25 000 km^2 of open and wooded grasslands along a mean annual rainfall gradient from less than 400 mm in the south-east to above 1100 mm in the north-west. Much of the herd's range is within Tanzania, but a portion of the critical dry season range extends across the international boundary into Kenya. The limits of the annual ranges of wildebeest and the other major migratory species, the plains zebra (*Equus burchelli*), were employed at the outset to define the bounds of the ecosystem (Lamprey, 1979). These bounds encompass the ranges of many resident large herbivores of over 25 species, a diverse mammalian carnivore fauna, and a spectacular avifauna.

The Serengeti National Park lies at latitudes of 1°–3° S within an elevation range of 1135–1800 m. It spans three of East Africa's 10 major geological regions from the fault-scarp of the Rift Valley in the east, across a northern extension of

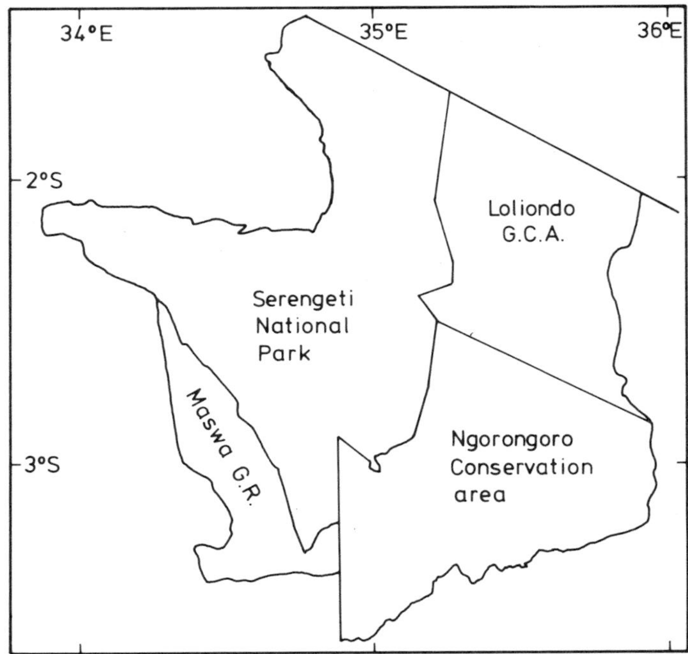

Figure 10.1 Map of the Serengeti National Park and associated wildlife areas in Tanzania that constitute core elements in the Serengeti ecosystem

Tanzania's huge central plateau in the middle, to the Lake Victoria basin in the west (Lundgren, 1975). The rainfall gradient is due to the rainshadow created by the Crater Highlands in the south-east, the Lake Victoria Convergence Zone in the west, and the seasonal movements of the Intertropical Convergence Zone (McNaughton, 1983). The large migratory herds of wildebeest and zebra that dominate the fauna spend the peak of the wet season (January to April) in the driest region in the south-east, and the peak of the dry season (July to October) in the wettest region in the north-west.

During the initial period of intensive research in the ecosystem, the Park was mapped geologically (Macfarlane, 1969), woodlands were assayed (Herlocker, 1976), soils were surveyed (deWit, 1978; Jager, 1982), animal populations were regularly censused (Sinclair and Norton-Griffiths, 1982), landscape classification systems were applied (Gerresheim, 1974; Epp, 1978), and autecological studies concentrated on all major mammalian herbivore and carnivore species (Sinclair and Norton-Griffiths, 1979). This period of comprehensive research was initiated as the Serengeti Research Project and was quickly converted into an institutional form as the Serengeti Research Programme.

10.3.2 DEVELOPMENT OF A LONG-TERM PROGRAM

Although the Serengeti Research Programme was not conceived initially as an ecosystem study, it did, nevertheless, rapidly evolve into one (Lamprey *et al.*, 1971; Lamprey, 1979), and the pioneering project research by early scientists (Pearsall, 1957; Grzimek and Grzimek, 1960; Talbot and Talbot, 1963) was important in laying a regional foundation for what was to follow. From the outset, research in the Serengeti was greatly influenced by potentially serious management problems associated with dense populations of large grazing and browsing mammals. Elephants, the major non-human agent of habitat change in Africa (Laws, 1970), were unknown in the ecosystem from its discovery at the turn of the century until 1951, when they invaded the Park due to displacement by growing human populations in the vicinity of Lake Victoria (Lamprey *et al.*, 1967).

Rinderpest, or cattle plague, an exotic disease introduced into Africa prior to the discovery of the Serengeti by Europeans, had a devastating effect on many susceptible ungulate species (Lydekker, 1908), particularly wildebeest and African buffalo (*Cyncerus caffer*). Conversion of the Serengeti region to a rinderpest-free zone by livestock inoculation in the early 1960s resulted in a resurgence of susceptible species, and the wildebeest, buffalo, and topi (*Damaliscus korrigum*) populations increased several-fold over succeeding years. Ongoing research sponsored by the European Economic Community has documented continued presence of the virus in Serengeti animals.

Regular, extended censuses of major animal populations has been a fundamental contribution of ecological monitoring (Figure 10.2) that is unlikely to have been achieved in the absence of a formal, supported monitoring program. Justification

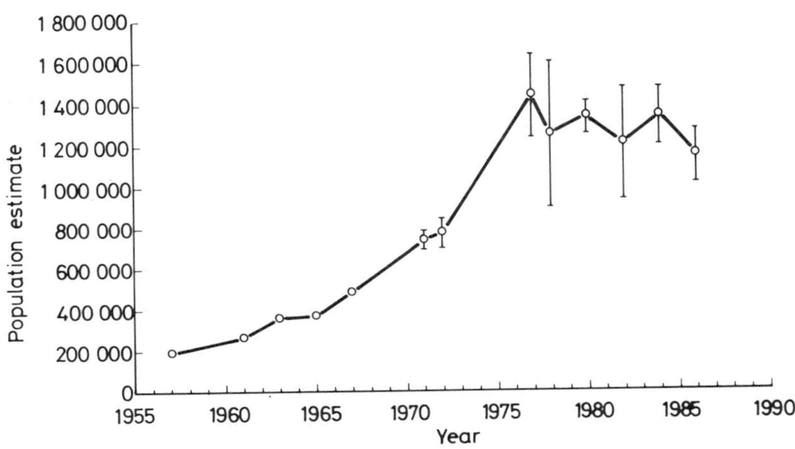

Figure 10.2 Thirty years of census records for the Serengeti migratory wildebeest population. Data through 1967 were 'total' counts and afterwards were photographic sample counts with error bars showing standard errors of the estimates. Data are part of the archives of the Serengeti Ecological Monitoring Programme, and were published in Swynnerton (1958), Sinclair (1973), Sinclair and Norton-Griffiths (1982), Sinclair *et al.* (1985), and Sinclair (1987)

of the expense of these censuses and integrity of record maintenance is unlikely in more project-oriented research programs. However, such population records can be essential to much project research, and can contribute to scientific progress in their own right through the demographic data that they provide.

The invasion of the Park by elephants and the resurgence of the ungulate populations created two serious management problems at which research was directed: loss of tree cover by the combined effects of elephants, browsing, fire, and overgrazing (Norton-Griffiths, 1979). The vision and efforts of the trustees of Tanzania National Parks led to establishment of the Serengeti Research Project in 1963 and creation of the Serengeti Research Institute (SRI) in 1966. A comprehensive ecosystem study was envisioned, but that was prevented by limited funds and facilities. Instead, a scheme of research priorities was designed by the SRI and its international Scientific Council. Priorities were based largely on the existing requirements of management staff, and as a matter of policy, a purely academic approach was avoided. An early major priority was the establishment of a monitoring program, and in 1969 the Serengeti Conservation and Research Programme was initiated. This program became the Serengeti Ecological Monitoring Programme (SEMP) in 1970, and was operational for eight years until withdrawal of international funding led to serious curtailment of its activities.

A broad range of research was carried out at SRI from the 1960s to the late 1970s, and SEMP (Norton-Griffiths, 1972) was integral to all of it. A comprehensive rain gauge system was established, among the most dense in Africa, and a complete weather station was constructed at SRI headquarters. The number of ungulates was monitored, and monthly reconnaissance of distributions of animals visible from the air was implemented on a grid system. Studies of major herbivore and carnivore species were commenced, some of which have continued throughout the intervening 25 years.

Over an approximate 15-year period of intense activity, data were collected on rainfall, animal distributions, population sizes and demographics of major large herbivores, ecology and behavior of the major large carnivores, the region's abundant scavenging avifauna, primary productivity, soils, geology, tree cover, fire frequency and extent, human settlement outside the Park and encroachment within the Park, hydrological properties of the major river system, and veterinary aspects of the animals. A technical and support staff was an integral component of the long-term research program. This staff included a botany technician, a photographic technician, a librarian, soils technicians, veterinary and wildlife assistants, vehicle drivers, mechanics, and a general laborer staff.

10.3.3 ORGANIZATION OF LONG-TERM RESEARCH

Research was organized in a typical administrative fashion as the program evolved. The Director of SRI was responsible for principal liaison with the national government, the Director and Trustees of National parks, the Chief Park Warden of the Serengeti National Park, international funding organizations, and the Scientific Council. The Scientific Council was an international body, including several scientists of major stature, created to evaluate ongoing research and to advise on program orientation. The director of SEMP was responsible for executing and maintaining the records of the principal monitoring functions: rainfall and weather records, population estimates of principal herbivore species, reconnaissance flight data on animal distributions, aerial photography for determining trends in tree cover, fire mapping, and base maps of regional geology, soils, and vegetation. The administrator was responsible for day-to-day operation and disposition of responsibilities among the technical and labor staff. An assistant director was responsible for co-ordinating the Institute's overall research objectives and meeting the needs of project research. A clear ecosystem approach was evident in the permissions allowed for this project research, largely funded outside the SRI budget. Some projects dealt with soils, primary production, and vegetation; others with each of the major herbivore species; others with major carnivores, including the longest continuous study, of over 20 years, on lions (*Panthera leo*); and other important ecosystem components, little studied in most long-term ecological research, such as small mammals, the scavenging avifauna, parasites and pathogens, termites, and dung beetles.

10.3.4 THE CONSEQUENCES OF HIATUS

Long-term ecological research, by definition, requires reliable, continuous funding.

With the exception of Kruger and Amboseli National Parks and Lamto Research Station, continuous funding has been only partially characteristic of research in Africa. Political, economic, and, in some places, military turmoil have hindered the continuity required by long-term research. Most project grants have been subject to significant funding hiatuses throughout the period of active research in the Serengeti.

Because its activities are central to long-term research, the disappearance of SEMP was symptomatic of an interruption of broad, cross-project approaches to Serengeti research. International economic problems, associated with oil prices and exacerbated by Tanzania's war with Uganda, closure of the international border between Tanzania and Kenya, and various political factors made international funding bodies wary, and funding for SEMP outside of Tanzania was no longer available after 1977. Internal funds were used to maintain rainfall records from rain gauges accessible by ground transport, but most of the region became inaccessible in the absence of light aircraft and maintained bush landing strips. The weather station at headquarters fell into disrepair, and crucial reference collections of mammal skins, bird skins, and insects were destroyed by pests. Integrity of the herbarium was maintained by project funds. Research in the Serengeti became largely confined to individual project programs, each with independent sources of funding, and the original research concepts and resultant priorities no longer applied. For eight years, long-term research in the Serengeti continued only in the form of those individual projects.

10.3.5 MONITORING: RE-INITIATION IN THE SERENGETI

During the funding hiatus for SEMP, ecological research in Tanzania was reorganized. SRI became the Serengeti Wildlife Research Centre, a subsidiary of the Serengeti Wildlife Research Institute (SWRI), headquartered in Arusha and no longer a component of Tanzania National Parks. SWRI assumed responsibility for all wildlife-related research throughout Tanzania and no longer focused exclusively or largely on the Serengeti. The need for continuous ecological monitoring of the Serengeti was emphasized at a workshop (IUCN, 1986), and SEMP was re-established in 1986, with a full- and part-time project officer, with funding from the Frankfurt Zoological Society, World Wide Fund for Nature, and the SWRI. In 1987, the Wildlife Division seconded a project officer to SEMP, and Tanzania National parks created the post of Serengeti National Park Ecologist under a newly formed Ecology Department. This ecologist is attached to SEMP, providing an important direct link to park management.

The principal objectives of the monitoring program have remained much the same over the last 20 years (Norton-Griffiths, 1972):

(1) To provide information for park management staff on the status and trends of animal and plant communities within the ecosystem.

(2) To provide information for park staff and land-use planners on land-use patterns and trends around the ecosystem.
(3) To provide timely warning of potential management problems.
(4) To provide baseline information for the formulation of management plans and policies.
(5) To provide a means of evaluating the effectiveness of management policies.

A specific requirement now exists for monitoring throughout Tanzania. As in the Serengeti, monitoring presently takes place at two levels: ground-based sampling by monitoring staff, functioning as part of the management of various conservation areas; and aerial survey. SEMP and the Serengeti Wildlife Research Institute currently meet requests for aerial monitoring from Tanzania National Parks, the Wildlife Division, and the President's Office. Training of Tanzanian personnel is an important component of this country-wide effort.

10.4 LONG-TERM ECOLOGICAL RESEARCH

10.4.1 PROJECT PERSPECTIVES

The longest term African research projects, aside from the Kruger burning plots, have been animal centered and behaviorally oriented. Primate research, concentrating on chimpanzees (*Pan troglodytes*), has been underway in Gombe Stream National Park since 1960 (Goodall, 1987). The lion research project in the Serengeti was begun by George Schaller in 1966 (Schaller, 1972) and has continued to the present, with some discontinuities, through four sets of prime investigators. The Amboseli elephant behavior project is 15 years old and has been largely supervised by a single individual with many assisting participants (Moss, 1988). Principal components of these long-term studies are individual animal recognition, detailed behavioral records, and lineage determinations. Although the focus of these autecological studies has been sociobiological, they nevertheless provide substantial long-term information on basic properties of the systems that the animals occupy.

The Serengeti Grazing Ecosystem Project has been ongoing for 15 years, and combines field studies (McNaughton, 1976, 1983, 1985) with simulation modeling (Coughenour *et al.*, 1984a,b) and laboratory experiments that are infeasible in the field on parameters necessary for model development and for understanding underlying mechanisms (McNaughton *et al.*, 1983; Coughenour *et al.*, 1985; Ruess and McNaughton, 1987). Although the goals of SEMP, as outlined above, are oriented toward management, the accumulated data bands and previous research at SRI were fundamental to rapid development of the grazing ecosystem project and to its progress. The rainfall and soil survey data provided a basis for rapid initial selection of a study site network, giving a broad span of coverage of the region's principal ecological variables: rainfall, edaphic properties, grazing intensities, and vegetation types. Although most study sites had to be individually assayed for rainfall during the hiatus in SEMP, site placement otherwise has taken advantage of the SEMP rain gauge network. Aerial observations of animal distributions allowed temporary study sites to be placed in a timely fashion to take advantage of specific experimental opportunities (McNaughton, 1985). All in all, experience with the Serengeti Grazing Ecosystem Project reveals a central role for long-term monitoring in the establishment, execution, and progress of long-term project research.

10.4.2 OBJECTIVES AND PRIORITIES

Long-term monitoring either is goal-specific or follows a more general course. In the former case, goals are determined before initiating monitoring activities, and methods established at the outset must be sufficiently accurate and precise to meet the stated objectives. In general, this requires both prior knowledge of the ecosystem and substantially greater personnel and funding than the second approach, in which monitoring activities follow a less-detailed, data-gathering path. The two approaches are usually complementary, with the more general, less specific approach often a necessary precursor to the goal-specific approach. An important aspect of less goal-oriented monitoring, however, is its ability to respond and adjust to explicitly formulated goals that arise out of the monitoring process, and those which management personnel may require.

Long-term ecological research in African ecosystems is, in our view, infused with five, not always explicit, goals. First, it is designed to establish baseline data on prevailing environmental conditions; this focuses primarily upon rainfall records at a representative range of locations. Second, it maintains and updates records on such fundamental underlying properties as geology, geomorphology and landscape regions, soils, and vegetation types. Third, it considers variance about baseline data, either as fluctuations about a mean or as directional trends. Fourth, it studies patterns of state and dynamics that cannot be revealed by short-term research. Finally, whether sociobiological or ecological in primary focus, it is comprehensive and integrative in scope. Animal behavior is placed in a system context, and ecosystem studies include the animals as an integral component. These are neither strictly monitoring nor strictly project or management goals, but a combination of the three.

What is long-term in the African context? For a multivoltine insect, a couple of years of research can encompass many generations and constitute a long-term study. But for ecosystems like those in Africa, whose most conspicuous feature is populations of long-lived mammals, a more extended temporal scale is relevant. For monitoring functions and most project programs, a decade seems a reasonable time span to establish prevailing patterns and general trends. For long-lived animals, behavioral studies with kinship and learning as fundamental elements may require somewhat longer time periods. Similarly, determining changes in the balances among an arborescent overstory, an intermediate bush canopy, and the underlying grass canopy may require periods longer than the decade range.

It is clear that a successful monitoring program cannot work as an independent unit, but must maintain a constant exchange of ideas and data between the ecologists working on specific projects and the management. Only through a flow of information from all directions can management design realistic policies, the monitoring program achieve its objectives, and the project research be efficiently prosecuted. The SEMP currently combines data collection with an information management system to update baseline records, advise management personnel, and provide ongoing data for project research in the following fields:

(1) Climate, principally rainfall;
(2) Wildlife numbers, population structure, and distribution;
(3) Fire and fire control;
(4) Vegetation distribution and changes; and
(5) Poaching distribution and intensity.

Beyond aerial measurement of tree cover during the active period of SRI, little vegetation monitoring has been accomplished, in spite of the central role that it could play in assessing long-term ecosystem trends. Two types of permanent vegetation plots were established early in the SRI program, permanent plots in the south-east and the north-east for aerial and terrestrial censuses, and experimental plots in several locations combining fencing and firebreaking. The former have rarely been quantitatively censused and the latter were dismantled or fell into disrepair in the mid 1970s. Quantitative baseline data on plant community composition were not collected on either type of plot when they were first established.

SEMP is now establishing permanent vegetation plots throughout the Serengeti. Twenty of a planned forty have been delimited with markers visible from the air. These will be sampled regularly from the ground with standardized vegetation sampling methods. In addition, about 2200 sample aerial photos have been taken on a systematic basis for monitoring tree–bush–grass balances throughout the Serengeti National Park and adjacent areas.

SEMP, when fully implemented under its current goals, should provide the fundamental data bases required for long-term project research. Implementation of a good vegetation monitoring program on fixed plots at a representative range of locations will strengthen the baseline considerably.

10.4.3 FUNDAMENTALS

Successful long-term ecological research, as practiced in the Serengeti specifically, and throughout Africa generally, has several fundamental properties. Those were largely evident from the outset, but have also gradually evolved in implementation.

First, there must be a clear goal-oriented vision of what the research is to accomplish. Although it is arguable whether long-term research in Africa has contributed more to basic science or to providing guidelines for management policies, the latter has been the central explicit objective. The inception of long-term research was coincident with recognition of a number of practical problems associated with political, economic, and demographic changes in Africa. For example, within the game reserves and national parks, the 'elephant problem' was a central concern throughout the continent. Increases in elephant populations due to pressures from human populations outside the parks and protection within them led to a rapid decline in tree communities in most parks (Laws *et al.*, 1975). While in recent years this has become a moot point due to wholesale poaching, nevertheless, advising management remains an explicit charge of most long-term ecological research.

Second, monitoring plays a critical role in long-term research. Monitoring programs commonly concentrate on accumulating baseline data on four factors: (1) weather, particularly rainfall; (2) numerical censuses of the principal animal species, together with some attention to such demographic generalities as cow/calf ratios and death rates; (3) the balance between tree and grass cover; and (4) extent and frequency of fires. In addition, basic geological, soil, and vegetation mapping often forms part of the central data archives of monitoring programs. Experimentation is rarely a component.

Third, project research focused on specific research areas is often an integral element of long-term research. The studies with the greatest continuity and detail have been primarily behavior oriented. But a clear delineation of three types of long-term project research is evident:

(1) Projects such as the Serengeti Grazing Ecosystem Project that concentrate on such fundamental ecological processes as energy and nutrient flow, emphasizing primary productivity and its consumption by the herbivore community;
(2) Autecological studies of the major herbivore species concentrating on demography, food requirements, and the degree and causes of population regulation; and
(3) Predator studies focusing in part on predator–prey relations but with a strong sociobiological component.

No one has yet accomplished a functional integration of the trophic web from the soil-related to the top predator and scavenger trophic levels. Rather, the long-term project research as been less comprehensive in scope and, because of the nature of its funding, oriented toward short-term, achievable goals. Experimental manipulations are often a major component of long-term project research.

Certain areas of research were only sporadically implemented, but they have continued over a sufficient period of time to become long-term research. For example, veterinary research was an integral component of the initial SRI program because of the importance of rinderpest as a regulator of the Serengeti's game animals. The research fell into abeyance with the inoculation of cattle and suppression of the disease in the wild animals, but reappearance of the disease (Rossiter *et al.*, 1983) led to its reinstatement, with EEC support.

10.4.4 A CRITICAL APPRAISAL

Our evaluation of long-term research in African ecosystems suggests that such research is uniquely or usefully appropriate when the phenomena being studied take place over long periods of time, like the population demographics of long-lived animals (McNaughton and Georgiadis, 1986), or when the phenomena are functionally and structurally complex, like the relationships among plants, herbivores, and the physical environment (McNaughton *et al.*, 1988). In the first case, a certain amount of time is required merely to document and determine any change of direction, and a variety of specific studies are needed to unravel the causes of the change or of constancy. In the second case, the complexity of the study requires a step-by-step experimental approach, involving a variety of research methods from field experimentation to computer modeling in order to identify the organizing linkages and determine their effects.

Long-term research has been fundamental in the progress of scientific understanding of African ecosystems and in the planning of management policies. The research has contributed to the establishment of principles of stability and resilience of the savanna ecosystems (McNaughton, 1977, 1985; Walker *et al.*, 1981; Ellis and Swift, 1988); to the determination of patterns and causes of change in animal and plant populations (Sinclair and Norton-Griffiths, 1979); to the development of management plans related to park goals, for example, fire plans for arresting the deterioration of woodland cover (Stronach, 1988); to the development of tools, such as simulation modeling, to provide scientific insight (McNaughton *et al.*, 1988) and serve as quantitative, predictive management tools; and to the identification of potentially important limiting factors through long-term information and sophisticated technologies (McNaughton, 1988).

The development and refinement of wildlife census methods was a particularly important contribution of long-term Serengeti research to basic ecological methodology (Jolly, 1969a,b). An ultimate outcome of the research was the production of five handbooks on techniques in wildlife ecology focusing on Africa (Norton-Griffiths, 1978; Bertram, 1979; Grimsdell, 1979; Western and Grimsdell, 1979; Sinclair and Grimsdell, 1982), and fundamental contributions to general approaches to ecological monitoring in long-term research (Clarke, 1986). General descriptions (Lamprey, 1979) and detailed descriptions (Sinclair and Norton-Griffiths, 1979) of the Serengeti research also provide ample evidence of the value of long-term ecological research and its applications to management policies.

Nevertheless, it would be disingenuous to suggest that research has contributed to solutions of some of the most critical practical problems that confront African resource managers. Chief among those problems has been the blitzkrieg of commercial poaching sweeping over Africa. It is correct, we believe, that regular censuses of animal populations have provided early information about the onset of intense poaching almost everywhere that it has occurred; the solutions to the problem, however, lie in economics and law enforcement. In spite of the fact that data-based simulation studies document that the greatest economic return is derived from harvesting mature ivory from dead animals (Pilgram and Western, 1986a,b), the most rapid economic yield to poachers is the complete harvest that is currently underway.

Long-term research can be most useful to management in three ways. First, it can provide records of subtle, chronic changes in ecosystems, such as changes in the balances among woodland, bushland, and grassland. Second, it can provide an early warning of the onset of acute changes, such as those due to poaching or disease outbreak. Third, it can suggest management strategies for dealing with changes. Dealing with the acute changes, such as poaching and disease, is complex and costly, but is well understood in the hypothetical sense.

10.4.5 PROSPECTS

Initial funding of long-term research in Africa came primarily from international funding bodies with a major conservation orientation. Later, project grants obtained funds from national scientific funding bodies such as the US National Science Foundation and the British Royal Society and Research Councils. But international conservation organizations later shifted funding away from research and into direct management activities and educational programs. This led to such consequences as the almost decade-long interruption of SEMP. Project research has continually been disrupted by the vagaries of funding.

The main impediment to long-term ecological research, in fact, has been funding discontinuity. Project proposals, by their nature, move in a generally linear fashion from one specific aspect to another in a general problem. This curtails the opportunity to take a broad perspective. Monitoring programs are so goal oriented and non-experimental in character that they are unable to provide the mechanistic insight into ecosystem function and population regulation that arises out of project research. Simulation modeling, an integral element of mechanistic understanding of large-scale ecological problems, has rarely been a component of research anywhere in Africa.

Mechanistic, quantitative simulation modeling, we believe, should be a core component of long-term ecological research, but only the Grassland Research and Serengeti Systems Model (GRASS) fulfils this requirement in African ecosystems (Coughenour, 1984; Coughenour *et al.*, 1984a,b; McNaughton *et al.*, 1986). Simulation modeling is important in long-term ecological research because it provides an organizing framework for research objectives, can point to areas of desirable information currently unknown, allows model manipulation to substitute for experiments that are impractical, impossible, or dangerous, and provides the only basis for extrapolations beyond the lifetimes of individual projects or scientists. Properly constructed and parameterized, simulation modeling could provide the basis for formulating long-term policies necessary in management. In addition, good models could allow the assessment of different management options and provide a basis for estimating the consequences of such trend deviations as drought, runs of wet years, severe poaching, and epidemic disease in some animal populations. There is a clear need for more thoroughly developed simulation modeling of ecosystems in Africa and in the tropics as a whole.

The United States' Long-Term Ecological Research Program (Callahan, 1984; Chapter 2 in this volume) could serve as a model for funding long-term research in Africa and throughout the world. The program funds multi-investigator teams for five-year, renewable intervals, concentrating on site-specific ecological phenomena. International funding with a similar scope, concentrating, most logically, on World Heritage sites representing a characteristic range of ecosystems, could provide the funding continuity necessary for long-term, integrated research. Combining the functions of monitoring with the experimental approaches characteristic of project research, including a strong modeling component, would do much to place long-term ecological research in the global scientific mainstream.

10.5 ACKNOWLEDGMENTS

We are grateful to the Trustees of Tanzania National Parks and the National Scientific Research Council of Tanzania for permission to live and do research in the Serengeti National Park. We express our appreciation to Mr David Babu, Director of Tanzania National Parks, Professor Karim Hirji, Director of the Serengeti Wildlife Research Institute, and Mr B. Maragesi, Principal Park Warden of the Serengeti National Park. Our research is supported by US National Science Foundation grants BSR 8505862 and NSF BSR-8817934 to SJM, and funds from the Frankfurt Zoological Society and World Wide Fund for Nature to KLC. Appreciation is expressed to Professor A.R.E Sinclair for organizing much of the wide census work conducted during 1986.

10.6 REFERENCES

Bertram, C.R. (1979). *Studying Predators*. Afr. Wildl. Leadership Fdn., Nairobi, Kenya.

Callahan, J.T. (1984). Long-term ecological research. *BioScience*, **34**, 363–367.

Clarke, R. (1986). *The Handbook of Ecological Monitoring*. Oxford University Press, Oxford.

Coughenour, M.B. (1984). A mechanistic simulation analysis of water use, leaf angles, and grazing in East African graminoids. *Ecol. Model.*, **26**, 203–220.

Coughenour, M.B., McNaughton, S.J. and Wallace, L.L. (1984a). Modelling primary production of perennial graminoids—uniting physiological processes and morphometric traits. *Ecol. Model.*, **23**, 101–134.

Coughenour, M.B., McNaughton, S.J. and Wallace. L.L. (1984b). Simulation study of East African perennial graminoid responses to defoliation. *Ecol. Model.*, **26**, 177–201.

Coughenour, M.B., McNaughton, S.J. and Wallace, L.L. (1985). Responses of an African graminoid (*Themeda triandra Forsk.*) to frequent defoliation, nitrogen, and water: a limit of adaptation to herbivory. *Oecologia* (Berlin), **68**, 105–110

deWit, H.A. (1978). Soils and grassland types of the Serengeti Plains (Tanzania). PhD Dissertation. University of Wangeningen, Wageningen, The Netherlands.

Ellis, J.E. and Swift, D.M. (1988). Stability of African pastoral ecosystems: alternate-paradigms and implications for development. *J. Range Manage.*, **41**, 450–459.
Epp, H. (1978). A natural resource survey of Serengeti National Park, Tanzania. *Ser. Res. Inst. Publ.*, 237, Arusha, Tanzania.
Gerresheim, K. (1974). *The Serengeti Landscape Classification—Map and Manuscript*. Afr. Wildl. Leadership Fdn., Nairobi, Kenya.
Gertenbach, W.P.D. and Potgieter, A.L.F. (1979). Veldbrandnavorsing in die struikmopan-ievel van die Nasionale Krugerwilduin. *Koedoe*, **22**, 1–28.
Goodall, J. (1987). *The Chimpanzees of Gombe*. Harvard University Press, Cambridge, MA.
Grimsdell, J.J.R. (1978). *Ecological Monitoring*. Afr. Wildl. Leadership Fdn., Nairobi, Kenya.
Grzimek, M. and Grzimek, B. (1960). A study of the game of the Serengeti Plains. *Z. Saugertierk*, **25**, 1–61.
Herlocker, D.J. (1974). *Woody Vegetation of the Serengeti National Park*. Texas A&M University, College Station, TX.
IUCN. (1986). *Toward a Regional Conservation Strategy for the Serengeti: Report of a Workshop held at the Serengeti Wildlife Research Centre*. IUCN, Nairobi, Kenya.
Jager, T. (1982). Soils of the Serengeti Woodlands, Tanzania. *Agr. Res. Rept.*, 912. Cen. Agr. Publ. Doc., Wangeningen, Netherlands.
Jolly, G.M. (1969a). Sampling methods for aerial censuses of wildlife population. *E. Afr. Agric. For. J.*, **34**, 46–49.
Jolly, G.M. (1969b). The treatment of errors in aerial counts of wildlife populations. *E. Afr. Agric. For. J.*, **34**, 50–55.
Lamprey, H.F. (1979). Structure and functioning of the semi-arid grazing land ecosystem of the Serengeti region (Tanzania). In *Tropical Grazing Land Ecosystems. A State of Knowledge Report*. UNESCO, Paris, 526–601.
Lamprey, H.F., Glover, P.E., Turner, M.I.M. and Bell.R.H.V. (1967). Invasion of the Serengeti National Park by elephants. *E. Afr. Wildl. J.*, **5**, 151–161.
Lamprey, H.F., Kruuk, J. and Norton-Griffiths, M., (1971). Research in the Serengeti. *Nature*, **230**, 497–499.
Laws, R.M. (1970). Elephants as agents of habitat change in East Africa. *Oikos*, **21**, 1–15.
Laws, R.M., Parker, I.S.C. and Johnstone, R.C.B. (1975). *Elephants and Their Habitats*. Clarendon, Oxford.
Lundgren, B. (1975). *Land Use in Kenya and Tanzania*. Roy Col. For., Int. Rur. Dev. Div., Stockholm.
Lydekker, R. (1908). *The Game Animals of Africa*. Rowland, Ward, London.
Macfarland, A. (1969). Preliminary report on the geology of the Central Serengeti, northwest Tanzania. *13th Ann. Rep. Afr. Geol.*, University of Leeds, 13, 14–15.
McNaughton, S.J. (1976). Serengeti migratory wildebeest: facilitation of energy flow by grazing. *Science*, **191**, 92–94.
McNaughton, S.J. (1977). Diversity and stability of ecological communities: a comment on the role of empiricism in ecology. *Am. Nat.*, **111**, 515–525.
McNaughton, S.J. (1983). Serengeti grassland ecology: the role of composite environmental factors and contingency in community organization. *Ecol. Monogr.*, **53**, 291–320.
McNaughton, S.J. (1985). Ecology of a grazing ecosystem: the Serengeti. *Ecol. Monogr.*, **55**, 259–294.
McNaughton, S.J. (1988). Mineral nutrition and spatial concentrations of African ungulates. *Nature*, **334**, 343–345.
McNaughton, S.J. and Georgiadis, N.J. (1986). Ecology of African grazing and browsing mammals. *Ann. Rev. Ecol. Syst.*, **17**, 39–65.
McNaughton, S.J. and Ruess, R.W. and Coughenour, M.B. (1986). Ecological consequences of nuclear war. *Nature*, **321**, 483–487.

McNaughton, S.J., Ruess, R.W. and Seagle, S.W. (1988). Large mammals and process dynamics in African ecosystems. *BioScience*, **38**, 794–800.

McNaughton, S.J., Wallace, L.L. and Coughenour.M.B. (1983). Plant adaptation in an ecosystem context: effects of defoliation, nitrogen, and water on growth of an African C4 sedge. *Ecology*, **64**, 307–318.

Moss, C. (1988). Elephant Memories. *Thirteen Years in the Life of an Elephant Family*. W. Morrow, New York.

Norton-Griffiths, M. (1972). *Serengeti Ecological Monitoring Program*. Afr. Wildl. Leadership Fdn., Washington, DC.

Norton-Griffiths, M. (1978). *Counting Animals*. Afr. Wildl. Leadership Fdn., Nairobi, Kenya.

Norton-Griffiths, M. (1979). The influence of grazing, browsing, and fire on the vegetation dynamics of the Serengeti. In Sinclair, A.R.E. and Norton-Griffiths, M. (Eds) *Serengeti. Dynamics of an Ecosystem*. University of Chicago Press, Chicago.

Pearsall, W.H. (1957). Report on an ecological survey of Serengeti National Park, Tanganyika. *Oryx*, **4**, 71–136.

Pilgrim, T. and Western, D. (1986a). Inferring hunting patterns of African elephants from tusks in the international ivory trade. *J. Appl. Ecol.*, **23**, 503–514

Pilgrim, T. and Western, D. (1986b). Managing African elephants for ivory production through ivory trade regulations. *J. Appl. Ecol.*, **23**, 515–529.

Rossiter, P.B., Hessett, D.M., Wafult, J.S., Harstad, L., Chemwa, S., Taylor, W.P., Rowe, L., Nyange, J.C., Otaru, M., Mumbala, M. and Scott, G.R. (1983). Re-emergence of rinderpest as a threat in East Africa since 1979. *Vet. Rec.*, **113**, 459–461.

Ruess, R.W. and McNaughton, S.J. (1987). Grazing and the dynamics of nutrient and energy regulated microbial biomasses in the Serengeti grasslands. *Oikos*, **49**, 101–110.

Schaller, G.B. (1972). *The Serengeti Lion*. University of Chicago Press, Chicago.

Sinclair, A.R.E. (1973). Population increases in buffalo and wildebeest in the Serengeti. *E. Afr. Wildl. J.*, **11**, 93–107.

Sinclair, A.R.E. (1987). Long-term monitoring in the Serengeti-Mara: trends in wildebeest and gazelle populations. Report No. 5, Serengeti Ecological Monitoring Programme, Arusha, Tanzania.

Sinclair, A.R.E., Dublin, H. and Borner, M. (1985). Population regulation: a test of the food hypothesis. *Oecologia* (Berlin), **53**, 266–268.

Sinclair, A.R.E. and Grimsdell, J.J.R. (1982). *Population Dynamics of Large Mammals*. Afr. Wildl. Leadership Fdn., Nairobi, Kenya.

Sinclair, A.R.E. and Norton-Griffiths, M. (1982). Does competition or facilitation regulate migrate ungulate populations in the Serengeti? A test of hypotheses. *Oecologia* (Berlin), **53**, 364–369.

Sinclair, A.R.E. and Norton-Griffiths, M. (Eds) (1979). *Serengeti, Dynamics of an Ecosystem*. University of Chicago Press, Chicago.

Stronach, N.R.H. (1988). *Fire Management Plan for the Serengeti National Park. A Report to Tanzania National Parks*. Serengeti Wildl. Res. Inst., Arusha, Tanzania.

Swynnerton, G.H. (1958). Fauna of the Serengeti National Park. *Mammallia*, **22**, 435–450.

Talbot, L.M. and Talbot, M.H. (1963). *The Wildebeest in Western Masailand*. Wildl. Monogr., Washington, DC.

Tucker, C.J., Townshend, J.R.G. and Goh, T.E. (1985). African landcover classification using satellite data. *Science*, **227**, 369–375.

Walker, B.H., Ludwig, D., Holling, C.S. and Peterman, R.M. (1981). Stability of semi-arid grazing systems. *J. Ecol.*, **69**, 473–498.

Western, D. and Grimsdell, J.J.R. (1979). *Measuring the Distribution of Animals in Relation to their Environment*. Afr. Wildl. Leadership Fdn., Nairobi, Kenya.

11 On Long-term Ecological Research in Australia

MARK WESTOBY
School of Biological Sciences, Macquarie University, NSW 2109, Australia

11.1 INTRODUCTION

Impetus for international collaboration in long-term ecological research (LTER) presently comes from two forces. The United States National Science Foundation has funded big-team research using the so-called 'ecosystem' approach through the

Long-term Ecological Research. Edited by Paul G. Risser

1980s. These large projects in the LTER program are intended to be continuing, and not closed after any particular time. Comparable programs have begun more recently in the Federal Republic of Germany, under the Man and Biosphere program. These big-team projects already benefit by networking among each other. Extension of the network to international collaboration would no doubt bring further benefits.

The other impetus for international collaboration comes from a new awareness among scientists that some ecological problems are global, and from a new idealism in tackling those problems. During the next century human life will be drastically affected by global atmospheric and climate change, and by alterations in land use which will stem from both further changes in agricultural politics and technology and from the atmospheric change. As scientists, we have faith that understanding these processes will improve human response to them. Many researchers now believe that if even quite simple processes can be measured simultaneously all over the world, new insights into global processes will emerge.

Australian ecologists certainly wish to make their contribution to the emerging global effort. Australia is one of the major landmasses in the southern hemisphere, and important to global processes for that reason alone. In addition, Australian soils and landforms are quite similar to those of important parts of the developing world, such as Brazil and Africa south of the Sahara. Much of Australia's ecological knowledge is more relevant to these tropical and subtropical areas than the ecological knowledge gained in the USA, Europe, or the USSR. At the September 1988 annual conference of the Ecological Society of Australia a resolution was passed which concluded: 'In particular, the Australian government is urged to support the establishment of a system of long-term study and monitoring sites throughout Australia. Such a system of sites is likely to be proposed within the International Geosphere-Biosphere Programme as one approach to the forecast of global change.'

At present any potential Australian contribution is stillborn for lack of resources. Although Australian science, including ecology, has an excellent international reputation and contributes substantially to the international peer-reviewed literature, it is funded at less than half the level, as a percentage of gross national product (GNP), of science in the United States or Federal Republic of Germany. Total expenditure by the Australian Research Council (ARC) on all aspects of ecology is in the range of one to two million dollars per year. Within present resources, most current Australian ecological research would have to be closed down in order to institute a single long-term site on the model of the United States or Federal Republic of Germany.

This chapter has four sections. First, the administrative situation for support of long-term ecological research in Australia is summarized. Second, a sample of Australian studies is described. Third, possible objectives of long-term ecological research are considered. Finally, possible scenarios for an Australian long-term ecological research program are considered in light of these objectives. Readers interested mainly in the science policy issues could begin with the final two sections.

11.2 ADMINISTRATIVE SITUATION

First, consider research in the sense of work aimed primarily at generating publishable advances in knowledge. In Australia there is no systematic support for long-term ecological research in this sense. Rather, long-term projects have been associated with individual scientists who have worked consistently at a particular site. Many studies have lasted for 10 to 30 years, in the hands of a single scientist; a very few have passed to a second generation. Koonamore (see below) has passed through several academic generations over 63 years; so far as I know, it is unique in this respect.

When Dr R. W. Johnson (now Director, Queensland Herbarium) was President of the Ecological Society of Australia during 1985 to 1986, he established a register of continuing ecological research sites, which he still holds. Forms giving basic information for this register were returned for 78 sites, 24 of these being forest growth plots of CSIRO Division of Forest Industry. The register is very far from giving complete coverage of continuing research sites in Australia, but nevertheless is a useful resource.

The CSIRO Division of Wildlife and Ecology held a workshop in November 1986 to work towards networks of research sites with a view to global change research. Different authors discussed possible sites and measurements in each of 12 major biomes (Shaughnessy *et al.*, 1988). No further action has been taken. It seems unlikely CSIRO will be able to establish substantial new programs, considering it continues to suffer annual funding cuts, and no funds have yet been allocated for global change research. During 1989 the Australian government has made a special allocation of A$8M to CSIRO for new global change research. Most of this went to atmospheric modeling. In terms of enhanced effort in the terrestrial ecology of global change, the infusion of funds is roughly equivalent to three scientists.

The Australian Research Grants Scheme (ARGS) has been the government's agency for competitive, peer-reviewed science funding in most areas of science, including ecology. ARGS did not actively encourage work on particular topics. It received proposals, had them reviewed, ranked them, and funded some. ARGS attempted to assure 3 years for funded proposals, but it was not uncommon for funding to be taken from a project after 1 or 2 years because new proposals had achieved higher ranking. ARGS did not sequester any funds for long-term ecology, and so far as I can discover never considered the possibility of doing so.

During 1988 ARGS was incorporated into a new Australian Research Council (ARC). The government intends the new ARC to guide research effort more actively into areas of national need. It is possible, therefore, that future Australian ecological research will have a larger element of central control. ARC's formulation of research priorities is still evolving.

Various agencies responsible for land management have monitoring schemes. By monitoring I mean the information collected is not intended mainly for publication, but to be used directly in management. Forest management agencies have forest growth plots. In recent years legislators have asked for the impact of livestock grazing to be monitored. (Comparisons between places thought to have different grazing histories have not produced a very clear consensus about grazing impact, so it has proved difficult for government agencies to demonstrate with certainty that individual leaseholders or landowners are managing badly enough for action to be taken against them.) The Western Australian Rangeland Monitoring System (WARMS) (Holm *et al.*, 1987) included 867 monitoring sites on 92 stations by 1984 (Holm, 1986). Of these, 254 included a photopoint, a belt transect for shrub numbers and dimensions, and observations of soil surface features. The large number of sites is necessary because of the aim of collecting significant amounts of evidence about individual stations. It is intended that sites should be remeasured at 4- to 5-year intervals, but it remains to be seen whether the Western Australia Department of Agriculture can muster the resources to achieve this for such a large number of sites. In general, agencies responsible for land management operate at the level of States or Territories, so there are no Australia-wide schemes of this sort.

11.3 SOME AUSTRALIAN STUDIES

Five Australian long-term sites (Figure 11.1) have been chosen to describe briefly here. They are rather varied, and serve to illustrate different virtues, weaknesses, and difficulties of long-term research.

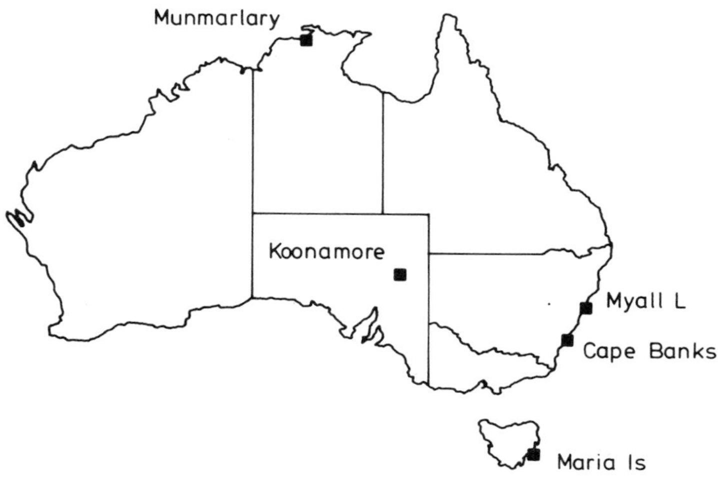

Figure 11.1 Locations of the five long-term study areas briefly described here

11.3.1 THE T.G.B. OSBORN VEGETATION RESERVE AT KOONAMORE

The reserve of 390 ha was established on a pastoral station in South Australia by Osborn in 1925, and has always been operated by the Botany Department of the University of Adelaide (Sinclair, 1986). The area is arid, with mean annual rainfall about 180 mm. The station had been heavily grazed by sheep for about 50 years of European occupation, and the intention was to follow processes of vegetation recovery. The reserve was fenced against sheep and rabbits. There are permanent quadrats, charted from time to time, and photopoints.

Research at Koonamore has fallen roughly into four phases. During the first 10 years much intensive work was done, reported by Osborn *et al.* (1935) and Wood (1936). There followed a long period when no research was reported. Intermittent recording was maintained, reputedly due to the personal commitment of Constance Eardley. Continued regeneration to 1962 was reported by Hall *et al.* (1964). Then during the 1970s two students of R.T. Lange made creative use of the long-run records. Noble found ways to estimate biomasses from photos of vegetation, and modeled germination and growth of both ephemerals and perennials in the chenopod shrublands (Noble, 1977; Noble and Crisp, 1980). Crisp examined demography of woody perennials from the records and from size frequency distributions (Crisp and Lange, 1976; Crisp, 1978), showing *inter alia* that sheep and rabbits had prevented establishment of *Acacia aneura* for nearly a century in this part of Australia. While short-term experiments can show such a grazing effect, they could not have indicated that it had applied to all germination episodes over an extended time.

Sinclair is currently in charge of the reserve. He is analyzing events of the past 15 years, particularly the consequences of the heavy summer rains of the mid 1970s, which led to widespread establishment of woody plants. It should be noted that Graetz, in the CSIRO draft report on long-term sites referred to earlier, regards Koonamore as not representative of the generality of either saltbush or bluebush types in Australia.

11.3.2 MARIA ISLAND

At Maria Island, off the south-east coast of Tasmania, surface water has been collected fortnightly since 1945. Physical and chemical characteristics of the water have been measured. Maria Island is now part of a network of coastal water sampling stations, managed by Dr Denis Mackey of CSIRO Division of Oceanography.

This long run of simple data showing between-year variation in seasonal cycles at one place has become valuable in interpreting wide-area processes. Satellite imagery and hydrographic cruise information have shown that the dominant feature of south-east Australian waters is the subtropical convergence (STC), running across from the southern end of Tasmania to New Zealand (Harris *et al.*, 1987). The broad zone where subtropical and subantarctic water masses meet has deep mixing, high phytoplankton productivity, and concentrated carbon export to deep water. Water temperature patterns and the timing of spring bloom at Maria Island are sensitive to penetration of subantarctic water up the east coast of Tasmania. This penetration in turn is driven by the zonal westerly winds at Tasmanian latitudes. These winds are cyclical with a mean periodicity of 11 years (Harris *et al.*, 1988). These phenomena in turn are correlated, positively or negatively, with atmospheric pressure at Darwin in the tropics and at Macquarie Island in the subantarctic, with rainfall, lake levels, and trout catches in the Tasmanian highlands, with seal abundance at Macquarie Island, and with catches of spiny lobsters not only off Tasmania but also off New South Wales and New Zealand. The correlations with lobster catches show an immediate effect and also an effect time-lagged by 5 to 7 years, consistent with known life history (Harris *et al.*, 1988). However, the biological basis of any effect on lobster recruitment is not known.

This is an example where correlations among different variables at one location showed up few relationships beyond those that were already well understood. It was the intercorrelations across space that led to substantial advances in understanding.

11.3.3 CAPE BANKS

Cape Banks is the north headland of Botany Bay, within the Sydney metropolitan area. Being so close to the universities in Sydney it has been used intermittently for research for a long time. Since 1972, under the leadership of A.J. Underwood of the University of Sydney, it has become a major center for the experimental ecology of rocky shores. For example, it has been shown that the boundary between the upper limit of algal turf and the lower limit of the barnacle and limpet zone is not determined by predation on molluscan herbivores, as had been expected on the basis of North American work. Rather, limpets are unable to establish lower on the shore where algal productivity is greater, and herbivores are unable to keep plant biomass low (Underwood and Jernakoff, 1981). Since 1972 a total of 98 publications and 32 theses have emerged from Cape Banks.

Most of this work has been experimentation lasting 1 to 3 years (reviewed in Underwood, in press). However, long-term monitoring (mostly not yet published) has been an important complement. For example, severe storms in 1983 stripped the tubeworm *Galeolaria* off rocks, which were then recolonized by algae and tunicates, and this change of state appears persistent. Another example is that during 1980 to 1984 the barnacle *Chamaesipho columna* declined due to a failure of recruitment. This failure was general throughout the 1000 km of the NSW coast.

We need to know not only how frequent extreme events are, but also how widespread. It will be recalled that by the late 1970s, much emphasis was being put on the role of stochastic effects in recruitment in controlling the structure of assemblages. An important recent trend in Australian ecology is to go beyond marveling at the variability of phenomena, and to use nested ANOVA designs to estimate percentages of the variance attributable to different scales in space and time. The first example of this dealt with barnacle recruitment, and was centered at Cape Banks (Caffey, 1985).

At present a program investigating succession in artificially-created tidepools is under way. A peculiarity of this site is that it lies above mean low water, but below mean high water. In consequence it falls into a legislative vacuum, is not anyone's property, and apparently no agency has the power to prevent the public from visiting the site as they please.

11.3.4 MUNMARLARY

The Munmarlary site was established by R.J. Hooper in 1972, and is operated by the Conservation Commission of the Northern Territory. It is in the monsoonal tropics, and the vegetation has a grassy understory beneath a variable canopy of eucalypts. Four different fire regimes have been applied, each to three replicate sites 100 × 100 m, this design being repeated in both forest (conventionally defined in Australia as projective foliage cover of trees 30–70%) and woodland (tree cover 10–30%). Both these vegetation types would be called savanna in tropical Africa or America.

The first 12 years of this study have recently been reported by Bowman *et al.* (1988). While 12 years may not seem very long term, in this environment it represents 12 fire-cycles of the prevailing fire regime. The four treatments have been yearly fires in early and in late dry season, fires every two years in early dry, and no fires.

To understand the significance of the work it is necessary to appreciate that also present in these landscapes are closed forests or monsoon rainforests (canopy cover >70%). These are dominated by species with rainforest rather than sclerophyll affinities. They are sensitive to fire, also less flammable, and at present are patchily distributed in fire-shadow locations. Some experimentation in Africa has indicated that if savannas are protected against fire they tend to be invaded by fire-sensitive species of closed forest (Rose-Innes, 1972). On the strength of this, Stocker and Mott (1981) suggested that much of the monsoonal north Australian landscape was occupied by monsoonal rainforest until an increase in fire frequency about 40 000 years BP, when people are thought to have arrived. A vegetation shift somewhat analogous can be seen in the 190 000-year pollen core from Lynch's Crater in north Queensland (Kershaw, 1986).

On the other hand, other evidence from Africa (Geldenhuys, 1977) and America (Kellman and Miyanishi, 1982; Sarmiento and Monasterio, 1975) suggests changes in fire regime affect vegetation structure considerably, but floristics are mainly controlled by soils.

The Munmarlary experiment has not shown any evidence that substantial shifts in floristics follow fire exclusion (Bowman *et al.*, 1988). However, a considerable woody understory develops, from species that were present before, but with suppressed height development. There are substantial differences between replicates, possibly attributable to soils effects.

Munmarlary shows the importance of manipulated treatments, with proper replication, in long-term studies. It also exposes the limitations of studies concentrated at one place in a landscape. The evidence now available from three continents strongly suggests that fire exclusion can allow closed forest to develop in some locations, but not in others. What we need now is to learn rules to identify parts of landscapes which can potentially become closed forest if protected against fire, and studies at single sites will not achieve this.

11.3.5 MYALL LAKES

In Myall Lakes National Park coastal sclerophyll forest and heath is found on a sand mass created about 6000 years BP The coastal dunes are mined for heavy minerals. Sand-mining involves bulldozing away the vegetation and stockpiling the topsoil. The dune is then excavated to a depth of tens of meters, making a pond. A slurry of sand and water is sucked from one edge of the pond, and the small percentage of rutile grains is separated by centrifugal force. The remainder of the sand is then pumped back onto the other side of the pond. In this way the pond moves progressively along the dune, creating a 'mining path' from which the vegetation has been removed, and the substrate reconstructed to considerable depth. Under current regeneration practices, the dune is first restored to approximately its original shape. The topsoil, including soil seed reserves, is replaced. The dune is stabilized with a crop of hybrid sorghum during the first year after mining, and natural vegetation develops over ensuing years. In consequence, a sequence of vegetation ages is found along each mining path.

Barry and Marilyn Fox recognized the potential of this situation for studying succession while still research students at Macquarie University, and established permanent plots during 1975 to 1976 on sites then aged 5 to 11 years after mining. They have studied successional processes in vegetation, small mammals, and ant assemblages. This research gains special strength from the complementary use of four types of information: comparisons among sites of different ages, longitudinal studies at each site, comparisons between regeneration after mining and regeneration after fire (permanent sites for fire studies were also set up during 1974 to 1977 (Fox, 1988)), and experimental manipulations of interactions between species.

For example, consider small mammals. *Mus musculus* abundance peaks at sites mined 3 to 5 years previously, while *Pseudomys novaehollandiae* abundance overtakes it at sites mined 5 to 7 years previously (Fox and Fox, 1984). The same pattern was present both when the mine path was sampled in 1982, and when it was sampled in 1987 (Twigg *et al.*, 1989). The actual sites representing a given time since mining were, of course, different in the two sampling years. On sites regenerating after fire, the first 5 to 6 years of the succession after mining are compressed into a single year, and *P. novaehollandiae* becomes more abundant during the second year (Fox and McKay, 1981; Fox and Fox, 1984). Probably this is because many plant species regenerate vegetatively after fire, whereas their rhizomes and lignotubers are destroyed during mining, so that understory vegetation develops much more quickly after fire. *P. novaehollandiae* abundance is correlated with vegetation cover below 50 cm (Fox and Fox, 1978). Indirect evidence indicated *M. musculus* abundance declines because they behaviorally avoid *P. novaehollandiae*. This hypothesis has been confirmed by experimental removals of *P. novaehollandiae*, and also by experimental additions (Fox and Pople, 1984), which are less often carried out.

A comparable situation is found among ants. *Iridomyrmex* species C dominates for the first 8 to 9 years after mining, and is then replaced by *Iridomyrmex* species A (Fox and Fox, 1982). Within the two-species mosaic of colonies found around the time one species replaces the other, experimental deletion of colonies of either species produced increases in the other species (Fox *et al.*, 1985; Haering and Fox, 1987), with the relationship mediated by sharp behavioral competition at territory boundaries. Because *Iridomyrmex* species are so active and behaviorally dominant (Greenslade, 1976), different assemblages of other ants are associated with dominance by the two different *Iridomyrmex* species (Fox and Fox, 1982).

11.4 OBJECTIVES OF LONG-TERM ECOLOGICAL RESEARCH

This section discusses what sorts of objectives might be achievable by long-term research but not by short-term research. The discussion does not specify variables to be measured. In any research, the variables worth measuring depend on which questions are thought interesting. The intention here is to consider the virtues of long-term ecological research generally, across the wide variety of possible questions and variables.

For the purpose of discussion, objectives of long-term measurement, or networking across space, or both, can be split into five categories:

(1) Enough replicate years to allow statistical analysis (e.g. correlation) of between-year variation;
(2) Perspective on rare events, estimating their return times;
(3) Studying second-phase effects of experimental manipulations. By second-phase effects are meant those which involve species arriving subsequently, or effects via consequences for slow variables; in contrast to first-phase effects, which involve those on species present at the outset of the experiments and on interactions among them;
(4) Estimating the proportions of variation in an ecological measurement which are attributable to differences between years or between places, or to place–year interactions, such as patterns which occur every year, but are shifted in space.

(5) Measuring changes over time in ecological variables, on the premise that we need to begin collecting a coherent record now, in order to have data against which to assess any understanding or models of global change we may develop in 10, 30, or 50 years' time.

11.4.1 PREVIEW

Briefly, the opinion we will put forward is that the highest priority for Australia is to contribute to global networks of many sites, making simple measurements, with a view to objectives 4 and (especially) 5. Extending intensive research to 10- to 50-year runs at a few individual sites, with a view to objectives 1-3, would not be such a high priority. Within a limited budget, there will be a conflict between having many sites and studying fewer sites more intensively. It will be important to restrain the natural instinct to measure more variables more often, lest spatial coverage be endangered.

This opinion is based both on a modest assessment of what can be achieved with respect to objectives 1 to 3 by runs of data 10 to 50 years long at any one site, and on a very high assessment of the importance of objective 5.

11.4.2 MODEST VALUE OF DATA RUNS OF 10 TO 50 YEARS FOR OBJECTIVES 1 TO 3

11.4.2.1 Objective 1: Years as replicates for analyzing correlation between two variables

Suppose two variables are in fact related by an ecological process. For example, the earliness of the wet season (*A*) might affect the peak population density achieved by a herbivorous insect (*B*). Now suppose *A* is capable of accounting for 20% of the variance in *B* (r^2=0.20); this is a decidedly strong relationship by ecological standards. Fifty replicate years would be needed to have a 90% chance of detecting such a relationship at $P<0.05$ (Figure 11.2). Thus runs of data in the 10- to 50-year range can only be relied on to detect very strong relationships. In order to reliably detect relationships with r^2 in the range 0.01 to 0.10 (where many interesting and important ecological relationships lie), data runs of 100 to 1000 years are needed. This suggests that monitoring ecological variables in one place for 10- to 50-year periods, while by no means valueless, is a relatively weak approach to detecting such correlations.

Two suggestions can be made for stronger approaches to detecting or assessing relationships between ecological variables at one place. First, variation in driving variables should be created experimentally where possible. Second, we should be turning stronger attention to the possibilities for obtaining long data runs from records of the past, whether historical or paleoecological. New technologies are

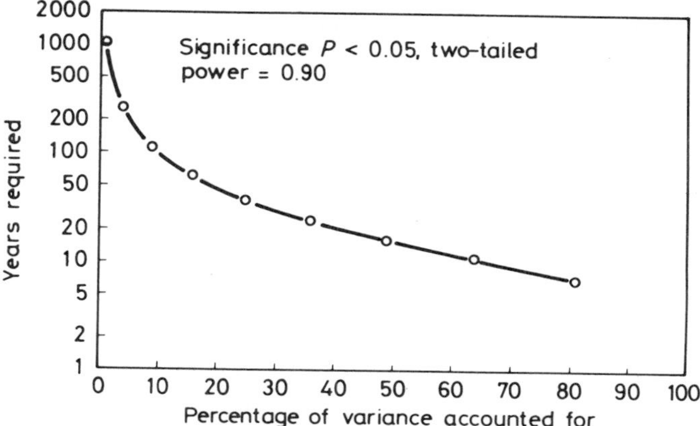

Figure 11.2 Number of replicate years required to detect with 90% probability (power = 0.90), at 5% significance, a correlation (between variables, across years) accounting for a given percentage of variation. Derived from a table in Cohen (1969)

opening up many possibilities for acquiring high-resolution environmental records from the past. There are many situations where only funding is restricting the availability of valuable data (Chappell 1988; Wasson, 1988). However, this should not be seen as a recommendation to simply hand funds to paleobiologists. Funding should be made conditional on paleobiologists collaborating closely with theoretical and experimental ecologists in defining questions and data collection designs which will allow the questions to be answered.

11.4.2.2 Objective 2: Perspective on the incidence of unusual but biologically important events

The domination of ecology by studies of 1 to 3-year duration has led to many misinterpretations of the incidence of unusual events (Weatherhead, 1986). However, even 20-year studies are only capable of assessing the return times of events which occur about 10 times per century or more often. Further, these return times are changing, so that past return times can not be simply extrapolated to the future.

Further still, it should be considered that if unusual events are biologically important, merely monitoring them will not advance our knowledge fast enough. We need to find ways to organize science funding such that hypotheses are tested experimentally during unusual events. Some rare events can be anticipated, e.g. the 18.6-year tidal cycle which Clark (1986) hypothesized reset Gramineae and Cyperaceae populations in a Long Island high marsh to a new exponential growth phase. Other experiments could in principle be designed in advance for implementation at an unknown future time (Westoby *et al.*, 1989). For example, a widespread rangelands problem is establishment of shrubs in semi-arid grasslands after they have germinated during two successive years of heavy summer rains. It is possible that destocking during a critical period would allow grass growth to suppress shrub seedling establishment. Experiments to test this hypothesis need to be funded in such a way that they can be implemented at whatever future time mass shrub germination comes about.

While a long study is always better than a short one, this should not lead us to concentrate our efforts into a relatively few research sites studied at length. On the contrary, my own experience has been that problems in generalizing from one or a few studies arise at least as often because they are unrepresentative in space as because they are unrepresentative in time.

At present, short-term studies at a single site often report that some event they think unusual had a significant effect on the outcome of their study (Weatherhead, 1986) (for example, mass kill of shrubs by drought or of echinoids by disease). It would be very valuable to have more perspective on how widespread such events are in space. In a reasonable proportion of cases some evidence (e.g. dead shrubs) is to be seen for a period after the event, so that a single visit to a site can assess whether the event happened there. It would be cost effective for agencies to make modest funds available so that when an unusual event is encountered during a 3- to 6-year study, funding can quickly be sought for travel, research assistance, satellite imagery, or whatever was necessary to allow a rapid wide-area assessment.

11.4.2.3 Objective 3: Second-phase effects of experimental manipulations

It is undoubtedly very important that second-phase effects should be recognized and studied in experimental ecology. This objective justifies some 10- to 50-year studies. However, it would be impossibly expensive to extend all field experiments for 10 to 50 years on the grounds that second-phase effects needed to be studied. Other approaches to second-phase effects should also be considered. If the nature of the effects can be anticipated, they can be factored into the initial design of experiments. For example, if effects involving species not yet present are anticipated, treatments in which they are introduced could be included in the design. For effects involving slow litter accumulation, litter could be added; and so forth.

Another possibility is to make better use of the many sites where intensive experimental studies have been undertaken over 3 to 6 years. Suppose investigators were to submit proposals to revisit their previously funded experimental sites from time to time. In effect, they would be saying they had no particular hypothesis, but would like to keep a watching brief on their sites to see if anything interesting happened. Under the criteria used by most competitive research funding agencies at present, such proposals would be ranked very low. However, there is a defensible case that by making modest funding (e.g. a week's fieldwork every 2 years) available for such proposals, funding agencies could procure insight into second-phase effects more cheaply than by any other means.

11.4.3 HIGH VALUE OF DATA RUNS OF 10 TO 50 YEARS FOR OBJECTIVES 4 AND 5

11.4.3.1 Objective 4: Assessing variation across space in relation to variation through time

Most of our present understanding of ecosystems is based on information which is restricted in space (often from plots of a hectare or less) as well as in time (studies extending over 3 years or less). This is especially true of that part of our understanding which is underpinned by field experimentation. Arguably the most important problem for ecological science over the next few decades is scaling up—sorting out situations where this narrow-scale understanding can be simply extrapolated to wider areas and longer times from situations where new concepts and approaches are needed.

Two of the Australian studies summarized above illustrate the strong relationship between wide spatial coverage and long-term research. Work centered on Cape Banks has shown how formal ANOVA design can be used to assess the relative strengths of variation at different scales in space and time. Such information allows more intensive narrow-scale studies to be put in context, and also permits estimates of the space and time scales over which aggregated variables become relatively predictable because narrow-scale variation is averaged out. The study centered on Maria Island detected a variety of interesting correlations between variables measured at sites hundreds or even thousands of kilometers apart. Given <50 year-replicates, only correlations with r^2 at least 0.20 can reliably be detected (Figure 11.2). Correlations this strong arising from processes at one place will often prove to be well understood already. For example, a relationship between surface water temperature and phytoplankton production at the Maria Island site, mediated by nutrients in surface water, would have been no surprise to marine ecologists. But correlations between spiny lobster catches off Tasmania and New Zealand and (in the opposite direction) off New South Wales would have been much more surprising. Such wide-area correlations may contribute strongly to understanding relationships between ground-truth ecological measurements and regional or global scale climatology.

In summary, much is to be gained in a research sense from having networks of many sites, even if only a few simple measurements are taken at each. The arrangement of sampling across space and through time should be designed with a view to partitioning variance into components attributable to different scales in space and time, and to interactions between space and time.

11.4.3.2 Objective 5: Data base on global change against which to assess future understanding

Global change—all its aspects, not just atmospheric change—is plainly the most important applied problem for this and the next generation of ecologists. It is not possible to predict what sort of understanding of global change processes will have been generated by research 30 to 50 years from now. However, it is possible to predict with certainty that confidence in that future understanding will be crippled unless we have in place, at that time, a coherent data base describing the actual course of change from the 1990s forward. Initiating such a data base should be regarded as easily the most important objective of long-term ecological studies, and the work should be designed with this objective dominant.

There is one desideratum of long-term ecological study which arises from this objective. One thing an ecological data base will need to do is to connect with general circulation models (GCMs) of global climate. Current GCMs work with cells about 500 × 500 km. These will become smaller with increased computing power, but it is quite likely that cells will still be on the order of 100 × 100 km in 30 years' time. That is, models of climate–vegetation relationships to be incorporated in GCMs need to deal with variables which represent averages over at least 100 × 100 km. A minimal requirement for a data base oriented to global change research is that it should provide descriptions of vegetation and land surface characteristics averaged over no less than 100 × 100 km. At present most ecological research measures variables at much smaller scales than this. It is, of course, possible that future research will develop rules which will allow us to scale up from small-scale measurements to wide-area estimates. A great deal of research is going into self-similarity of spatial patterns at different scales, for example; and into nesting regional-scale climate models within GCMs. However, it would be wrong to rely on methods being developed in the future to scale up from local-scale measurements to averages over wider areas. The data base should be designed from the outset with a view to providing reliable averages over wide areas. This could be done either through remote sensing, or through proper procedures of stratification and randomization in choosing sites for ground measurement, depending on the variable. We should not fall into the errors of using a small number of sites, choosing them arbitrarily on grounds of convenience, and relying on future developments to allow reliable scaling up.

11.5 SCENARIOS FOR AN AUSTRALIAN EFFORT IN LONG-TERM ECOLOGICAL RESEARCH IN THE CONTEXT OF GLOBAL CHANGE

There are many complexities in assessing what form Australian efforts should best take. I will summarize the most important issues by putting forward and commenting on three scenarios. Of these, I would regard the first as a disastrous outcome (though there is a real risk it might happen), the second as having some good and some bad features, and the third scenario as the path we ought to take if the government can be persuaded.

Under scenario 1, no extra funding would flow to Australian ecology for purposes of global change research. Instead, some of the funds currently spent on research in all areas of ecology would be redirected towards long-term studies. This would be most likely to come about through ARC shifting its funding criteria to put greater emphasis on long-term studies in the context of global change, and relatively less emphasis on peer-review rating of the excellence of the proposal and the investigators.

Under this scenario Australia's existing research effort in ecology would be seriously damaged. A significant development of long-term studies would require much or all of the funds currently spent by ARC on ecology in general. The advance of our fundamental understanding of ecosystem organization would slow or stop. The long-term studies would not substitute for the previous ecology research effort in this respect, because they would not be asking questions of the same incisiveness.

Further, it needs to be considered that a policy of funding long-term projects automatically means that projects once begun must be favoured for continued funding. The most important factor in research progress is the quality of the principal investigators and the intensity with which they commit their time and energy to the research. The best method yet found for assuring this quality and commitment is free and open competition among all applicants, on the basis of peer review, at frequent intervals. The hard truth must be faced that if some projects are given continuity for periods upwards of 10 years, this will be done at least partly at the expense of other projects and other investigators which would be rated better by peer review. There is a cost of continuity, which takes the form of foregoing the opportunity to switch funds to better investigators, and which increases over time. This cost means that long-term research cannot be expected to substitute satisfactorily for open-competition research. We need long-term studies to complement research on a shorter funding cycle, not to replace it.

Under scenario 2, the Australian government would commit substantial new funds to global change research, and a reasonable proportion of it, say A$5 to 10 million per year, would be allocated to ecology. Because of the pressure of precedent from the style of long-term research adopted in the United States and Federal Republic of Germany, combined with science politics within Australia, these funds would be directed into about five big-team projects. Depending on the strength of influence from overseas, these projects might have an 'ecosystem' orientation, or might be oriented towards approaches more highly regarded in Australia, such as vegetation dynamics or experimental ecology. Under this scenario, understanding of Australian ecosystems would be advanced by a burst of fieldwork, representing at least a doubling of current ARC expenditure. This would certainly be an excellent thing. However, the data base which resulted would come from relatively few sites, perhaps five to 20. Its geographical coverage would be far too thin to give a satisfactory picture of the spatially patchy processes expected in response to global change (for example, processes of certain elements of some vegetation types colonizing into new locations, or of increased incidence of outbreaks of some herbivorous insects within some vegetation types). For these reasons, scenario 2 would not be making a very good contribution to the data base for global change, which is the most important objective for long-term ecological studies in Australia.

Under scenario 3, as under scenario 2, the Australian government would find significant new funds for global change research, and A$5 to 10 million per year of these would be spent on ecological aspects. Under scenario 3, however, these extra funds would be spent on a large number (more than 200) of sites. Relatively simple measurements would be taken at most of these sites, not necessarily every year.

This would be the best of the three scenarios in that it would be directed at the most important objective for global change research, which is to set in place a data base recording the time-course of global change with complete geographical coverage and relatively high spatial resolution. In saying that it would be the best of the three scenarios, I do not at all mean that it would cover all the ecological research that needs to be done in relation to global change; much more than A$5 to 10 million would be needed for that. I mean simply that it would be better than scenarios 1 and 2.

There would certainly be problems with scenario 3. Notably, means would have to be found to assure effective quality control of the data and management of the data base. Normally, researchers can be expected to pay close attention to their data because they examine and analyze them themselves and publish from them within a few years. But in the case of simple measurements collected from less than 100 locations, many people would be involved in taking the measurements, and many of the data would have no immediate use in testing hypotheses. Comparability among different sites and over time would need to be assured, and the data base would need to be organized to allow ready access by any interested parties.

Protocols for data management can, no doubt, be organized. However, the key to quality research is the sort of people who are involved. When choosing sites for a monitoring network, we face the enervating prospect of a series of workshops during which everyone advocates the sites they work at, naturally for entirely objective reasons of the site's interest, representativeness, and backlog of data. There is a risk that we could emerge from such a process without the country's best researchers involved with the selected sites. We should start out instead by selecting first-rate researchers, using the usual criteria of continuing publication in international peer-reviewed journals, etc. These researchers could then be put in a position to run networks of sites, whether in person or by subcontracting.

The use of new funds to set in place such a data base should parallel a steady enhancement of fundamental research in ecology, through the medium of short-term grants given competitively through ARC. Some minor modifications could usefully be made to current funding practices, with a view to gaining better perspective on the representativeness of results in time and space. Specifically, paleobiology could be better funded provided proposals are devised in collaboration with experimental and theoretical ecologists; means could be found to fund experiments to be implemented at unknown future times, when the relevant configuration of circumstances arises; when unusual events occur during a 3- to 6-year study at a single site, some funds could be available for rapid assessments of how widespread the events were; and after intensive studies have been completed involving experimental manipulations at a site, modest funds could be provided for proposals to revisit the site occasionally to see whether any interesting longer-term effects have become apparent.

11.6 ACKNOWLEDGMENTS

My thanks for helpful discussion, comments and contributions of information from Paul Adam, Andy Beattie, David Bowman, Barry Fox, Frank Golley, Graeme Harris, Ian Hume, Bob May, Ron Pulliam, Graham Pyke, Barbara Rice, Paul Risser, Russell Sinclair, Tony Underwood, Dedee Woodside, Paul Zedler, an anonymous referee and from many people at the Berchtesgaden workshop. The conclusions are mine.

11.7 REFERENCES

Bowman, D.M.J.S., Wilson, B.A. and Hooper, R.J. (1988). Responses of *Eucalyptus* forest and woodland to four fire regimes at Munmarlary, Northern Territory, Australia. *Journal of Ecology*, **76**, 215–232.

Caffey, H.M. (1985). Spatial and temporal variation in settlement and recruitment of intertidal barnacles. *Ecological Monographs*, **55**, 313–332.

Chappell, J. (1988). The quaternary environmental record. In *Australian National Committee for the IGBP, Convenors. Global Change*. Australian Academy of Science, Canberra, 30–35.

Clark, J.S. (1986). Late-Holocene vegetation and coastal processes at a Long Island tidal marsh. *Journal of Ecology*, **74**, 561–578.

Cohen, J. (1969). *Statistical Power Analysis for the Behavioral Sciences*. Academic Press, New York.

Crisp, M.D. (1978). Demography and survival under grazing of three Australian semi-desert shrubs. *Oikos*, **30**, 520–528.

Crisp, M.D. and Lange, R.T. (1976). Age structure, distribution, and survival under grazing of the arid zone shrub, *Acacia burkittii*. *Oikos*, **27**, 86–92.

Fox, B.J. (1982). Fire and mammalian secondary succession in an Australian coastal heath. *Ecology*, **63**, 1415–1425.

Fox, M.D. (1988). Understory changes following fire at Myall Lakes, New South Wales. *Cunninghamia*, **2**, 85–95.

Fox, B.J. and Fox, M.D. (1978). Recolonization of coastal heath by *Pseudomys novaehollandiae* (Muridae) following sand-mining. *Australian Journal of Ecology*, **3**, 447–465.

Fox, M.D. and Fox, B.J. (1982). Evidence for interspecific competition influencing ant species diversity in a regenerating heathland. In Buckley, R.C. (Ed.) *Ant-Plant Interactions in Australia*. Dr W. Junk, The Hague, 99–110.

Fox, B.J. and Fox, M.D. (1984). Small-mammal recolonization of open-forest following sandmining. *Australian Journal of Ecology*, **9**, 241–252.

Fox, B.J., Fox, M.D. and Archer, E. (1985). Experimental confirmation of competition between two dominant species of *Iridomyrmex* (Hymenoptera: Formicidae). *Australian Journal of Ecology*, **10**, 105–110.

Fox, B.J. and McKay, G.M. (1981). Small-mammal responses to pyric successional changes in eucalypt forest. *Australian Journal of Ecology*, **6**, 29–41.

Fox, B.J. and Pople, A.R. (1984). Experimental confirmation of interspecific competition between native and introduced mice. *Australian Journal of Ecology*, **9**, 323–334.

Geldenhuys, C.J. (1977). The effect of different regimes of annual burning on two woodland communities in Kavango. *South African Forestry Journal*, **103**, 32–42.

Greenslade, P.J.M. (1976). The meat ant *Iridomyrmex purpureus* (Hymenoptera: Formicidae) as a dominant member of ant communities. *Journal of the Australian Entomological Society*, **15**, 237–240.

Haering, R. and Fox, B.J. (1987). Short-term coexistence and long-term competitive displacement of two dominant species of *Iridomyrmex*: the successional response of ants to regenerating habitats. *Journal of Animal Ecology*, **56**, 495–507.

Hall, E.A.A., Specht, R.L. and Eardley, C.M. (1964). Regeneration of the vegetation on Koonamore Vegetation Reserve, 1926–1962. *Australian Journal of Botany*, **12**, 205–264.

Harris, G.P., Davies, P., Nunez, M. and Meyers, G. (1988). Interannual variability in climate and fisheries in Tasmania. *Nature*, **333**, 754–757.

Harris, G., Nilsson, C., Clementson, L. and Thomas, D. (1987). The water masses of the east coast of Tasmania: seasonal and interannual variability and the influence of phytoplankton biomass and productivity. *Aust. J. Mar. Freshw. Res.*, **38**, 569–590.

Holm, A.McR. (1986). The assessment of range trend in Western Australian pastoral lands. In Joss, P.J., Lynch, P.W. and Williams, O.B. (Eds) *Rangelands: A Resource Under Siege. Proceedings 2nd International Rangeland Congress*. Australian Academy of Science, Canberra.

Holm, A.McR., Burnside, D.G. and Mitchell, A.A. (1987). The development of a system for monitoring trend in range condition in the arid shrublands of Western Australia. *Australian Rangeland Journal*, **9**, 14–20.

Kellman, M. and Miyanishi, K. (1982). Forest seedling establishment in neotropical savanna: observations and experiments in the Mountain Pine Ridge savanna, Belize. *Journal of Biogeography*, **9**, 193–206.

Kershaw, A.P. (1986). Climatic change and Aboriginal burning in north-east Australia during the last two glacial/interglacial cycles. *Nature*, **322**, 47–49.

Noble, I.R. (1977). Long-term biomass dynamics in an arid chenopod shrub community at Koonamore, South Australia. *Australian Journal of Botany*, **25**, 639–653.

Noble, I.R. and Crisp, M.D. (1980). Germination and growth models of short-lived grass and forb populations based on long-term photo-point data at Koonamore, South Australia. *Israel Journal of Botany*, **28**, 195–210.

Osborn, T.G.B., Wood, J.G. and Paltridge, T.B. (1935). On the climate and vegetation of the Koonamore Vegetation Reserve to 1931. *Proceedings of the Linnean Society of New South Wales*, **60**, 392–427.

Rose-Innes, R. (1972). Fire in West African savanna. *Proceedings of the Tall Timbers Fire Ecology Conference*, **11**, 147–173.

Sarmiento, G. and Monasterio, M. (1975). A critical consideration of the environmental conditions associated with the occurrence of savanna ecosystems in tropical America. In Golley, F.B. and Medina, E. (Eds) *Tropical Ecosystems: Trends in Terrestrial and Aquatic Research*. Springer-Verlag, New York, 223–250.

Shaughnessy, P.D., Leigh, J.H. and Walker, B.H. (1988). Preliminary proposals for a set of long-term ecological study and monitoring sites in Australia. Working Paper, available from CSIRO Division of Wildlife and Ecology, PO Box 84, Lyneham, ACT 2602 Australia.

Sinclair, R. (1986). The T.G.B. Osborne Vegetation Reserve at Koonamore: long-term observations of changes in arid vegetation. In Joss, P.J., Lynch, P.W. and Williams, O.B. (Eds) *Rangelands: A Resource Under Siege. Proceedings 2nd International Rangeland Congress*. Australian Academy of Science, Canberra, 67–68.

Stocker, G.C. and Mott, J.J. (1981). Fire in the tropical forests and woodlands of Northern Australia. In Gill, A.M., Groves, R.M. and Noble, I.R. (Eds) *Fire and the Australian Biota*. Australian Academy of Science, Canberra, 425–439.

Twigg L.E., Fox, B.J. and Luo Jia. The modified primary succession following sandmining: A validation of the use of chronosequence analysis. *Australian Journal of Ecology*, **14**, 441–448.

Underwood, A.J. Rocky intertidal shores. In Hammond, L.S. and Synnot, R. (Eds) *Australian Marine Biology*. Longman-Cheshire, Melbourne. In press.

Underwood, A.J. and Jernakoff, P. (1981). Interactions between algae and grazing gastropods in the structure of a low-shore algal community. *Oecologia*, **48**, 221–233.

Wasson, R.J. (1988). Landscape denudation in Australia in the late Holocene. In *Australian National Committee for the IGBP, Convenors. Global Change*. Australian Academy of Science, Canberra, 56–59.

Weatherhead, P.J. (1986). How unusual are unusual events? *American Naturalist*, **128**, 150–154.

Westoby, M., Walker, B.H. and Noy-Meir, I. (1989). Opportunistic management for rangelands not at equilibrium. *Journal of Range Management*, **42**, 266–274.

Wood, J.G. (1936). Regeneration of the vegetation of Koonamore Vegetation Reserve, 1926–36. *Transactions of the Royal Society of South Australia*, **60**, 96–111.

12 Computer Models and Long-term Ecological Research

H.H. SHUGART*, G.B. BONAN‡, D.L. URBAN*, W.K. LAUENROTH†, W.J. PARTON† and G.M. HORNBERGER*

**Department of Environmental Sciences, The University of Virginia, Charlottesville, VA 22903, USA*

†Range Science Department, Colorado State University, Fort Collins, CO 80523, USA

‡Earth Resources Branch/ Code 623, Laboratory for Terrestrial Physics, NASA / Goddard Space Flight Center, Greenbelt, MD 20771, USA

12.1 INTRODUCTION

Ecological modeling has a diversity of roles to play in the development and coordination of long-term ecological research. In particular, models can be used to augment data collection in a predictive mode, or to inspect and interact theories about ecological systems. The two cases that will be discussed in the present chapter involve the use of models to simulate system dynamics at differing spatial

Long-term Ecological Research. Edited by Paul G. Risser

scales. One example involves development of a model at a site with a considerable history of long-term ecological research (The US National Science Foundation's Long-Term Ecological Research (LTER) Boreal Forest Site near Fairbanks, Alaska). The testing of this model has been on continental and global scales. The second example involves an ongoing project in which a set of models, developed at several LTER sites, is applied to a single site (the Konza Prairie LTER site) in the transition zone between forest and prairie in the United States.

The rationale and justification for the use of simulation models in these two examples stems directly from the scope of the LTER objectives. Large-scale ecological studies are logistically difficult, and the LTER network represents an infrastructure to support such studies, and so to provide an empirical foundation for cross-site comparison and synthesis. This foundation can be built upon with the aid of ecosystem models.

Simulation models offer a means of incorporating a knowledge base at one level and extending this information to its higher level implications (Shugart, 1984). This approach is conceptually compatible with the empirical approach of establishing LTER sites in a variety of ecosystems, and synthesizing larger patterns from the site-specific data.

A simulation model is a dynamic (working) implementation of current understanding of the structure and functioning of a given system. The degree of realism in the models used in the present examples implicitly ensures model verification—the models must adequately reproduce the systems they represent. The linkage to LTER research sites ensures the availability of an empirical basis for model verification. We foresee that specific tests will emerge as theory is developed. Opportunities to test model predictions against independent data will be especially valuable. Collaboration with the LTER network and other established research sites will ensure opportunities for model validation.

12.2 CASE 1

12.2.1 SIMULATION OF PERMAFROST IN THE CIRCUMPOLAR BOREAL ZONE

Our current understanding of the ecology of boreal forests indicates that vegetation patterns within the circumpolar boreal forest reflect a complex interrelationship among climate, solar radiation, soil moisture, soil temperature, and the forest floor organic layer. Forest fires and insect outbreaks further shape the vegetation mosaic over large areas.

Research work at the Bonanza Creek LTER site has identified the soil thermal regime as the dominant controlling factor in boreal forest ecosystem production and element cycling (Van Cleve and Viereck, 1981; Viereck *et al.*, 1986). Bonan (1989) used a simulation model parameterized using the extensive data base from the Bonanza Creek LTER site to explore the interactions among solar radiation, soil moisture, soil freezing and thawing, the forest floor organic layer, and forest fires.

In the Bonan model, depth to permafrost is a direct function of heat load (solar radiation, air temperature). Mineral soil water content is also important because the thermal conductivity and the latent heat of fusion of the soil are functions of soil moisture. Shallow depth to permafrost, in turn, impedes soil drainage, creating saturated soil conditions. A thick moss-organic layer on the forest floor has a high water-absorbing capacity that also helps to maintain high soil moisture conditions. This layer also greatly impedes heat flow because of its low bulk density and low thermal conductivity. Likewise, by reducing the flux of heat into the soil profile, the forest canopy is an important factor in maintaining low soil temperatures and a shallow soil active layer. In this chapter, we present the results from simulations of the soil thermal conditions at the Bonanza Creek LTER research site and subsequent model applications on the global scale. Details of the model are provided in Bonan (1989).

Two general approaches to modeling the soil thermal regime exist. First, knowledge of the physical and thermal properties of a soil can be combined with principles of energy transfer and heat flow to formulate highly detailed, physically-based models. For example, the fundamental one-dimensional heat flow equations have been used to develop detailed simulation models of the soil thermal regime at Barrow, Alaska (Goodwin and Outcalt, 1975; Outcalt *et al.*, 1975a,b; Outcalt and Goodwin, 1980; Goodwin *et al.*, 1984), and in Sweden (Jansson and Halldin, 1980). However, the generality of these models is restricted by the detailed soils and meteorological parameters and the small time step required to solve the soil thermal calculations.

A second approach is to avoid the theoretical and computational problems involved in solving the fundamental heat flow equations by using empirical models that predict soil temperature from surface meteorological data. For example, the soil heat flux is often a linear function of net radiation (Idso *et al.*, 1975; DeHeer-Amissah *et al.*, 1981; Rosenberg *et al.*, 1983) and can be predicted in the winter from daily maximum and minimum air temperatures, solar radiation, and snow depth (Cary, 1982; Zuzel *et al.*, 1986). Soil temperatures at various depths can be predicted from above-ground climatic conditions, especially air temperature (Ouellet, 1973; Hasfurther and Burman, 1974; Bocock *et al.*, 1977; Toy *et al.*, 1978; MacLean and Ayres, 1985) or from the Julian date (Meikle and Treadway, 1979, 1981). This approach does require large data sets to develop and test the functions. Such data are available only from a research site with a relatively long-term record (such as an LTER site). The present model, which is parameterized largely from data developed at the Bonanza Creek LTER site, can be taken as an example of developing a model locally and then testing and applying it on an extensive (global) scale.

In the Bonan (1989) model, the depths of seasonal freezing and thawing in the soil profile are solved on a monthly basis using the Stefan formula for freezing and thawing in a multi-layered soil (Carlson, 1952; Jumikis, 1966; Lunardini, 1981). This avoids the small time steps required by physically based heat flow models and the large data requirements of empirical models. Nonetheless, the model takes into account the physical and thermal properties of the soil, the soil moisture content, the latent heat of fusion of water, and the duration of freezing or thawing (Jumikis, 1966; Lunardini, 1981). Like most of the simple theories of heat transfer in soils, the Stefan equation solves the fundamental, one-dimensional heat conduction equation while neglecting convective heat flow from precipitation, snow melt, and surface water which, though usually negligible (de Vries, 1975), can be an important factor in soil thawing (Moskvin, 1974; Ryden and Kostov, 1980; Kingsbury and Moore, 1987). In addition, the Stefan equation provides an approximate solution to heat conduction under the assumption that sensible heat effects are negligible. This is true if latent heat is much larger than sensible heat or if sensible heat is small—a condition that is true for soils with high water contents (Lunardini, 1981). The Stefan equation is also based on the assumption that temperature gradients in the soil are linear. When these conditions are met, Stefan's equation provides an accurate approximation of depths of freezing and thawing in soils (Carlson, 1952; Jumikis, 1966; Lunardini, 1981). When they are not met, the equation will yield depths of freeze that are too large (Lunardini, 1981).

In the Bonan model, the soil profile is treated as a one- or two-layered system composed of a forest floor moss-organic layer, if present, and an underlying mineral soil layer. The thermal conductivity of an organic layer is a linear function of moisture content (de Vries, 1975; Jansson and Halldin, 1980). These properties of northern mineral soils have been described for North America (Kersten, 1949; Lunardini, 1981), Sweden (Tamm, 1950), and the Soviet Union (Chudnovskii, 1962; Shul'gin, 1965; Dimo, 1969). Kersten's (1949) equations, which calculate mineral soil thermal conductivities from bulk density and water content, closely correspond to similar data from the Soviet Union (Lunardini, 1981). Estimates of surface temperatures are needed to calculate the amount of heat (°C/day) applied to the soil profile during thawing and freezing. One approach has been to derive the equilibrium surface temperature as the surface temperature that balances the energy budget over a specific period of time (Outcalt, 1972; Outcalt *et al.*, 1975a,b; Terjung and O'Rourke, 1982; Bristow, 1987). However, a simpler approach uses empirically derived correction factors to adjust air temperature sums for surface conditions (Carlson, 1952; Lunardini, 1981).

In the Bonan model, parameters for the surface conditions are used to adjust the heat load. Several such parameters have been measured directly for particular surfaces (e.g. blackened soil after a fire), and other surface conditions were estimated to correspond with these known surface parameters. Details regarding estimation of these parameters are listed in Bonan (1989).

Air temperatures also need to be corrected for the effect of different slopes and aspects on the surface energy budget and the depths of freezing and thawing (Tamm, 1950; Shul'gin, 1965; Dingman and Koutz, 1974; Ryden and Kostov, 1980; Rosenberg *et al.*, 1983; Rieger, 1983; Viereck *et al.*, 1983). Air temperature can be adjusted for slope and aspect by the ratio of solar radiation received on a sloped surface to that received on a horizontal surface at the same latitude (Riley *et al.*, 1973).

The depth of thaw algorithm in the model was validated locally by its ability to reproduce characteristics of the soil thermal regime for different forested sites at Fairbanks, Alaska (the Bonanza Creek LTER site). Climatic data were taken from Hare and Hay (1974). Air temperatures were adjusted for elevation by a dry adiabatic lapse rate of 1.0°C per 100 m (Rosenberg *et al.*, 1983). Monthly solar radiation and soil moisture estimates were obtained from the solar radiation and soil moisture algorithms.

The depth of seasonal thaw algorithm used in the Bonan model provides good estimates of the soil thermal regime in relation to topography, elevation, and the forest floor thickness (Table 12.1), though depth of seasonal thaw is not necessarily the same as depth to permafrost (Ryden and Kostov, 1980; Lunardini, 1981). Shallow seasonal thaw depths occurred in poorly drained upland and floodplain sites with a thick forest floor; deep seasonal thaw depths took place in well-drained, south-facing upland and floodplain soils. Simulated seasonal soil freezing and thawing (Figure 12.1) corresponded with observed seasonal patterns (Viereck and Dyrness, 1979; Ryden and Kostov, 1980; Viereck, 1982; Chindyaev, 1987; Kingsbury and Moore, 1987). The algorithm also reproduced the observed decrease in depth of thaw with increased organic layer thickness (Figure 12.2) (Dyrness, 1982; Viereck, 1982).

The depth of thaw algorithm was also validated by its ability to predict global patterns in the distribution of permafrost. Permafrost is likely to exist and be maintained if the depth of seasonal freeze exceeds that of seasonal thaw (Lunardini,

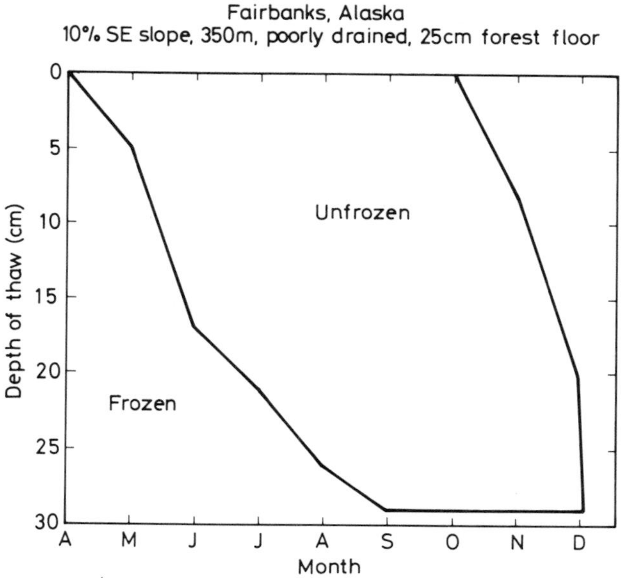

Figure 12.1 Predicted seasonal mineral soil freezing and thawing for a closed-canopy forest, Fairbanks, Alaska

Table 12.1 Observed depth to permafrost and predicted depth of seasonal thaw for several sites near Fairbanks, Alaska (Viereck *et al.*, 1983)

					Observed	Predicted
Slope (%)	Aspect	Elevation (m)	Drainage[a]	Forest floor (cm)	Permafrost[b] depth (cm)	Thaw[c] depth (cm)
Uplands						
30	N	427	PD	38	22	22
0	–	167	PD	14	55	59
10	SE	385	PD	23	35	31
0	–	468	MD	12	–	–
0	–	343	MD	19	–	–
15	NW	470	MD	17	–	–
12	S	747	MD	10	–	–
25	S	396	WD	9	–	129
18	SE	229	WD	6	–	154
Floodplain						
0	–	177	PD	25	16	30
0	–	122	PD	19	20	39
0	–	120	WD	18	–	116
0	–	177	WD	5	–	153
0	–	120	WD	15	–	125

[a] Drainage class
PD = poorly drained field capacity: m_V = 38%
MD = moderately drained field capacity: m_V = 29%
WD = well drained field capacity: m_V = 20%
[b] No permafrost
[c] Depth of thaw exceeded depth to bedrock (50 cm)

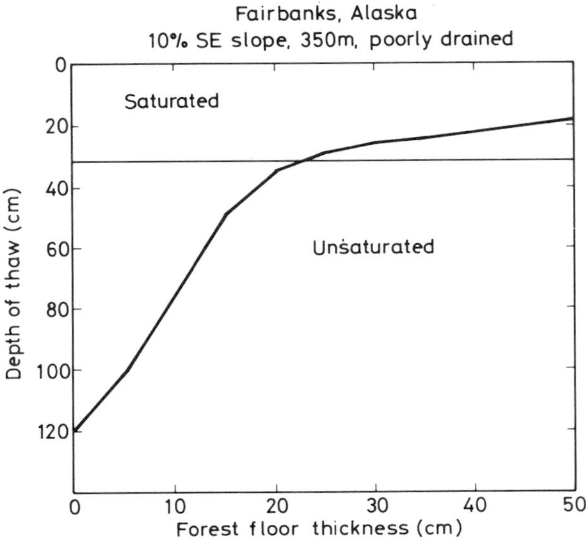

Figure 12.2 Predicted depth of seasonal thaw in mineral soil at Fairbanks, Alaska, in relation to forest floor thickness

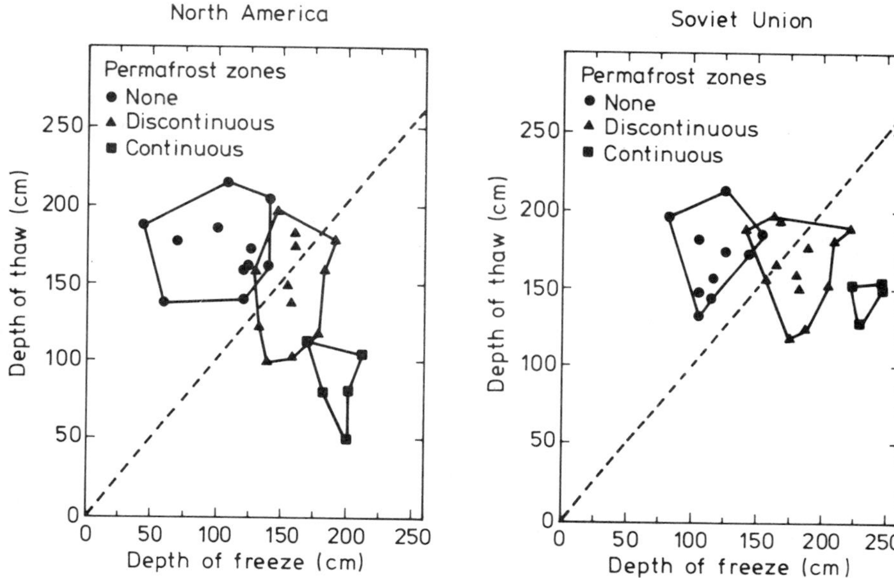

Figure 12.3 Depths of freeze and thaw penetration in a closed-canopy, forested, well-drained, fine-textured soil for climatic sites located throughout northern North America and the Soviet Union

1981). Depths of freeze and thaw simulated for a well-drained, fine-textured soil beneath a closed forest canopy with no forest floor were compared for 28 climatic sites in northern North America (Hare and Hay, 1974) located in zones of no permafrost, discontinuous permafrost, and continuous permafrost (FAO, 1975), and for 27 climatic sites in the Soviet Union (Lydolph, 1977) also located in zones classified as no permafrost, discontinuous permafrost, and continuous permafrost (Lydolph, 1977). Monthly solar radiation and soil moisture estimates were obtained from the solar radiation and soil moisture algorithms.

In both North America and the Soviet Union, the observed presence or absence of permafrost was related to thaw and freeze penetration depths (Figure 12.3). In both the North American and Soviet locations, depths of thaw exceeded depths of freeze for sites without permafrost. For sites in the continuous permafrost zone, depths of freeze exceeded depths of thaw. Sites located in the discontinuous permafrost zone were scattered to either side of the critical line of equal thaw and freeze depths. The clear discrimination of permafrost zones by the depths of freeze and thaw was somewhat surprising given the coarse resolution of the observed data (FAO, 1975; Lydolph, 1977). This indicates that thaw and freeze penetration depths in a soil provide a strong ability to predict regional climatic zones where permafrost can exist, cannot exist, or has the potential to exist. Of course, these methods cannot be used to predict zones of 'relic permafrost'—permafrost that is unstable under current climatic conditions but exists because of past conditions (Lunardini, 1981). In comparison to mean annual air temperature (Brown, 1969) or air thawing and freezing degree-day sums (Harris, 1983; Nelson and Outcalt, 1983, 1987), computation of thawing and freezing penetration depths for predicting the presence or absence of permafrost incorporates both soil heat transfer theory and vital soil properties.

12.3 CASE 2

12.3.1 MODELS SIMULATING ECOSYSTEM PATTERN IN FORESTS VERSUS GRASSLANDS

The climatic changes that are being predicted by computer models of the earth's 'weather machine' (General Circulation Models or GCMs) in response to a doubling of atmospheric carbon dioxide appear to be ecologically significant, at least at the higher northern latitudes (Shugart *et al.*, 1986). Nevertheless, 'uncertainty' is one of the key words that arises in any discussions evaluating the possibility and impact of anthropogenic global climate change. As reviewed in Bolin *et al.* (1986), the uncertainty of the magnitude of the increase in the earth's ambient CO_2, and the temporal pattern of the projected increase, is great, as it also is in the regional pattern of climatic change (particularly with regard to the climatic variables that involve water—precipitation, cloudiness, soil moisture, etc.).

From the point of view of an ecologist, the central question to be asked about the response of ecosystems to global climatic change is, 'Will the rate and magnitude of global change be ecologically significant?' A principal consideration involves the possible changes in the boundaries between major biomes under an altered climatic condition.

The dynamic modeling of the position and the nature of transition between grassland and forest is a challenging problem because the 'rules' for model construction may be very different from those in either forests or grasslands. In general, the formulation of parsimonious models of a given ecosystem involves identifying the ecosystem processes that are important in producing the patterns of interest at a given time or space scale. Even when considered at equivalent time and space scales, the important pattern-producing processes may be different in grasslands and forests. Further, in the transition zone, ecosystem processes that are not essential in either models of grasslands or forests may become important factors.

We are currently involved in research to develop models to simulate the transition from prairie to forest in the central part of the United States. This project is ongoing and the present discussion is intended to provide a documentation of the modeling approach being used in this project. We have chosen to merge two models that emphasize the biology of individuals (either grasses or trees), and two physico-chemical-based models of soil development and the chemistry of soils. The models have been developed and tested in the context of a single biome.

Our work will involve merging and modifying these models to understand better the communalities and differences in grasslands and forests, and the processes that emerge as important in the transition between these two major biome-types. The set of models consists of two ecological models and two soil process models developed at different sites which have all developed long-term data sets on ecosystems processes. The four models are (1) the CENTURY model of carbon and nutrient dynamics, (2) the MAGIC model of soil and water chemistry, (3) the FORET model of forest dynamics, and (4) the STEPPE model of succession in semi-arid grasslands. Each of these models is described below.

12.3.2 THE CENTURY MODEL

The CENTURY model (Parton *et al.*, 1987a) was developed to simulate soil organic matter dynamics and plant production in grazed grasslands and agroecosystems. The data used to develop the model came from long-term incubation studies of ^{14}C-labeled plant material in different soil types (e.g. Sorenson, 1983; Ladd *et al.*, 1981), soil carbon-dating (Martel and Paul, 1974), and soil particle size fractionation data (Tiessen *et al.*, 1982, Tiessen and Stewart, 1983). There have also been several model analyses at different levels of resolution (Cole *et al.*, 1977; Hunt, 1977; Parton *et al.*, 1983; Van Veen *et al.*, 1984). The model simulates the dynamics of C, N, and P in the soil–plant system using monthly time steps. The input data required for the model include soil texture, monthly precipitation, maximum and minimum air temperature, and plant lignin content. The CENTURY model has been used to simulate regional patterns of soil C, N, and P and plant production for the US central grasslands region (Parton *et al.*, 1987a, 1988) and the impact of management practices on agroecosystems (Parton *et al.*, 1987b). In the model (Figure 12.4), the soil organic matter is divided into (1) an active soil fraction consisting of live microbes and microbial products (1- to 2-year turnover time), (2) a protected fraction that is resistant to decomposition (20- to 40-year turnover time), and (3) a fraction that is physically or chemically isolated (800- to 1200-year turnover time).

The plant residue is divided into structural (2- to 5-year turnover time) and metabolic (0.1- to 1-year turnover time) pools as a function of the lignin to N ratio of the residue. Decomposition is calculated by multiplying the decay rate specified for each state variable by the combined effect of soil moisture and soil temperature on decomposition. The decay rate of the structural material is also a function of its lignin. The active soil organic matter decay rate changes as a function of the soil silt plus clay content (low values for high-silt and clay soils). The respiration loss for carbon is fixed, except for active soil organic matter, the respiration of which decreases with the soil silt plus clay content. Submodels for N and P parallel the carbon structure with modifications as appropriate.

The model also includes a plant production submodel, which simulates the monthly dynamics of C, N, P, and S in the live and dead above ground plant material, live roots, and resistant (structural) and labile (metabolic) surface and

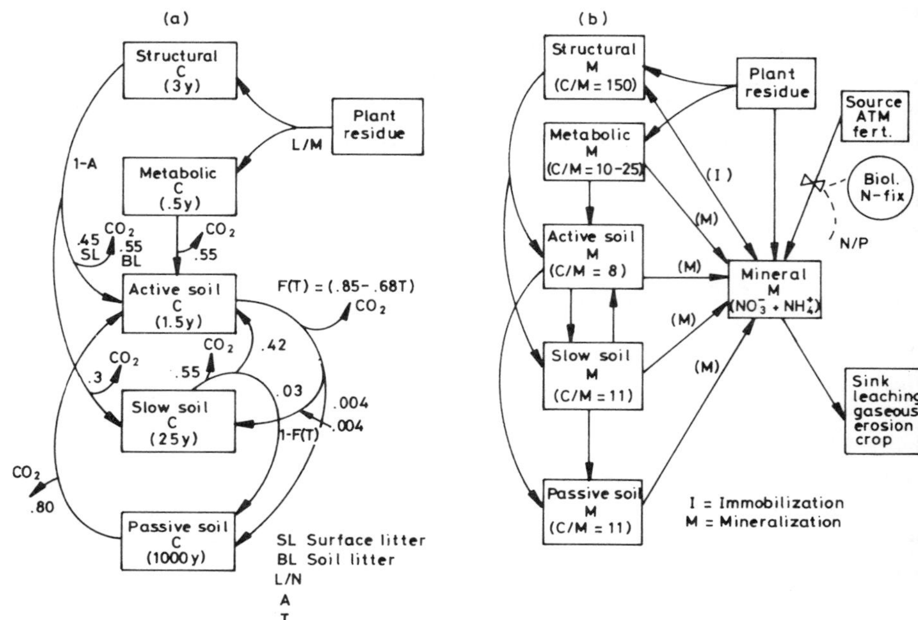

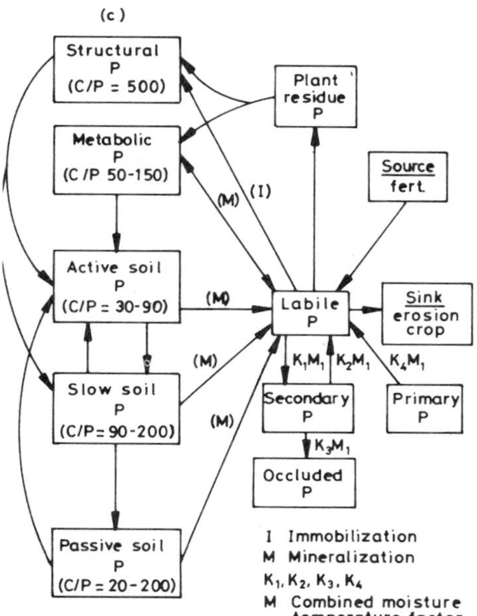

Figure 12.4 Flow diagram for carbon, nitrogen and phosphorus dynamics in the CENTURY model. (a) Carbon pools and fluxes; (b) nitrogen pools and fluxes; (c) phosphorus pools and fluxes

root detritus pools. Maximum potential plant growth is estimated as a function of the annual precipitation and is reduced if sufficient N, P, or S is not available.

In a regional-scale simulation, spatial patterns of soil organic carbon, nitrogen, and phosphorus were simulated for a fine-textured soil (25% sand, 30% clay, and 45% silt) and a sandy soil (75% sand, 10% clay, and 15% silt). The model was run for 10 000 years at 56 sites for each soil texture. The contours in Figure 12.5 represent the state of the system after 10 000 years of soil formation. At the beginning of the runs, all of the P (100 g m^{-2}) in the soil was in parent P material. Nitrogen inputs into the systems were assumed to be proportional to the annual precipitation. The climate is spatially variable in this region, with precipitation ranging from <30 cm in the western part to >120 cm in the south eastern part. The temperature ranges from 22°C in the south to less than 4°C in the northern part of the region. The predicted patterns of plant productivity and soil organic C, N, and P in the Central Grasslands Region are illustrated in Figure 12.5. These patterns have been validated by comparing simulated results with extensive regional data bases for the Central Grasslands Region (Parton *et al.*, 1988).

12.3.3 THE MAGIC MODEL

MAGIC (Cosby *et al.*, 1985a,b,c, 1986a, 1987a; Wright *et al.*, 1986) was developed to examine the effects of soil chemical reactions on drainage water quality from catchments and to simulate the effects of atmospheric inputs and biological processes on those soil reactions. The temporal scale of model simulations ranges from monthly to centennial. MAGIC is readily adaptable to a biological interface. For instance, variations in dissolved organic acids in response to soil genesis can be used to drive MAGIC, as can changes in inorganic cation or anion concentrations in response to nitrification or forest growth and uptake. MAGIC in turn can simulate changing soil chemical conditions that can be used as inputs to forest or grassland growth models.

The model has been applied to individual catchments in the United States, Scotland, Wales, Finland, Sweden, and Norway (Cosby *et al.*, 1985a,b,c, 1986a,b, 1987a; Neal *et al.*, 1986; Wright *et al.*, 1986; Wright and Cosby, 1986; Lepisto *et al.*, 1987; Whitehead *et al.*, 1987a,b). The model has also been used to simulate regional distributions of soil and water quality in the United States (Hornberger *et al.*, 1986a,b), Norway (Hornberger *et al.*, 1987a,b; Cosby *et al.*, 1987b) and Scotland (Musgrove *et al.*, 1987).

MAGIC postulates that a relatively small number of important soil processes can produce observed spatial and temporal variations in soil and surface water chemical properties. Reuss (1980, 1983) and Reuss and Johnson (1985) proposed a simple system of reactions describing the equilibrium between dissolved and adsorbed ions in the soil–soil water system. MAGIC has its roots in the Reuss–Johnson conceptual system, but has been expanded from their simple two-component (Ca–Al) system to include other cations and anions in catchment soil and surface waters (Cosby *et al.*, 1985a).

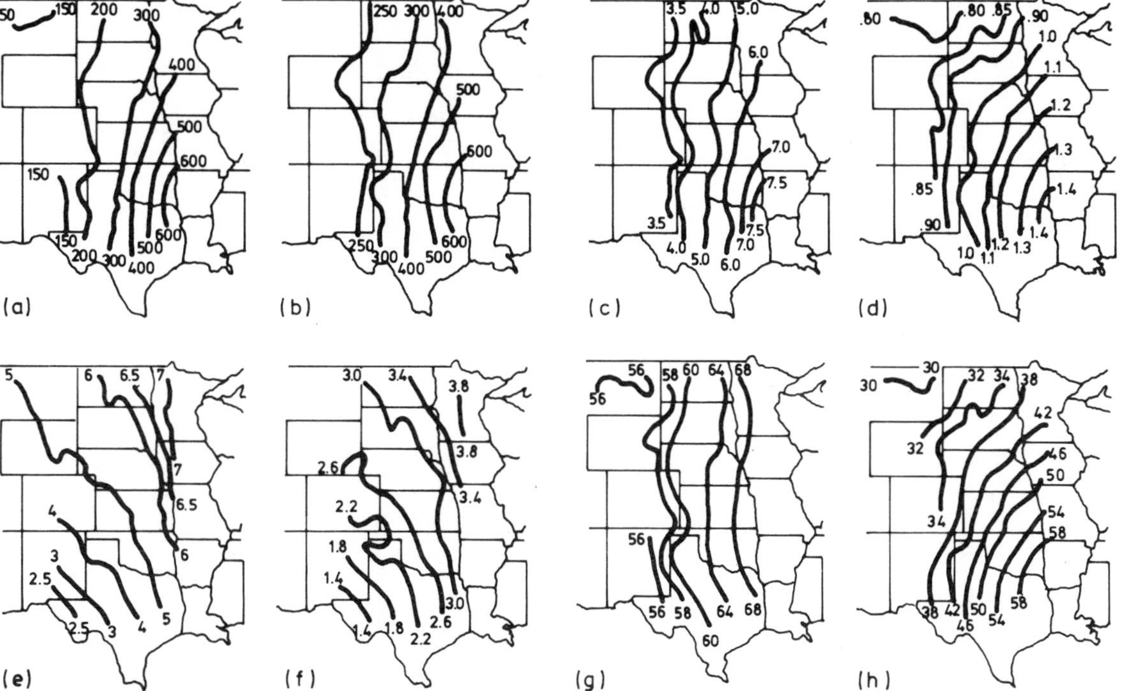

Figure 12.5 Regional patterns simulated by the CENTURY model for (a) above-ground production (gm^{-2}); (b) below-ground production (gm^{-2}); (c) nitrogen mineralization ($gm^{-2}y^{-1}$); (d) phosphorus mineralization ($gm^{-2}y^{-1}$); (e) soil carbon on fine-textured soils (gm^{-2}); (f) soil carbon on sandy soils (gm^{-1}); (g) organic phosphorus on fine-textured soils (gm^{-2}); and (h) organic phosphorus on sandy soils (gm^{-2}). Maps are developed by running the CENTURY model for 56 sites in the region and then using a contouring routine from the S package (Bell Labs, Murray Hill, NJ, USA) to generate the maps. The model was run for 10 000 years at each site for each soil texture, and the plotted variables represent the state of the system after 10 000 years of soil formation. At each site the model response was based on sites with both a fine-textured soil (25% sand, 30% clay, and 45% silt) and a sandy soil (75% sand, 10% clay, and 15% silt). Only the results from the sandy soil are shown in 4.c and 4.d.

MAGIC assumes that atmospheric deposition, mineral weathering, ion exchange, and biological processes in the soil control the soil water chemical composition. Dynamic changes in cation and anion concentrations in soil water result from buffering by organic acids and soil ion exchange; dissolution, precipitation, and exchange of aluminum; alkalinity generation driven by CO_2 variations; aqueous complexation and speciation reactions; and biological uptake, release, and transformation. The chemical status of the soil matrix is continuously simulated. Soil and drainage water chemistry are controlled in part by the rate at which the chemical status of the soil matrix changes. This in turn is affected by the weathering, leaching, and relative availability and mobility of cations and anions.

12.3.3.1 An example site-specific application

MAGIC was calibrated for Dargall Lane, a small moorland catchment situated in the Galloway Hills of south-western Scotland (Cosby *et al.*, 1986b), and used to reconstruct the water quality of the drainage stream since 1844 (Figure 12.6). Confidence intervals for the reconstructed water quality were estimated through a model sensitivity analysis. Model simulations were compared with an independently determined reconstruction of stream pH changes based on paleoecological studies of the sediments in the loch receiving the stream drainage (diatom–pH relationships, see Batterbee *et al.*, 1985). Similar comparisons for other sites in Europe and the United States (Wright *et al.*, 1986) also indicate that the temporal response of the model is correct on the order of decades to centuries.

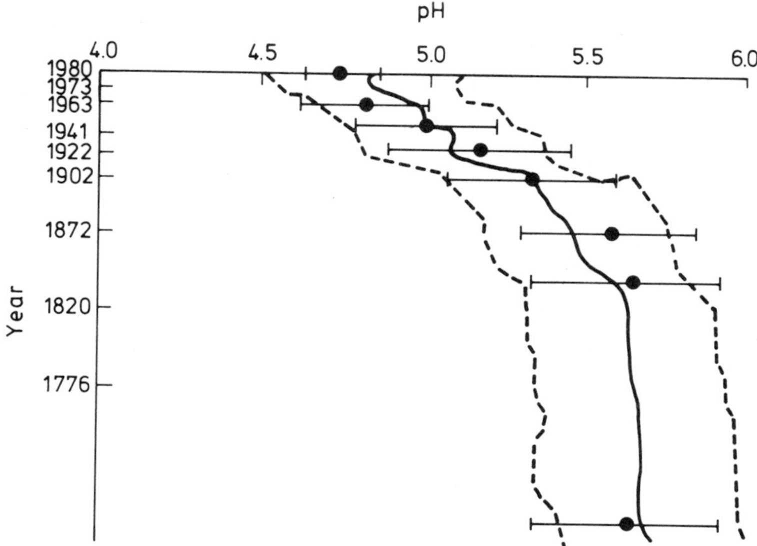

Figure 12.6 Reconstruction of historical pH variations for Dargall Lane. Solid line and dashed lines are median and 95% confidence bands for MAGIC. Circles and associated error bars are estimates from diatom records

12.3.3.2 An example regional application

MAGIC was applied to a 1974 survey of 715 lakes in southern Norway (Hornberger *et al.*, 1987a; Cosby *et al.*, 1987b). Regional distributions of water quality variables were simulated (with uncertainty estimates) for the survey year, for a pre-acidic deposition year (1844), and for a future year (2000). The simulated distribution bands overlapped the observed distribution function for several water quality variables in 1974 (Figures 12.7(a) and 12.7(b)), indicating that the calibrated regional model reproduced observed regional water quality characteristics. Simulations (Figure 12.7(c)) imply that significant changes in water quality have occurred over the past several decades; the results are in agreement with historical records. Forecast water quality distributions (in response to a 30% reduction of acidic deposition by the year 2000, Figure 12.7(d)) suggest that significant shifts in the spatial pattern of water quality may occur in the future.

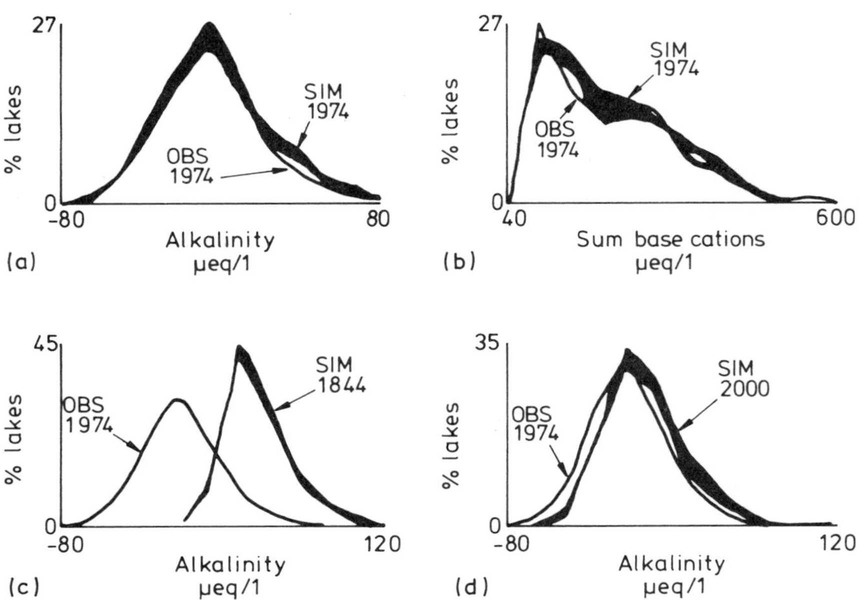

Figure 12.7 (a) and (b) Simulated and observed regional distribution of lake chemical composition for southernmost Norway in 1974; (c) MAGIC estimates of changes in the regional distribution of alkalinity that have occurred since 1844 in response to atmospheric acidic deposition; (d) MAGIC forecasts of changes in the regional distribution of alkalinity in response to a 30% decrease in atmospheric acidic deposition

12.3.4 THE FORET MODEL

The FORET model (Shugart and West, 1977; Shugart, 1984) is derived from the earlier JABOWA model (Botkin *et al.*, 1972). The model simulates the annual growth in diameter of each tree on a simulated study plot of a size between 0.01 and 0.1 ha. Each of the trees on this plot is interactive with the others by competition for light and for limiting nutrients. The model is well suited to simulating the dynamics of the 'gaps', the spatial elements that comprise a mature forest mosaic (Watt, 1925, 1947; Borman and Likens, 1979a,b; Shugart and West, 1981; Shugart, 1984).

The growth of trees in the models is based on species-specific growth functions relating growth increment to individual size (Botkin *et al.*, 1972; Shugart, 1984). Environmental constraints incorporated by FORET include available light, soil moisture, and soil fertility. The realized diameter increment of a tree in any given year thus reflects its maximum potential as constrained by available light, soil moisture, and soil fertility. Regeneration of trees on the simulated plot is a stochastic process in which the likelihood of establishment of new trees depends on species response to the forest floor environment. Mortality is also stochastic. Natural mortality is based on the expected longevity for each species, and is implemented as a constant (low) probability of mortality. A second form of mortality occurs as a result of stress (e.g. through suppression); stressed trees are subjected to a higher probability of mortality. Many versions of FORET also include a third source of mortality resulting from natural disturbances (e.g. fires) or forest management (e.g. logging).

Species parameters in the model are largely derived from natural-history or silvicultural accounts (Fowells, 1965). These model parameters are, in general, not freely 'fitted' to calibration data, because they cannot vary except in a prescribed nominal range. Implementing a version of the model for a particular study site (forest type) typically involves the incorporation of site-specific environmental factors (e.g. a disturbance regime), and a moderate amount of 'tuning' to a set of calibration data.

The FORET model has been used as a base model to develop a large set of forest simulators in several different parts of the world. A map of these locations (Figure 12.8) also indicates the range of environmental conditions to which this basic modeling approach has been extended.

The stochastic nature of the model makes it necessary to use Monte Carlo methods to obtain interpretations at the level of the forest stand. For this, a large number of plots (e.g. 100) are simulated, and the results are aggregated to represent the mean and the range of responses of the simulated forest. Model tests typically involve the comparison of modeled forest pattern with real forests sampled with a similar intensity. An elaboration of these tests for several FORET-derived models may be found in Shugart (1984) and in the sources documenting the various models (see legend of Figure 12.8).

A recent version of FORET, still under development and testing, provides the

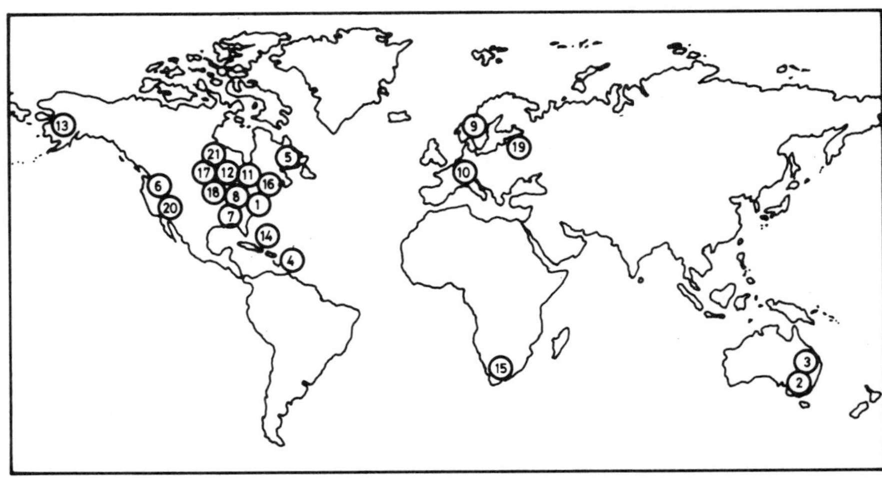

Figure 12.8 Geographical locations of published gap models and locations of ongoing applications of gap models that involve participants in the present proposal. The models have been applied in a wide variety of forests in different parts of the world and under different environment and disturbance regimes.

FORET-derived models

1. FORET (Shugart and West, 1977; Shugart, 1984). Southern Appalachian deciduous forest.
2. BRIND (Shugart and Noble, 1981). Australian eucalyptus forest.
3. KIABRAM (Shugart *et al*., 1981). Australian subtropical rain forest.
4. FORICO (Doyle, 1981). Puerto Rican Montane rain forest.
5. SMAFS (El Bayoumi *et al*., 1984). Eastern Canadian mixed-wood forest.
6. CLIMACS (Dale and Hemstrom, 1984). Pacific North-west coniferous forest.
7. FORFLO (Pearlstine *et al*., 1985). Southern floodplain forest.
8. FORCAT (Waldrop *et al*., 1986). Oak/hickory forest.
9. FORSKA (Leemans and Prentice, 1987) Scandinavian forest.
10. FORECE (Kienast, 1987). Central European forest.
11. FORANAK (Busing and Clebsch, 1987). Montane boreal forest.
12. ZELIG (Smith and Urban, 1988). Boreal-temperate forest transition.
13. LOKI (work in progress). Circumpolar boreal forest.
14. MANGRO (work in progress). Caribbean mangrove forest.
15. OUTENIQUE (work in progress). African temperate rain forest.

Other gap models

16. JABOWA (Botkin *et al*., 1972). Northern hardwood forest.
17. FORTNITE (Aber and Melillo, 1982). Wisconsin mixed-wood forest.
18. SWAMP (Phipps, 1979). Arkansas floodplain model.
19. SJABO (Tonu, 1983). Estonian conifer forest.
20. SILVA (Kercher and Axelrod, 1984). Mixed conifer forest.
21. LINKAGES (Pastor and Post, 1986). Temperate boreal transition.

capability to simulate spatial pattern explicitly in forests (Smith and Urban, 1988). This model, called ZELIG, is implemented on a grid of 100-m^2 cells. Trees on one cell may shade or crowd trees on adjacent cells. The model was designed to model environmental gradients, and can also simulate spatially contagious interactions (disturbances or seed dispersal) that are larger scaled than gap dynamics.

12.3.5 THE STEPPE MODEL

The STEPPE model was developed to simulate successional dynamics for a semi-arid grassland. The general structure of the model is similar to FORET in that it simulates annual recruitment, growth, and death of individual plants on a small plot representing the zone of influence of a single dominant individual. In STEPPE, plot size is 0.12 m^2, and is based on the resource space associated with individual *Bouteloua gracilis* plants, the dominant species in shortgrass plant communities. The size and age of each individual on the plot are followed through time. The input data required for STEPPE include annual precipitation and temperature, frequencies and intensities of disturbances, and life-history information pertaining to the recruitment, growth, and mortality of each species or species-group.

In semi-arid grasslands, such as the shortgrass steppe region of North America, soil water is the most frequent control on plant growth and community structure (Noy-Meir, 1973; Lauenroth *et al.*, 1978), and below-ground net primary production contributes approximately 85% to total primary production (Sims and Singh, 1978). Therefore, STEPPE is based on the importance of below-ground processes associated with the acquisition of soil water resources. In the model, resource use by plants is based on the overlap between the distribution of soil water and the distribution of roots with depth in the soil profile. Each life form simulated by the model has a specific distribution of roots with depth; STEPPE includes *Bouteloua*, succulent, annual grass and forb, other perennial grass (except *Bouteloua*), and shrub life forms. The distribution of soil water varies with the amount of annual precipitation, while the distribution of roots alters with species composition and plant demographics.

The recruitment of plants onto the plot occurs either through seedlings or vegetative propagation. The probability that a seedling from a particular species will become established is based either on the occurrence of suitable microenvironmental conditions or the relative abundance of seeds on the plot (Wilson and Briske, 1979; Briske and Wilson, 1977, 1878; Coffin *et al.*, 1987; Lauenroth *et al.*, 1987).

The annual increase in size of each plant on the plot is a function of its optimum growth rate, the effects of precipitation and temperature, and interactions with other plants for below-ground resources. The optimum growth rate is used to calculate the amount of resources required by each individual in each resource group. Precipitation and the effects of other plants are used to calculate the amount of resources available to each group. The actual growth rate for each individual is a function of the relationship between the resources required to sustain the optimum

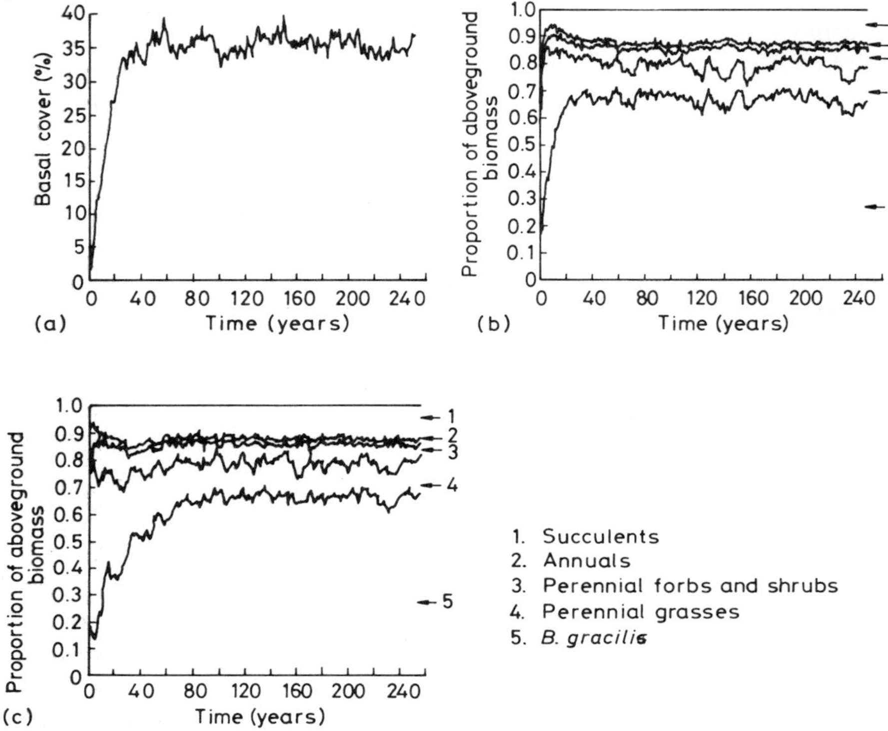

Figure 12.9 (a) Simulated basal area for *Bouteloua gracilis* for 50 plots simulated by the STEPPE model and averaged over 250 years; (b) Proportion of above-ground biomass occupied by *B. gracilis* and four other types of plants for 50 plots simulated by the STEPPE model and averaged over 250 years; (c) Proportion of above-ground biomass occupied by *B. gracilis* and four other types of plants with the seed availability made a function of precipitation for 50 plots simulated by the STEPPE model and averaged over 250 years. The share of the total above ground biomass associated with each group is indicated by the distance between the boundary lines in this cumulative proportion graph. All simulations were initiated at the bare-ground stage and parameterized for a moderately grazed swale under conditions appropriate to the Central Plains LTER site

growth rate and those available to the plant. This relationship is mediated by the effects of temperature.

Three sources of mortality in the STEPPE model are based on information from Shugart (1984): an age-independent intrinsic likelihood of mortality, a probability of mortality associated with slow-growing plants, and probabilities associated with three types of disturbances (cattle fecal pats, western harvester ant mounds, and burrows from small animals).

The STEPPE model was used to evaluate the successional dynamics of plants on small plots (gaps) and the spatial variability in gap dynamics for a shortgrass plant community (Coffin and Lauenroth, 1988). The recovery time after a disturbance of the dominant species, *Bouteloua gracilis*, and the effects of the availability of *Bouteloua* seeds on recovery time were also evaluated. The simulations were initiated as bare ground, and were parameterized for a moderately grazed lowland at the Central Plains LTER in north central Colorado.

The average basal cover of *Bouteloua* (Figure 12.9(a)) and the average proportion of total above-ground biomass for five species groups (grouped by similar life-history traits, Figure 12.9(b)) were comparable to the composition of shortgrass communities (Sims *et al.*, 1978). The recovery time of *B. gracilis* in the model is more rapid than has been reported from long-term studies of large-scale disturbances (Riegel, 1941; Hyder *et al.*, 1971; Briske and Wilson, 1977); this faster recovery occurs whether seed availability is modeled as constant or as varying with precipitation (Figure 12.9(c)). This result indicates a scale-dependency of processes important to the recovery of *B. gracilis* after disturbance and suggests that future modeling efforts may need to attend more explicitly to the spatial context of grassland succession.

12.3.6 MODEL COMMUNALITIES, INTERFACINGS, AND ANALYSIS

The four models share a number of variables and parameters that will facilitate their interfacing (Figure 12.10). For example, the CENTURY model requires as input an estimate of plant biomass, which can be provided by either the STEPPE or FORET models; CENTURY output (soil physical and chemical properties) can then serve as input for the MAGIC model; and soil moisture and fertility, estimated from CENTURY and MAGIC, can be used as input in FORET and STEPPE. These couplings permit an effective linkage of the physical and biological components of the modeled ecosystems. The compatible temporal and spatial domains of the models are important.

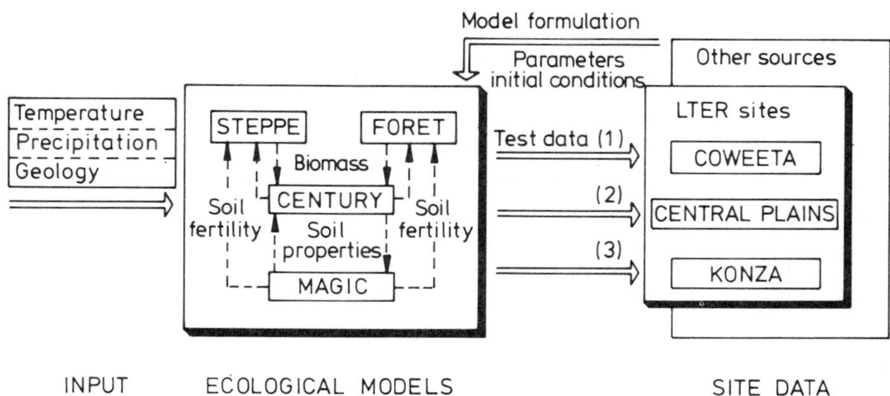

Figure 12.10 Shared attributes in the CENTURY, FORET, MAGIC and STEPPE models and the relationship of these models to the LTER network

Each model can run at an annual time step (some variables are treated at monthly steps and integrated to annual steps), and runs are typically on the order of 10^1 to 10^3 simulation years. Each model simulates a small base unit, and larger scale inferences or interpretations are derived from Monte Carlo simulations of a distribution of base units.

The Monte Carlo method of simulation allows one to account for spatial heterogeneity in environmental factors at three degrees of resolution. As a simplest case, heterogeneity can be implemented by simulating a distribution of parameters representing an appropriate domain of values for a site factor of interest (e.g. soil water-holding capacity). Model output then indicates the corresponding range of system behaviors. This approach does not account explicitly for space, but may be sufficient for many applications. At the other extreme, specific values for a study site could be entered as a distribution of parameters, with each value associated with a model plot, so that model output could then be mapped explicitly back onto the study site; this approach would predict a specific spatial pattern in system attributes (plant pattern, soil properties) for the study site. As an intermediate approach, one can simulate a distribution of parameters representing an abstract or hypothetical resource gradient (e.g. an idealized soil catena); this approach is amenable to theoretical purposes because resource gradients of interest can be represented 'cleanly' yet model predictions can still be applied to specific study sites.

Beyond the common variables shared by the four models, each model simulates a set of ecosystem-level attributes. These attributes are the interface between the model-generated theory and empirical records at LTER and other research sites (Figure 12.10). The vegetation models (FORET and STEPPE) can provide estimates of plant species composition, plant size (or age) distributions, plant biomass, vegetation pattern in space, and successional trends; these are comparable to vegetation data available from most empirical surveys. MAGIC estimates surface and groundwater chemistry, emphasizing major cations and anions in the soil solution; such data are diagnostic of the state of an ecosystem, and hence represent a potentially valuable tool for ecosystem monitoring. CENTURY simulates a soil profile, including soil organic matter and major nutrients, which can be verified by reference to readily available soil survey data.

Three of the major factors that control the nature and extent of terrestrial vegetation are light, soil moisture, and soil fertility. While each of these constraints may influence plant growth and regeneration in any given system, it is difficult (and may be impossible) to distinguish their relative contributions empirically. One can distinguish the relative contributions within simulation models, however. In fact, this is one of the primary rationales for undertaking the proposed work—to ascertain which of the many possible functional constraints are actually important (in the models) under a variety of conditions representative of an environmental gradient, and to determine how these constraints express themselves in patterns of soil properties and vegetation. To the extent that the models capture the essence of the behavior of the natural system, the inferences drawn from the model system should be transferable. At the very least the results should be suggestive of further tests (using field data) that might either refute or enhance the verisimilitude of the model.

The techniques that are used to distinguish how model responses are conditioned by inputs and parameter values are generally referred to as 'uncertainty analyses' (see Beck, 1987, for a recent review). O'Neill and Gardner and their colleagues have been particularly productive in the investigation of ecosystem models using Monte Carlo methods (O'Neill *et al.*, 1980; Gardner and O'Neill 1981, 1983). An adaptation of the procedure developed by Iman and Conover (1979, 1980) can be used to gauge model response over a wide range of variation. The method has been successfully used by Jaffe and Ferrara (1984) with a stream pollution model and by members of our group at the University of Virginia (Wolock *et al.*, 1988) with a hydrological model.

12.4 CONCLUSIONS

Within particular biomes, the factors governing structure and dynamics of ecosystems may vary considerably. For example, species composition and age structure in plant communities are largely controlled by asymmetric competition for an above-ground resource (light) and symmetric competition for below-ground resources (moisture and nutrients). In mesic forests, the principal constraint is the availability of light. As a forest environment tends from mesic to xeric conditions, the effective constraint shifts towards below-ground resources (Tilman, 1988; Smith and Huston, in review). In still drier environs, the forest yields to grassland, where the principal constraint is below-ground. This suggests that there are general trends or patterns in the relative influence of environmental constraints in structuring ecosystems.

In a similar vein, there is a need to detail the factors responsible for system scaling. The developing concern for environmental effects on a large scale reawakens a recurrent theme in ecological research (McIntosh, 1985). Scale and pattern studies date back at least to Watt (1925, 1947) with intellectual roots that go back to nineteenth-century biogeographers (O'Neill *et al.*, 1986). Interest in measuring vegetation pattern has led to a significant body of statistical and analytical theory (Cliff *et al.*, 1975; Pielou, 1977). It is perhaps the juxtaposition of new problems and new techniques that has resulted in a convergence of the ecological community on problems of scale. The current interest in scale is reflected in research on soil processes (Sollins *et al.*, 1983), paleoecology (Delcourt *et al.*, 1983), avian communities (Maurer, 1985; Weins, 1986; Urban and Shugart, 1986), ecotoxicology (Perez *et al.*, in press), and both marine (Levasseur *et al.*, 1984) and freshwater (Carpenter and Kitchell, 1987) ecosystems. A body of concepts has been spearheaded by field researchers (Weins *et al.*, 1985; Ricklefs, 1987; Addicott *et al.*, in press). This movement is evident in the new emphasis on landscape ecology (Risser *et al.*, 1984; Forman and Godron, 1986; Urban *et al.*, 1987). This same concern for scale is clear in aquatic ecosystems (Steele, 1985) where hydrodynamics impose a hierarchy of scales on the system (Harris, 1980).

Current discussions on biospheric dynamics and global ecology (Bolin *et al.*, 1986; Rosswall *et al.*, 1988) have sharpened the research interest in time- and space-scales in ecological systems, as have discussions of 'hierarchy theory' (Allen and Starr, 1983; Allen and Hoekstra, 1984; O'Neill *et al.*, 1986; Urban *et al.*, 1987). An appreciation of scale is a clear prerequisite to unifying the dynamics of atmospheric processes and the dynamics of terrestrial ecosystems. The categorization of controlling factors at different space- and time-scales in particular ecosystems has been the topic of several reviews (Delcourt, *et al.*, 1983, Pickett and White, 1985).

Several studies have attended the natural scales of ecosystems in a qualitative or semi-quantitative manner (Delcourt *et al.*, 1983; Urban *et al.*, 1987). By natural scales we refer to the temporal frequencies or patterns of spatial variability to which a particular system is responsive. Obversely, this implies the smaller scales to which the ecosystem cannot respond (i.e. higher frequencies and finer grained spatial patterns are witnessed only as average values), and also indicates the larger scales that the ecosystem witnesses as effectively constant values.

Clearly, natural patterns in environmental constraints contribute substantially to the spatial pattern and temporal dynamics of particular ecosystems. There is some evidence that these patterns, especially temporal ones, may resonate with natural frequencies of plant growth forms (i.e. phenology and longevity) to amplify environmental patterns (Neilson, 1986, 1987). Thus, there may be an interaction between abiotic environmental constraints and biotic natural-history phenomena that results in system-specific spatial and dynamic patterns. It seems likely that various ecosystems or biomes should vary systematically in their natural scaling.

The problems in dealing with the effects of scaling the known responses of ecological systems to larger spatial or temporal scales (an apparent prerequisite to the global-scale ecological studies) create a necessary role for ecological models. In the examples we presented above, the models are capable of simulating the dynamics and patterns of important ecological features at continental scales. It is significant that these modeling studies were developed in the context of long-term ecological study sites. The importance of properly designed long-term studies lies in the nature of the data sets that such studies ultimately produce, and in their ability to allow the development and testing of ecological models. The modeling cycle of construction, testing, and application is typically of a duration that requires long-term studies, and the associated logistical and intellectual support.

12.5 ACKNOWLEDGMENTS

This research was supported, in part, by the US National Science Foundation (Grants BSR-8702333 and BSR-8807882) and, in part, by the US National Aeronautics and Space Administration (Grant NAG-5-1018).

12.6 REFERENCES

Addicott, J.F., Aho, J.M., Antolin, M.F., Padilla, D.K., Richardson, J.S. and Doluk, D.A. Ecological neighborhoods: scaling environmental patterns. *Oikos* (in press).

Allen, T.F.H. and Hoekstra, T.W. (1984). Nested and non-nested hierarchies: a significant distinction for ecological systems. In Smith, A.W. (Ed.) *Proceedings of the Society for General Systems Research. I. Systems Methodologies and Isomorphies*. Intersystems Publ., Coutts Lib. Serv., Lewiston, New York, 175–180.

Allen, T.F.H. and Starr, T.B. (1983). *Hierarchy: Perspectives for Ecological Complexity*. University of Chicago Press, Chicago, Illinois.

Bocock, K.L., Jeffers, J.N.R., Lindley, D.K., Adamson, J.K. and Gill, C.A. (1977). Estimating woodland soil temperature from air temperature and other climatic variables. *Agricultural Meteorology*, **18**, 351–372

Bolin, B., Doos, B.R., Jager, J. and Warrick, R.A. (1986). *The Greenhouse Effect, Climatic Change and Ecosystems*. SCOPE 29, John Wiley, Chichester.

Bonan, G.B. (1988). *Environmental Processes and Vegetation Patterns in Boreal Forests*. Dissertation, University of Virginia.

Bonan, G.B. (1989). A computer model of the solar radiation, soil moisture, and soil thermal regimes in boreal forests. *Ecological Modelling*, **45**, 275–306.

Bonan, G.B. and Shugart, H.H. (1989). Environmental factors and ecological processes in boreal forests. *Annual Review of Ecology and Systematics*, **20**, 1–28.

Bormann, F.H. and Likens, G.E. (1979a). *Patterns and Process in a Forested Ecosystem*. Springer-Verlag, New York.

Bormann, F.H. and Likens, G.E. (1979b). Catastrophic disturbance and the steady state in northern hardwood forests. *Amer. Scientist*, **67**, 660–669.

Botkin, D.B., Janak, J.F. and Wallis, J.R. (1972). Some ecological consequences of a computer model of forest growth. *J. Ecol.*, **60**, 849–873.

Briske, D.D. and Wilson, A.M. (1977). Temperature effects on adventitious root development in blue grama seedlings. *Journal of Range Management*, **30**, 276–280.

Briske, D.D. and Wilson, A.M. (1978). Moisture and temperature requirements for adventitious root development in blue grama seedlings. *Journal of Range Management*, **31**, 174–178.

Bristow, K.L. (1987). On solving the surface energy balance equation for surface temperatures. *Agricultural and Forest Meteorology*, **39**, 49–54.

Brown, R.J.E. (1969). Factors influencing discontinuous permafrost in Canada. In Pewe, T.L. (Ed.) *The Periglacial Environment*. McGill-Queen's University Press, Montreal, 11–53.

Carlson, H. (1952). Calculation of depth of thaw in frozen ground. In *Frost Action in Soils: A Symposium*. United States National Research Council, Highway Research Board, 192–223.

Carpenter, S.R. and Kitchell, J.F. (1987). The temporal scale of variance in limnetic primary production. *American Naturalist*, **129**, 417–433.

Cary, J.W. (1982). Amount of soil ice predicted from weather observations. *Agricultural Meteorology*, **27**, 35–43.

Chindyaev, A.S. (1987). Freezing and thawing of drained peat soils in forests of the middle Urals. *Soviet Forest Sciences*, **1**, 68–72.

Chudnovskii, A.F. (1962). *Heat Transfer in the Soil*. Israel Program for Scientific Translations, Jerusalem.

Cliff, A.D., Haggett, P., Ord, J.K., Bassett, K. and Davies, R.B. (1975). *Elements of Spatial Structure*. Cambridge University Press, Cambridge.

Coffin, D.P., Dougherty, R.L. and Lauenroth, W.K. (1987). Influences of soil texture and vegetation structure on seed dynamics of a shortgrass site. *Abstracts of the 40th Annual Meeting*, The Society for Range Management, Boise, Idaho.

Coffin, D.P. and Lauenroth, W.K. (1988). A gap dynamics simulation model of succession in a semiarid grassland. Submitted to *Ecological Modelling*.

Cole, C.V., Innis, G.S. and Stewart, J.W.B. (1977). Simulation of phosphorus cycling in semiarid grassland. *Ecology*, **58**(1), 1–15.

Cosby, B.J., Hornberger, G.M., Galloway, J.N. and Wright, R.F. (1985c). Freshwater acidification from atmospheric deposition of sulfuric acid: a quantitative model. *Environ. Sci. Tech.*, **19**, 1144–1149.

Cosby, B.J., Hornberger, G.M., Wolock, D.M. and Ryan, P.F. (1987a). Calibration and coupling of conceptual rainfall-runoff/chemical flux models for long-term simulation of catchment response to acidic deposition. In Beck, M.B. (Ed.) *Systems Analysis In Water Quality Management*. Pergamon Press, New York, 151–160.

Cosby, B.J., Hornberger, G.M. and Wright, R.F. (1987b). A regional model of surface water acidification in southern Norway: calibration and validation using survey data. Submitted for: Kamari, J. (Ed.) *Proceedings of the IIASA-IMGW Task Force Meeting on Environmental Impact Models to Assess Regional Acidification*, Reidel, Dordrecht.

Cosby, B.J., Hornberger, G.M., Wright, R.F., Rastetter, E.B. and Galloway, J.N. (1986a). Estimating catchment water quality response to acid deposition using mathematical models of soil ion exchange processes. *Geoderma*, **38**, 77–95.

Cosby, B.J., Whitehead, P.G. and Neale, R. (1986b). A preliminary model of long-term changes in stream acidity in south-west Scotland. *J. Hydrol.*, **84**, 381–401.

Cosby, B.J., Wright, R. F., Hornberger, G.M. and Galloway, J.N. (1985a). Modelling the effects of acid deposition: assessment of a lumped parameter model of soil water and streamwater chemistry. *Wat. Resour. Res.*, **21**, 51–63.

Cosby, B.J., Wright, R.F., Hornberger, G.M. and Galloway, J.N. (1985b). Modelling the effects of acid deposition: estimation of long-term water quality responses in a small forested catchment. *Wat. Resour. Res.*, **21**, 1591–1601.

DeHeer-Amissah, A., Hogstrom, U. and Smedman-Hogstrom, A.S. (1981). Calculation of sensible and latent heat fluxes and surface resistance from profile data. *Boundary-Layer Meteorology*, **20**, 35–45.

Delcourt, H.R., Delcourt, P.A. and Webb, T. (1983). Dynamic plant ecology: The spectrum of vegetational change in space and time. *Quat. Sci. Rev.*, **1**, 153–175.

Dimo, V.N. (1969). Physical properties and elements of the heat regime of permafrost meadow-forest soils. In *Permafrost Soils and their Regime*. Indian National Scientific Documentation Center, New Delhi, 119–191.

Dingman, S.L. and Koutz, F.R. (1974). Relations among vegetation, permafrost, and potential insulation in Central Alaska. *Arctic and Alpine Research*, **6**, 37–42.

Dyrness, C.T. (1982). Control of depth to permafrost and soil temperature by the forest floor in black spruce/feathermoss communities. United States Forest Service Research Note PNW–396.

Food and Agriculture Organization (1975). *Soil Map of the World. II. North America*. Food and Agriculture Organization, United Nations Educational, Scientific and Cultural Organization, Paris.

Forman, R.T.T. and Godron, M. (1986). *Landscape Ecology*. John Wiley, New York.

Gardner, R.H. and O'Neill, R.V. (1981). A comparison of sensitivity analysis and error analysis based on a stream ecosystem model. *Ecol. Modell.*, **12**, 173–190.

Gardner, R. H. and O'Neill, R.V. (1983). Parameter uncertainty and model predictions: a review of Monte Carlo results. In Beck, M.B. and van Straten, G. (Eds) *Uncertainty and Forecasting of Water Quality*. Springer-Verlag, New York, 245–257.

Goodwin, C.W., Brown, J. and Outcalt, S.I. (1984). Potential responses of permafrost to climatic warming. In McBeath, J.H. (Ed.) *The Potential Effects of Carbon Dioxide-Induced Climatic Changes in Alaska*. School of Agriculture and Land Resources Management, University of Alaska, Fairbanks. Miscellaneous Publication 83-1, 92–105.

Goodwin, C.W. and Outcalt, S.I. (1975). The development of a computer model of the annual snow-soil thermal regime in Arctic Tundra Terrain. In *Proceedings of the AAAS-AMS conference: Climate of the Arctic*. Geophysical Institute, University of Alaska, Fairbanks, 227–229.

Hare, F.K. and Hay, J.E. (1974). The Climate of Canada and Alaska. In Bryson, R.A. and Hare, F.K. (Eds) *World Survey of Climatology*, Volume 11 *Climates of North America*. Elsevier Scientific Publishing Company, New York, 49–192.

Harris, G.P. (1980). Temporal and spatial scales in phytoplankton ecology. *Can. J. Fish Aquat. Sci.*, **37**, 877–900.

Harris, S.A. (1983). Comparison of the climatic and geomorphic methods of predicting permafrost distribution in western Yukon Territory. In *Permafrost: Proceedings Fourth International Conference*. National Academy Press, Washington, DC, 450–455.

Hasfurther, V.R. and R.D. Burman. (1974). Soil temperature modeling using air temperature as a driving mechanism. *Transactions of the American Society of Agricultural Engineers*, **17**, 78–81.

Hornberger, G.M., Cosby, B.J. and Rastetter, E.B. (1986b). Regionalization of predictions of effects of atmospheric deposition on surface waters. In *Proceedings of the International Conference on Water Quality Modelling in the Inland Natural Environment*. Bournemouth, England, 10–13 June 1986. BHRA, The Fluid Engineering Centre, Bedford, England, 535–550.

Hornberger, G.M., Cosby, B.J. and Galloway, J.N. (1986a). Modeling the effects of acid deposition: Uncertainty and spatial variability in estimation of long-term sulfate dynamics in a region. *Wat. Resour. Res.*, **22**, 1293–1302.

Hornberger, G.M., Cosby, B.J. and Wright, R.F. (1987a). Analysis of historical surface water acidification in southern Norway using a regionalized conceptual model (MAGIC). In Beck, M.B. (Ed.) *Systems Analysis In Water Quality Management*, Pergamon Press, New York, 127–132.

Hornberger, G.M., Cosby, B.J. and Wright, R.F. (1987b). A regional model of surface water acidification in southern Norway: uncertainty in long-term hindcasts and forecasts. Submitted for: Kamari, J. (Ed.) *Proceedings of the IIASA-IMGW Task Force Meeting on Environmental Impact Models to Assess Regional Acidification*, Reidel, Dordrecht.

Hunt, H.W. (1977). A simulation model for decomposition in grasslands. *Ecology*, **58**, 469–484.

Hyder, D.N., Everson, A.C. and Bement, R.E. (1971). Seedling morphology and seeding failures with blue grama. *Journal of Range Management*, **24**, 287–292.

Idso, S.B., Aase, J.K. and Jackson, R.D. (1975). Net radiation-soil heat flux relations as influenced by soil water content variations. *Boundary-Layer Meteorology*, **9**, 113–122.

Iman, R.L. and Conover, W.J. (1979). Small sample sensitivity analysis techniques for computer models, with an application to risk assessment. *Commun. Statist. Theor. Meth.*, **49**, 1749–1842.

Iman, R.L. and Conover, W.J. (1980). The use of rank transform in regression. *Technometrics*, **21**, 499–509.

Jaffee, P.R. and Ferrara, R.A. (1984). Modeling sediment and water column interactions for hydrophobic pollutants. *Water Res.*, **18**, 1169–1174.

Jansson, P.-E. and Halldin, S. (1980). Soil water and heat model: technical description. *Swedish Coniferous Forest Project*, Swedish University of Agricultural Sciences, Uppsala. Technical Report Number 26.

Jumikis, A.R. (1966). *Thermal Soil Mechanics*. Rutgers University Press, New Brunswick, New Jersey.

Kersten, M.S. (1949). *Thermal Properties of Soils*. University of Minnesota, Institute of Technology Bulletin No. 28.

Kingsbury, C.M. and Moore, T.R. (1987). The freeze–thaw cycle of a subarctic fen, northern Quebec, Canada. *Arctic and Alpine Research*, **19**, 289–295.

Ladd, J.H., Oades, J.M. and Amato, M. (1981). Microbial biomass formed from ^{14}C and ^{15}N-labeled plant material decomposition in soils in the field. *Soil Biol. Biochem.*, **13**, 119–126.

Lauenroth, W.K., Dodd, J.L. and Sims, P.L. (1978). The effects of water- and nitrogen-induced stresses on plant community structure in a semiarid grassland. *Oecologia* (Berlin), **36**, 211–222.

Lauenroth, W.K., Sala, O.E. and Kirchner, T.B. (1987). Soil water dynamics and the establishment of seedlings of *Bouteloua gracilis* in the shortgrass steppe: A simulation analysis. *Abstracts of the 40th Annual Meeting, The Society for Range Management*, Boise, Idaho.

Levasseur, M., Therriault, J.C. and Legendre, L. (1984). Hierarchical control of phytoplankton succession by physical factors. *Mar. Ecol. Prog. Ser.*, **19**, 211–222.

Lunardini, V.J. (1981). *Heat Transfer in Cold Climates*. Van Nostrand Reinhold, New York.

Lydolph, P.E. (1977). *World Survey of Climatology*, Volume 7 *Climates of the Soviet Union*. Elsevier Scientific, New York.

MacLean, S.F. and Ayres, M.P. (1985). Estimation of soil temperature from climatic variables at Barrow, Alaska, U.S.A. *Arctic and Alpine Research*, **17**, 425–432.

Martel, Y.A. and Paul, E.A. (1974). Effects of cultivation on the organic matter of grassland soils as determined by fractionation and radiocarbon dating. *Can. J. Soil Sci.*, **54**, 419–426.

Maurer, B.A. (1985). Avian community dynamics in desert grasslands: observational scale and hierarchical structure. *Ecol. Monogr.*, **55**, 295–312.

McIntosh, R.P. (1985). *The Background of Ecology*. Cambridge University Press, Cambridge.

Meikle, R.W. and Treadway, T.R. (1979). A mathematical method for estimating soil temperatures. *Soil Science*, **128**, 226–242.

Meikle, R.W. and Treadway, T.R. (1981). A mathematical method for estimating soil temperatures in Canada. *Soil Science*, **131**, 320–326.

Moskvin, Yu. P. (1974). Investigations of the thawing of the active soil layer in the permafrost zone. *Soviet Hydrology*, **5**, 323–328.

Musgrove, T.J., Whitehead, P.G., Cosby, B.J, Hornberger, G.M. and Wright, R.F. (1987). Regional modelling of acidity in the Galloway region in southwest Scotland. Submitted for: Kamari, J. (Ed.) *Proceedings of the IIASA-IMGW Task Force Meeting on Environmental Impact Models to Assess Regional Acidification*. Reidel, Dordrecht.

Neal, C., Whitehead, P.G., Neale, R. and Cosby, B.J. (1986). Modelling the effects of acidic deposition and conifer afforestation on stream acidity in the British uplands. *J. Hydrol.*, **86**, 15–26.

Neilson, R.P. (1986). High resolution climatic analysis and southwest biogeography. *Science*, **232**, 27–34.

Neilson, R.P. (1987). Biotic regionalization and climatic controls in western North America. *Vegetatio*, **70** (in press)

Nelson, F. and Outcalt, S.I. (1983). A frost index number for spatial prediction of ground-frost zones. In *Permafrost: Proceedings, Fourth International Conference*. National Academy Press, Washington, DC, 907–911.

Nelson, F.E. and Outcalt, S.I. (1987). A computational method for prediction and regionalization of permafrost. *Arctic and Alpine Research*, **19**, 279–288.

Noy-Meir, I. (1973). Desert ecosystems: Environment and producers. *Annual Review of Ecology and Systematics*, **4**, 25–51.

O'Neill, R.V., DeAngelis, D.L., Waide, J.B. and Allen, T.F.H. (1986). A Hierarchical Concept of Ecosystems. Princeton University Press, Princeton, NJ.

O'Neill, R.V., Gardner, R.H. and Mankin, J.B. (1980). Analysis of parameter error in a nonlinear model. *Ecol. Odell.*, **8**, 297–311.

Ouellet, C.E. (1973). Macroclimatic model for estimating monthly soil temperatures under short-grass cover in Canada. *Canadian Journal of Soil Science*, **53**, 263–274.

Outcalt, S.I. and C. Goodwin. (1980). A climatological model of surface modification effects on the thermal regime of the active layer at Barrow, Alaska. In *Proceedings American Society of Mechanical Engineers, Petroleum Division, Energy Technology Conference*, New Orleans, Louisiana. ASME Publication 80-Pet-20.

Outcalt, S.I., Goodwin, C. Weller, G. and Brown, J. (1975a). Computer simulation of snowmelt and soil thermal regime at Barrow, Alaska. *Wat. Resour. Res.*, **11**, 709–715.

Outcalt, S.I., Goodwin, C., Weller, G. and Brown, J. (1975b) A digital computer simulation of the annual snow and soil thermal regimes at Barrow, Alaska. *Cold Regions Research and Engineering Laboratory*, Hanover, New Hampshire. Report Number 331.

Parton, W.J., Cole, C.V., Stewart, J.W.B. and Schimel, D.S. (1987a). Regional modeling of C, N and P in temperate grassland soils. *Plant and Soil* (in press).

Parton, W.J., Persson, J. and Anderson, D.W. (1983). Simulation of soil organic matter changes in Swedish soils. In Lauenroth, W.K., Skogerboe, G.V. and Flug, M. (Eds) *Analysis of Ecological Systems: State–of–the–Art in Ecological Systems*. Elsevier, New York, 511–516.

Parton, W.J., Schimel, D.S., Cole, C.V. and Ojima, D. (1987b). Analysis of factors controlling soil organic levels of grasslands in the Great Plains. *Soil. Sci. Soc. Am. J.*, **51**, 1173–1179.

Parton, W.J., Stewart, J.W.B. and Cole, C.V. (1988). Dynamics of C, N, P and S in grassland soils: model. *Biogeochemistry*, **5**, 109–131.

Perez, K.T., Morrison, G.E., Davey, E.W., Lackie, N.F., Soper, A.E., Blasco, R.J., Winslow, D.L., Johnson, R.L., Marino, S.A. and Heltshe, J.F. Is bigger better? Physical simulation models of a marine system perturbed by a contaminant. *Ecology* (in press).

Pickett, S.T.A. and White, P.S. (Eds) (1985). *The Ecology of Natural Disturbance and Patch Dynamics*. Academic Press, New York.

Pielou, E.C. (1977). *Mathematical Ecology*. Wiley Interscience, New York.

Reuss, J.O. (1980). Simulations of soil nutrient losses resulting from rainfall acidity. *Ecol. Mod.*, 11, 15–38.

Reuss, J.O. (1983). Implications of the Ca–A1 exchange system for the effect of acid precipitation on soils. *J. Environ. Qual.*, **12**, 591–595.

Reuss, J.O. and Johnson, D.W. (1985). Effect of soil processes on the acidification of water by acid deposition. *J. Environ. Qual.*, **14**, 26–31.

Ricklefs, R.E. (1987). Community diversity: relative roles of local and regional processes. *Science*, **235**, 167–171.

Riegel, A. (1941). Life history habits of blue grama. *Kansas Academy of Sciences Transactions*, **44**, 76–83.

Rieger, S. (1983). *The Genesis and Classification of Cold Soils*. Academic Press, New York.

Riley, J.P., Israelsen, E.K. and Eggleston, K.O. (1973). Some approaches to snowmelt prediction. In *The Role of Snowmelt and Ice in Hydrology*. International Association of Hydrological Sciences. Publication Number 107, Volume 2, 956–971.

Risser, P.G., Karr, J.R. and Forman, R.T.T. (1984). Landscape Ecology. *Illinois Nat. Hist. Survey*, Special Publication 2. Champaign, Illinois.

Rosenberg, N.J., Blad, B.L. and Verma, S.B. (1983). *Microclimate*. John Wiley, New York.

Rosswall, T., Woodmansee, R.G. and Risser, P.G. (1988). *Scales and Global Change*. Scope 35, John Wiley, Chichester.

Ryden, B.E. and Kostov, L. (1980). Thawing and freezing in tundra soils. In Sonesson, M. (Ed.) *Ecology of a subarctic mire*. Ecological Bulletin (Stockholm), 30, 251–281.

Shugart, H.H. (1984). *A Theory of Forest Dynamics: The Ecological Implications of Forest Succession Models*. Springer-Verlag, New York.

Shugart, H.H. and West, D.C. (1977). Development and application of an Appalachian deciduous forest succession model. *J. Environ. Manage.*, **5**, 161–179.

Shugart, H.H. and West, D.C. (1980). Forest succession models. *BioScience*, **30**, 308–313.

Shugart, H.H. and West, D.C. (1981). Long-term dynamics of forest ecosystems. *Am. Sci.*, **69**, 647–652.

Shul'gin, A.M. (1965). *The Temperature Regime of Soils*. Israel Program for Scientific Translations, Jerusalem.

Sims, P.L. and Singh, J.S. (1978). The structure and function of ten western North American grasslands. III. Net primary production, turnover and efficiencies of energy capture and water use. *Journal of Ecology*, **66**, 573–597.

Sims, P.L., Singh, J.S. and Lauenroth, W.K. (1978). The structure and function of ten western North American grasslands. I. Abiotic and vegetational characteristics. *Journal of Ecology*, **66**, 251–285.

Slaughter, C.W. (1983). Summer shortwave radiation at a subarctic forest site. *Canadian Journal of Forest Research*, **13**, 740–746.

Smith, T.M. and Huston, M. (1988). A theory of spatial and temporal dynamics of plant communities. *Vegetatio* (in review).

Smith, T.M. and Urban, D.L. (1988). Scale and resolution of forest structural pattern. **Vegetatio**, **74**, 143–150.

Sollins, P., Spycher, G. and Topik, C. (1983). Processes of soil organic matter accretion at a mudflow chronosequence, Mt. Shasta, California. *Ecology*, **64**, 1273–1282.

Sorenson, L.H. (1981). Carbon–nitrogen relationships during the humification of cellulose in soils containing different amounts of clay. *Soil Biology and Biochemistry*, **13**, 313–321.

Steele, J.H. (1985). A comparison of terrestrial and marine ecological systems. *Nature*, **313**, 355–358.

Tamm, O. (1950). *Northern Coniferous Forest Soils*. Scrivener Press, Oxford.

Terjung, W.H. and O'Rourke, P.A. (1982). The effect of solar altitude on surface temperatures and energy budget components on two contrasting landscapes. *Boundary-Layer Meteorology*, **24**, 67–76.

Tiessen, H. and Stewart, J.W.B. (1983). Parcticle-size fractions and use in studies of soil organic matter: II Cultivation effects on organic matter composition in size fractions. *Soil Sci. Soc. Am. J.*, **47**, 509–514.

Tiessen, H., Stewart, J.W.B. and Bettany, J.R. (1982). Cultivation effects on the amounts and concentration of carbon, nitrogen, and phosphorus in grassland soils. *Agron. J.*, **74**, 831–835.

Tilman, D. (1988). *Plant strategies and the structure and dynamics of plant communities*. Princeton University Press, Princeton, NJ.

Toy, T.J., Kuhaida, A.J. and Munson, B.E. (1978). The prediction of mean monthly soil temperature from mean monthly air temperature. *Soil Science*, **126**, 181–189.

Urban, D.L., O'Neill, R.V. and Shugart, H.H. (1987). Landscape ecology. *Bioscience*, **37**, 119–127.

Urban, D. and Shugart, H.H. (1986). Avian demography in mosaic landscapes. In Verner, J., Morrison, M.L. and Ralph, C.J. (Eds) *Modeling Habitat Relationships in Terrestrial Vertebrates*. University of Wisconsin Press, Madison.

Van Veen, J.A., Ladd, J.H. and Frissel, M.J. (1984). Modeling C and N turnover through the microbial biomass in soil. *Plant Soil*, **76**, 257–274.

Viereck, L.A. (1982). Effects of fire and firelines on active layer thickness and soil temperatures in interior Alaska. In *Proceedings Fourth Canadian Permafrost Conference*. National Research Council of Canada, Ottawa, 123–135.

Viereck, L.A. and Dyrness, C.T. (1979) Ecological effects of the Wickersham Dome fire near Fairbanks, Alaska. *United States Forest Service General Technical Report* PNW-90.

Viereck, L.A., Dyrness, C.T., Van Cleve, K. and Foote, M.J. (1983). Vegetation, soils, and forest productivity in selected forest types in interior Alaska. *Canadian Journal of Forest Research*, **13**, 703–720.

de Vries, D.A. (1975). Heat transfer in soils. In de Vries, D.A. and Afgan, N.H. (Eds) *Heat and Mass Transfer in the Biosphere*. I. *Transfer Processes in Plant Environment*. Scripta Book Company, Washington, DC, 5–28.

Watt, A.S. (1925). On the ecology of British beechwoods with special reference to their regeneration. Part 2, Sections II and III. The development of the beech communities on the Sussex Downs. *J. Ecol.*, **13**, 27–73.

Watt, A.S. (1947). Pattern and process in the plant community. *J. Ecol.*, **35**, 1–22.

Weins, J.A. (1986). Spatial scale and temporal variation in studies of shrubsteppe birds. In Diamond, J. and Cale, T.J. (Eds) *Community Ecology*. Harper & Row, New York.

Whitehead, P.G., Bird, S., Hornung, M., Cosby, B.J., Neal, C. and Paricos, P. (1987a). Stream acidification trends in the Welsh Uplands: a modelling study of the Llyn Brianne catchments. Submitted to *J. Hydrol*.

Whitehead, P.G., Reynolds, B., Hornung, M. Neal, C., Cosby, B.J. and Paricos, P. (1987b). Modelling long term stream acidification trends in upland Wales at Plynlimon. Submitted to *Process Hydrology*.

Wilson, A.M., and Briske, D.D. (1979). Seminal and adventitious root growth of blue grama seedlings on the Central Plains. *Journal of Range Management*, **32**, 205–213.

Wolock, D.M., Hornberger, G.M., Beven, K.J. and Campbell, W.G. (1988). Topographic and edaphic control of soil contact times and flow paths: a regional analysis of hydrochemical catchment response. Submitted to *Water Resources Research*.

Wright, R.F. and Cosby, B.J. (1987). Use of a process-oriented model to predict acidification at manipulated catchments in Norway. *Atmospheric Environment*, **21**, 727–730.

Wright, R.F., Cosby, B.J., Hornberger, G.M. and Galloway, J.N. (1986). Comparison of paleolimnological with MAGIC model reconstructions of water acidification. *J. Water Air Soil Poll.*, **30**, 367–380.

Zuzel, J.F., Pikul, J.L. and Greenwalt, R.N. (1986). Point probability distributions of frozen soil. *Journal of Climate and Applied Meteorology*, **25**, 1681–1686.

13 Model- and Strategy-driven Geographical Maps for Ecological Research and Management

WOLF-DIETER GROSSMANN
Institut für Sozio-Okonomische Entwicklungsforschung, Österreichische Akademie der Wissenschaften, A-1030 Wien, Kegelgasse 27, Austria

Long-term Ecological Research. Edited by Paul G. Risser

13.1 A HIERARCHICAL DESCRIPTION OF SYSTEMS

13.1.1 THREE DIMENSIONS OR ASPECTS OF COMPLEX SYSTEMS

Problems in ecosystems research and management can have three dimensions:

(1) A dimension of details (for example, the spatial distribution of characteristics such as soil types);
(2) A dynamic dimension, of development and change; and
(3) A strategic dimension, concerning, for example, the role of heterogeneity or disruptions and surprise for sustaining life.

For many problems all of these dimensions have to be considered together. These dimensions are used for a three-layer view of systems, and different methods are appropriate on each layer. Results from one layer provide inputs for the other layers, and a combined, 'multifaceted' method is thus defined to deal with 'multifaceted' problems in complex systems. The word 'multifaceted' is due to Zeigler (1979). 'Ecosystem' is used to include human ecosystems.

13.1.2 THE BOTTOM LAYER IN THE THREE-LAYER VIEW OF A SYSTEM

The bottom layer in the three-layer view of a system described above is the layer of simple data and other details within one system. These data are often precise, usually numerous, and simple to measure and evaluate. One important class of such data are spatial distributions of features. Examples of such features are the soil type or orographic features, such as aspect, slope, and elevation, or other factors such as microclimate, location of rivers, boundaries of properties, and administrative boundaries. Other details may be those due to diversity of species, differences within populations, and other forms of heterogeneity. Most of these latter details do not pose scientific or management problems other than the fact that they are numerous.

The bottom layer structures are simple, obvious, and often linear. Problems on this layer are well defined, and lend themselves to fast reactions to problems and the solution of problems, frequent needs in industrial production or in metabolism. Dynamics of the system on this layer are also simple.

13.1.3 INTERMEDIATE LAYER OR LAYER OF COMPLEX DYNAMICS

The intermediate layer in the system is the layer of such important aspects of ecosystems as their structures and their dynamics. Some examples of dynamics are seasonal changes, successions, actions by man, migratory patterns, or adaptations. Ecosystems both generate dynamics and react to changes of their environment.

Problems on this intermediate layer deal with aggregated characteristics, such as the complex structures which generate dynamics in, for example, predator–prey relationships, the structure of a company and its customers in determining the liquidity of the company, or the global systems structures which drive weather and climate.

Uncertainty is often high on this layer because of variable impacts from the system's environment. Climate variations and human behavior are both examples of variable impacts, and are both important causes of uncertainty. The preservation of liquidity in the company is an example of many factors as a cause of increased uncertainty. New orders may create the need for prefinancing, and thus temporarily decrease liquidity; customers may disappear from the market, and new customers may emerge.

Feedback loops are appropriate as reactions to many such changes. As long as a feedback reaction is appropriate to deal with a change, the structure of the system can remain unchanged.

13.1.4 HIGHEST OR STRATEGIC LAYER

Structures can and do change. Since models on the intermediate layer depict structures, they usually become invalid if the real structure changes. Change can be caused by rare events, by destruction, by learning, or by chance; an ecosystem seems to use as many different strategies to bring about change as it uses to cope with it. Some of the coping strategies are diversity, variability, system architecture, resilience, combinations and decoupling of systems, and replacement of systems (Ashby, 1959, 1960; Simon, 1962; Grossmann, 1978; Holling, 1978).

The results of the likelihood and form of changes from the strategic layer are important for evaluations on the intermediate layer. If the structure is changed, the resulting dynamics will usually change also. However, dynamics can change without change of the structure, as in the case of change in modes of behavior (Thom, 1975; Haken, 1978), and conditions which lead to change of the behavioral mode are, therefore, evaluated on the highest layer. Not all structural changes result in changes of the system dynamics, as a changed, and therefore different, structure can continue to fulfil the same purpose. Factors important for the evaluation of the behavioral mode and for the assessment of structural changes are called 'strategic factors' or 'strategic criteria'. These criteria are usually qualitative, imprecise, often subjective (as, for example, in portfolio-analysis), and highly aggregated. Changes of structure and of mode, and the reasons for change, characterize a third aspect of ecosystems. On this layer, the uncertainty is very high.

13.1.5 HISTORY OF THE THREE-LAYERED DESCRIPTION OF SYSTEMS APPLIED HERE

The systems view outlined above also has hierarchical features. Hierarchical or stratified perceptions of complex systems have a long tradition. Simon

(1962) provided examples demonstrating why a hierarchical architecture increases the probability of reliable functioning; Bertallanffy (1969) gave a hierarchical description of the cosmos; Mesarovic *et al.* (1971) gave a detailed mathematical description of specific hierarchical systems; Bossel and Strobel (1977) combined a dynamic model with strategic criteria; Vester and von Hesler (1980) suggested a six-level scheme for regional planning; hierarchical descriptions were applied to ecosystems by Allen and Starr (1982); and implications of cybernetics for socio-ecological systems have been investigated by Rappaport (1979).

The view of systems presented here is partially based on this material, but foremost, it is based on experience derived from problems in manifold projects (MAB 1 project, Adisoemarto and Brunig, 1978; Grossmann, 1978; MAB 11 project, Vester and von Hesler, 1980; ARP-Project IIASA, Grossmann, 1983; MAB 1 project, Brunig *et al.*, 1986; pollution abatement projects, Grossmann and Grossmann, 1985; Grossmann and Orthofer, 1987; and projects on forest die-back, Grossmann, 1988a). The view was developed to applicability by the author for the MAB 6 Project Berchtesgaden (Grossmann *et al.*, 1983), and was refined in several case studies within this project (Grossmann *et al.*, 1984; Grossmann and Schaller, 1986; Haber *et al.*, 1984; Haber, 1989a, b) and other projects.

13.1.6 A RELATIVISTIC VIEW OF SYSTEMS

The hierarchical view of systems is relativistic. A specific ecosystem usually is part of another hierarchy, the hierarchy of its lower and higher ecosystems, and, eventually, of its environment, even if this is no longer an ecosystem. The highest layer of one system may represent a simple datum on the lowest layer of a higher system in another hierarchy.

13.2 APPROPRIATE METHODS TO DEAL WITH THE DIFFERENT PROBLEMS ON THE DIFFERENT LAYERS

In the three-layer view of systems given above, there are differences between the layers with respect to

Problems,
Type of data,
Type of structure,
Type of appropriate research,
Time horizons,
Advisors, local experts, and
Audience for problems and solutions.

Methods chosen to deal with problems must be appropriate for these characteristics.

The decision of which methods are appropriate can be based on the characteristics of the data.

13.2.1 METHODS FOR THE BOTTOM LAYER

Geographical Information Systems (GIS) are an important tool for storing, evaluating, depicting, updating, and processing spatial data. A GIS should store the topology of spatial relationships, i.e. which polygon is adjacent to which others, adjacent to which line features, and so on. This task is difficult if maps have to be combined (so-called map overlay, for example, of maps of the soil, microclimate, slope, and aspect, to determine the ecological conditions of the site) and is offered by only a few GISs. In our projects, we use ARC/INFO, which combines software for spatial evaluations with a relational data bank to store the topology. Data banks are, in general, appropriate on this layer, as are statistics, spread sheets, and real-time control for management.

13.2.2 METHODS FOR THE INTERMEDIATE LAYER

On this layer, the complex structures and their dynamics must be depicted and evaluated. Multiloop feedback models are one appropriate method for this task, and often exhibit dynamics comparable to those of reality if they mirror real structures. However, all methods that either model real complex structures or lead to complex or aggregated dynamics are appropriate. Object-oriented programming can be used to model dynamic processes. Some AI methods can be used to model and evaluate structures of decision making.

13.2.3 METHODS FOR THE HIGHEST LAYER

General or 'strategic' criteria are appropriate to evaluate reasons and forms of structural changes, or to develop policies on how to make systems more viable or how to replace them with systems that are more suitable to prevailing or emerging conditions. Several such criteria were mentioned above. *Markowitz's Portfolio Analysis* (Markowitz, 1959) and its extensions help in the economics to determine a tradeoff between risk and profit by appropriate diversification of shares, resources, customers, etc. The subjectivity in decisions on this layer was mentioned above. In natural ecosystems no conscious decision making is done; however, similar criteria seem to be applicable to natural ecosystems. Scenarios may be developed from the evaluation of strategic criteria; for example, the 'let it burn' policy in some United States forests is based upon strategic evaluations of the consequences of too-rigid fire control.

Long-term ecological research and long-term monitoring are adequate approaches to problems on the strategic layer. Table 13.1 is a summary of systems characteristics and appropriate methods.

Table 13.1 Summary of characteristics of the three layers and of appropriate methods

Characteristic	Layer of details		Intermediate layer	Strategic layer
	Numbers	Logical variables		
dominant relationships	causal	rules	structures	valuations
uncertainties				
data	low	low–middle	middle	high
structure	low	low	low	considerable
outside influence	low	low	intermediate	high
methods	many, precise,	deduction	feedback, holistic, structural	strategic
data	numbers	knowledge	functions	concepts
precision	high	middle	middle	low
number	very high	high	middle	low
importance in overall system	low	middle	middle	high
character	simple	literal	composed	colligative
aggregation	low	low	middle	high
variables				
precision	high	middle	middle	low
number	very high	high	middle	low
importance in overall system	low	low-middle	middle	high
character	simple	simple, literal	aggregated	colligative
aggregation	low	low–middle	middle	high
time horizon	short	short–inter mediate	intermediate	long

13.2.4 COMBINATION OF METHODS IN THE THREE-LAYER APPROACH

Implementation of the results of a strategic evaluation is usually difficult; this is because the analysis is 'soft' and the implementation often has to be done on the layer of details. This became very obvious in the Man and the Biosphere (MAB 11) project 'Lower Main' (MAB program area 11: 'Human Settlements'), where the planners could not translate the results of strategic evaluations into maps. The planners agreed with the outcomes of the strategic analysis (for example, to increase the diversity of the means of transport) but could not derive from this recommendation where to build what. Regional plans often have to be precise to the centimeter, as there is a considerable gap between the outcome of strategic analyses and the details of implementation. The method outlined here helps to narrow or even bridge this gap.

13.2.4.1 Combination between the strategic layer and the layer of complex dynamics

The evaluation of strategic criteria usually allows the derivation of different scenarios on change, including structural change, of the system on all layers. These scenarios are then evaluated with models on the intermediate layers. The structure of the models may have to be adapted to accommodate the changes, or their mode of operation may have to be changed. Models may even have to be replaced. The results of strategic analyses thus translate into corresponding model outcomes. The models are in turn used to check the strategic analyses, because on the intermediate layer far more and more precise knowledge is available. One well-known feature of models is counterintuitive behavior (Forrester 1969), which is highly effective in testing and refining scenarios. This top-down and bottom-up linkage between the strategic and the intermediate layer helps to build reasonable scenarios that are more suitable to implementation, and to translate them into model structures and model dynamics. This is one step to implementation of strategies, but not a sufficient one.

13.2.4.2 Combination of the layer of complex dynamics and the bottom layer

In a combination of the intermediate and bottom layers, the resulting dynamics are translated into maps using the GIS. The precise and numerous data on the bottom layer help in testing the model dynamics and the underlying strategic analyses, and in improving analyses and models. The whole procedure can be iterative. The resulting maps can be used to guide actions, allowing planners to depict where to do what and when. The combination was applied to projects on forest die-back (Grossmann *et al.*, 1983, 1984; Grossmann and Schaller, 1986; Grossmann, 1988a) and to the development of new agricultural strategies (Haber *et al.*, 1989a). In forest die-back the procedure was

(1) To select one hypothesis on forest die-back and the spatial variables that are important to evaluate this specific hypothesis. Important variables for all hypotheses are the age and species of trees and quality of the soils. Important additional variables for the soil acidification hypothesis are at least the pH-value and buffering capacity of the soils and isolines of H-deposition. Important additional variables for an ozone hypothesis are isolines of durations of the different ozone doses. Usually a map is composed of different polygons such that within one polygon the value of the specific variable is roughly constant (for example, in a soil map there is one type of soil within one polygon).
(2) To overlay all maps selected for one specific hypothesis. This is done with the Geographical Information System (GIS). Overlay proceeds like having all maps on transparencies and putting them on top of each other. The resulting map has many more polygons than each individual map. It is called the 'smallest common geometry', SCG. A polygon in the SCG as derived here has the property that all variables of the maps that are combined in the SCG have roughly a constant value within this polygon: within one polygon there is only one tree species (or one type of stand), the trees are in the same age class, the pH-value (in the case of the soil-acidification hypothesis) is the same, and so on. The GIS automatically creates a new data bank for the SCG.
(3) To attribute an ordering to the polygons of the SCG based on the evaluation of the risk according to the different variables. This ordering is qualitative. It gives the liability for something to happen (for example, the risk that the corresponding areas in a forest are damaged). In regional planning the outcome is the probability of use of the different polygons for housing or other construction, if suitable criteria such as availability, accessibility, steepness, etc., are used. Such a SCG is therefore called a 'risk map'.
(4) To connect the risk map (bottom layer) with the dynamic model (intermediate layer). If for a specific year forest area containing 3% of severely damaged trees (damage class 5) has to be located, take the polygons with the highest risk until 3% of the forest area is found and attribute damage class 5 to these polygons. If 7% of damage class 4 is to be found, take the polygons with the next lower risks until another 7% of the forest area is found and so on.
(5) To plot the resulting map using the GIS.
(6) To repeat this procedure for all points in time for which maps are wanted.

Analysis on the strategic layer

The diversities of forest sites affected by forest damage and of damaged tree species are both high. This led the group in 1983 (Grossman *et al.*, 1983) to believe that biological factors such as diseases or pests are not the main factor of the damage. Due to the same reason, climate factors should also be only a part of the problem. The concentration of most pollutants, such as SO_2 or NO_x, is also highly diverse. Only a few pollutants in particular, ozone pose continental problems during extended periods of time. However, the observed symptoms of tree damage are usually different from ozone damage. Therefore, other secondary substances beside ozone (e.g. hydrogen peroxide, formaldehyde, formic acid, and nitric acid) were also more closely investigated because their concentrations correspond on the average to ozone concentrations and should hence also temporarily pose a continental problem (Grossmann, 1988).

Many other combinations of methods are possible. They seem to work best if they are made according to the three-layer description of complex systems given above. Several projects that adopted different, promising approaches sooner or later ran into difficulties and gave disappointing results. Analysis of the shortcomings will be reported in Haber (1989b).

13.2.4.3 Soft coupling; the local expert

Some variables are not always available. However, certain variables can be calculated; for example, the distribution of fog using elevation of the sites and temperature, humidity, and other climate data from a climate station. Most of the calculations on the distribution and frequency of fog are correct; although some are not, and the map of calculated fog distribution must be discussed with people who know the area, the so-called local experts, who in this particular case are farmers and foresters. The map supports communication.

An understanding of data is important. One value of a variable may have a totally different meaning depending on the context in which it is used. For example, one subspecies of spruce has naturally downward-bent branches. But this shape may also be one of the early symptoms of forest damage. The local people must be asked what the data mean in their area. Large central data banks often lead to disaster in understanding when the source and the definition or meaning of the data are not known (Jeffers 1978). The numbers must be translated into information or knowledge which gives the actual meaning. These corrected data can then become the contents of maps from which time series of maps can be derived. This translation process, using knowledge and judgment of local experts, is called 'soft coupling' (Grossmann, 1983).

13.2.4.4 Software for combining maps and models

Software (the DYS-ARC package, devised by the author) is now available to evaluate models and translate their dynamics into dynamic maps, using a SCG. This software asks which risk map to choose, which time series of data to read, and which type of combination of the time series and the risk map to select. The software then asks for the points in time for which to make the maps, checks the times, and then starts ARC/INFO to produce the corresponding time series of maps. This software can also handle more complex types of risk maps with different classes of risk, where no transition between classes is allowed. It can also process dynamics from sources other than models.

13.3 APPLICATIONS OF DYNAMIC MAPS

The maps on forest die-back were produced without knowledge of the actual distribution of the damage, and thus could be compared with the actual situation. In this comparison, the forest scientists acknowledged the shortcomings of their maps showing the distribution of damage based on on-site evaluations, and rated the dynamic map as superior to their evaluation. Actually, both types of map had weaknesses. The two methods, of on-site inspection and of dynamic maps, can be used in conjunction with each other, to complement and check each other.

Dynamics are not only derived from dynamic models. Experts can provide projections based on their (subjective) expectations. As the resulting maps are very precise, and represent the obvious expression of expert expectations, they provide a graphic display by which to judge the validity of the expert opinions.

Other applications of dynamic maps are

(1) To test scientific hypotheses, as described for forest die-back. Different hypotheses on the cause of forest die-back lead to different dynamics in the models, and to different SCGs in the GIS. Hence different dynamic maps result from different hypotheses. There exist about 200 different hypotheses on the causes of forest die-back. Comparison with the present real situation and its past and future development is a very effective test of the different hypotheses.

(2) To update information in the GIS. This means progress towards Automated Geographical Data Banks. As a GIS usually stores large amounts of data, it is time consuming and expensive to update these data. But priorities for update can be determined if dynamic models are combined with the static maps to find out where changes are most likely or most important.

(3) To improve classification of remote sensing data. Dynamic maps can be produced to show the same area for the same time as the remote sensing data, but based on different sources of information. Therefore, a comparison of differences can help to classify the remote sensing data, in particular to reveal new and unexpected developments. Pattern evaluation often works even when statistics fail.

(4) To provide support for environmental monitoring through the comparison of computed maps to maps based on remote sensing data. The expected development (from any vehicle of anticipation) is translated into dynamic maps and compared with the actual ongoing development (from on-site inspection, false-color infrared photographs, or remote sensing data). Differences may reveal unexpected developments at a very early stage.

(5) To give support for environmental early warning. An unexpected development is more easily revealed if the expected development is presented as explicitly and visibly as possible. Dynamic maps are excellent for this purpose.

(6) To give support in detection of rare events, slow dynamics, and other somewhat hideous developments, thereby showing that dynamic maps can effectively support long-term ecological research.
(7) To guide and schedule management actions in ecosystems (e.g. where to plant or to harvest what at a specific time).
(8) To allow evaluation and implementation of strategic criteria.

The combined method is also helpful in modeling to

(9) Assess the applicability of models, and to find errors in structures and data.
(10) Calibrate, (in-)validate, or improve models. One data base, on the intermediate layer, allows calibration; the other, quite different and on the bottom layer, allows (in-)validation. Besides historical 'validation', about 30 more tests exist for feedback models (Forrester and Senge, 1980); many of these are more powerful if combined with dynamic maps.

13.4 TEST CASE: REASONS FOR THE CHANGES OF AGRICULTURAL LAND USE AND TESTING OF OPTIONS FOR CHANGE

One case study in a five-year project (MAB Project 6, Berchtesgaden) was on the development of new agricultural policies (Haber, 1989a). The study is briefly summarized below.

The problem concerned farming in Berchtesgaden, Germany, which is especially unprofitable due to unfavorable climate and poor soils. The majority of the farmers have second jobs to subsidize their own farming, and many offer 'holidays on the farm'. Tourists enjoy the landscape, which is characterized by high mountains, lakes, and farms with mountainous pastures. Farming in the area is slowly decreasing as farmers concentrate on tourism. As a consequence, the attractiveness of the landscape is decreasing as the number of cow herds decrease, and natural forest growth takes over abandoned mountain pastures.

An aim of the study was the development of new options for the region, and an indication of whether concentration on tourism will pay or will be counterproductive. Alternative sources of income must be compatible with the climate and the ecological conditions of the region, and to its very low accessibility. Options should be environmentally beneficial.

Strategic analyses

The project group evaluated data on the natural potential, accessibility for traffic, numbers of inhabitants and composition of the economy, new technologies, and other sources of possible innovations in order to assess the options. The history and attitude of the local population were also evaluated. The outcome of the evaluation showed that neither farmers nor the local economy have choices other than tourism. However, since agriculture seems to be important for the viability of tourism in the area, one option is the production of high-quality agricultural products ('organic farming') and direct marketing by the farmers to the tourists. The outcomes of the strategic analyses were used to define three scenarios: abandoning of agriculture and concentration on tourism, continuation of the present behavior, and development of agricultural high-value production and internal marketing.

Layer of complex dynamics

Extensive research on socio-economic factors had been done by one group taking into account the history of the area during the last centuries. Time series for the years 1970 to 1986 of actual data were available for the number of tourists, accommodations, farms, and cattle, the land use in all categories, and several economic parameters. Several models were constructed in interdisciplinary workshops to evaluate the scenarios. These models depict relationships between farmers and their available labor, extent and type of farming, number of cattle, land use in eight categories, aesthetic value of the landscape, supply and demand in tourism, number of tourists in three categories, crowding by tourists, income from farming and tourism, consequences for the regional economy, etc. Extreme values were determined with a linear programming model; dynamics were evaluated with a model using difference equations (Grossmann, 1988b).

It was easy to develop and evaluate this latter model due to new software, the DYS-ARC package, described earlier in this chapter. A run for the years 1970 to 2020 with time steps of 0.1 years with 300 output variables took 1 minute on an AT compatible PC; in earlier case studies a smaller model needed several hours on a VAX using a DYNAMO dialect. The DYS-ARC package allows the user to view interactively, graphically, and numerically all variables in any composition, so that the consequences of changes can be traced through the whole system.

Validation of models is done with up to 11 methods; for example, the models had to reproduce the known time series starting from 1970. Other tests were extreme value tests, discussion of model structures in workshops, etc. Translation of model results into dynamic maps provided further tests (see below).

Results showed that abandoning agriculture was the most unfavorable scenario because it increased competition with nonagricultural offers in tourism and caused slow deterioration of important characteristics of the landscape. In this scenario, the demand for and the quality of the tourism decreased, and supply and competition increased. Establishment of agricultural high-value production and internal marketing needed far less investment than was anticipated, and turned out to be the most profitable option for the whole region. One unexpected result of this latter option was the too-great increase in intensity of land use, which would cause degradation of the land and agricultural pollution. Restrictions on the most intense types of land use would be needed.

Bottom layer

Geographical information was entered into the GIS using scales down to 1:5000 where necessary. Different map scales were used complimentarily. Sources of information were the available maps, aerial photography, and additional field data collection. Maps now exist within the ARC/INFO GIS on soil types, humidity of the soils, detailed land use including roads, other line information such as railroads, polygon information on settlements, forests, location of farms, type of farming area and present use (from aerial photography and on-site visits), vegetation, and other information.

With this information, the agricultural areas were ordered according to suitability for grazing so that any change in land use suggested by the dynamic model could be translated into time series of maps where these changes are most likely to occur. Important information for this was the soil type, steepness, aspect, size, present state, present use, vegetation, and elevation. This translation is similar to the one explained for forest damage, but slightly more complex.

This translation allows checks on whether predicted changes are possible and reasonable.

Evaluation of patterns

Inhabitants of the area and local experts instantly perceived remarkable developments in time series of maps, and started heated discussions if the developments seemed reasonable and loudly rejected unlikely or wrong maps.

Statistical analyses

Complex maps are composed of many polygons, and for each polygon all data are available in the data bank of the GIS. Hence statistical analyses could be done with reasonable and homogeneous data.

In several studies, statistics showed a poor correlation whereas there was marked similarity in actual and computer-produced patterns. In one of these cases the prevailing wind direction was specified erroneously so that predicted and actual patterns differed by about 45°; in another case an anemometer had malfunctioned during the last 4 years. The team had not used these specific data from the last 4 years because they gave an unlikely pattern, and a long-term time series and checking of this device (one year later) helped to establish that the team had been right.

With these data and methods it is possible to predict further changes of abandoned areas, as detailed information on ecological conditions of all sites and succession types of sites exists. Tourists and inhabitants had been asked about their preferences for different landscapes using photographs. Evaluation of anticipated landscape changes with these preferences allowed judgments on the validity of model outcomes on the layer of dynamics. With a detailed balancing model, possible increased levels of fertilization have been evaluated, taking into account characteristics of the soils, orographic features (slope), vegetation, and location and intensity of land use. These parameters determine the consequences of agricultural pollution of groundwater.

The study was made possible by the availability of long-term time series and an understanding of the history of the region and the likely behavior of the local population. Nearly all predictions, model construction, and, in all probability, model results would have been wrong without this knowledge of the history of the region and its ecosystems. For example, the inhabitants tend to stay in their area in spite of crowded housing, lack of jobs, and poor climate. Model parameters quite different from those suitable for other regions in Germany had to be chosen. Forests were nearly completely destroyed in the past in the production of salt, and reforestation was often done without consideration of which subspecies are suitable for climate and soil. Catastrophes in forests are partially due to poor choice of subspecies.

On several occasions the models and historical time series were at odds, but the model was more often correct than the data. In some cases, typing errors in the time series had caused the discrepancies, in other cases established knowledge about the area was deliberately ignored due to scientific prejudices, but had to be included in the model so that it could pass the tests, in particular the historical validation. For this, long-term time series were necessary.

The resulting dynamic maps are effective for testing hypotheses of behavior, successional changes, changes in nitrate content of groundwater, etc. These maps are also effective for support of ongoing environmental monitoring. All of these are important problems in long-term ecological research.

13.5 CONCLUSIONS

Dynamic maps have been successful in testing and improving models, in evaluating scientific hypotheses, in finding errors in data and assumptions, and in checking and developing strategies. Many more applications of model- and strategy-driven maps seem possible, and new options for ecosystem research and management seem to emerge through this method.

13.6 REFERENCES

Adisoemarto and Brunig, E.F. (Eds) (1978). *Transactions of the Second International MAB-IUFRO Workshop on Tropical Rainforest Ecosystem Research*, 21–25 October 1978, Special Report No. 2, Chair of World Forestry, University of Hamburg.

Allen, T.F.H. and Starr, T.B. (1982). *Hierarchy. Perspectives for Ecological Complexity*. The University of Chicago Press, Chicago, Illinois.

Ashby, W.R. (1959). *An Introduction to Cybernetics*. Chapman and Hall, London.

Ashby, W.R. (1960). *Design for a Brain*. John Wiley, New York.

Bertalanffy, L.V. (1969) (rev. edition). *General Systems Theory: Essays on its Foundations and its Development*. Braziller, New York.

Bossel, H. and Strobel, M. (1977). Experiments with an intelligent world model. *Futures*, **10**, No. 3, 191–212.

Forrester, J.W. (1969). *Urban Dynamics*. MIT Press, Cambridge, Massachusetts.

Forrester, J.W. and Senge, P. (1980). Tests for building confidence in system dynamics models. In Legasto, A.A. Jr, Forrester, J.W., Lyneis, J.M. (Eds) *System Dynamics*. North-Holland, Amsterdam, 209–228.

Grossmann, W.D. (1978). The dynamic meta-model—a tentative suggestion how to approach the solution of a pressing problem. In Adisoemarto, S. and Brunig, E.F. (Eds) *Transactions of the Second International MAB-IUFRO Workshop on Tropical Rainforest Ecosystem Research*. Special Report No. 2. Chair of World Forestry, University of Hamburg, 166–185.

Grossmann, W.D. (1983). Systems approaches towards complex systems. In Messerli, P. and Stucki, E. (Eds) *Fachbeiträge der schweizerischen MAB-Information*, Vol. 19. Bundesamt für Umweltschutz, Bern.

Grossmann, W.D. (1987). *Systemprojekt Lehrforst Rosalia: Untersuchungen zum Problem der neuartigen Waldschäden*. Final report. Federal Ministery for Science and Research, Vienna.

Grossmann, W.D. (1988a). Products of photo-oxidation as a decisive factor of the new forest decline? Results and considerations. *Ecological Modelling*, **42**, 281–305.

Grossmann, W.D. (1988b). *Regio: Ein Modellbausatz zur Bearbeitung ökonomisch-ökologischer Probleme im Rahmen eines hierarchischen Systemkonzeptes* (REGIO: A Model Kit for Socio-Economic Systems). Final Report Research Area 99, MAB Project 6 Berchtesgaden, Parts 1 and 2. National Park Administration, Berchtesgaden.

Grossmann, W.D. and Brunig, E.F. (1983). Concepts of hierarchy in forest sector modeling. In Seppala, R., Row, C. and Morgan, A. (Eds) *Forest Sector Modelling*. AB Academic Publishers, 105–117.

Grossmann, W.D., Haber, W., Kerner, F., Richter, U., Schaller, J. and Sittard, M. (1983). Ziele, Fragestellungen und Methoden. Ökosystemforschung Berchtesgaden. *MAB-Mitteilungen* Nr. 16. Deutsches Nationalkomitee MAB, Bonn.

Grossmann, W.D. and Orthofer, R. (1987). Grobbeurteilung von Kohlenwasserstoffemissionen im Hinblick auf ihre möglichen Beiträge zum Waldsterben. *Report OEF2S-A—1121*. Austrian Research Center, Seibersdorf.

Grossmann, W.D. and Schaller, J. (1986). Geographical maps on forest die-off, driven by dynamic models. *Ecological Modelling*, **31**, 341–353.

Grossmann, W.D., Schaller, J. and Sittard, M. (1984). 'Zeitkarten': eine neue Methode zum Test von Hypothesen und Gegenmassnahmen beim Waldsterben. *Allgemeine Forstzeitschrift*, München.

Haber, W. (Ed.) (1986). Mögliche Auswirkungen der geplanten Olympischen Winterspiele 1992 auf das Regionale System Berchtesgaden. *Lehrstuhl für Landschaftsökologie*. TU München /Weihenstephan.

Haber, W. (Ed.) (1989a). (1): Report on the consequences of different agricultural policies in Berchtesgaden, using a combination of dynamic models and a GIS. (2): An update of Grossmann *et al.* 1983, taking into account the experiences from 5 years research in the MAB 6 Project Berchtesgaden. In press.

Haber, W. (Ed.) (1989b). Report on the project on forest die back investigated by combining dynamic feedback models and area related balancing. *Lehrstuhl für Landschaftsökologie*. TU München /Weihenstephan. In press.

Haber, W., Grossmann, W.D., Schaller, J. (1984). Integrated evaluation and synthesis of data by connection dynamic feedback models with a geographic information system. In Brandt, J. and Agger, P. (Eds) *Methodology in Landscape Ecological Research and*

Planning. Proc. 1st International Seminar. Int. Assoc. of Landscape Ecol. Roskilde Univ. Center, Roskilde.

Haken, H. (1978). *Synergetics*, 2nd edition. Springer-Verlag, Berlin.

Holling, C.S. (Ed.) (1978). *Adaptive Environmental Assessment and Management*. John Wiley, New York.

Jeffers, J.N.R. (1978). *An Introduction to Systems Analysis: With Ecological Applications*. Edward Arnold, London.

Markowitz, H.M. (1959). *Portfolio Selection; Efficient Diversification of Investments*. John Wiley, New York.

Mesarovic, M., Macko, M. and Takahara, Y. (1971). *Theory of Hierarchical, Multilevel Systems*. Academic Press, New York.

Rappaport, R.A. (1979). *Ecology, Meaning and Religion*. North Atlantic Books, Richmond.

Simon, H. (1962). The architecture of complexity. *Proc. American Philos. Soc.*, **106**, December.

Thom, R. (1975). *Structural Stability and Morphogenesis*. Benjamin, Reading, Massachusetts.

Vester, F. and von Hesler, A. (1980). *The Sensitivity Model*. RPU (Regionale Planungsgemeinschaft Untermain), Frankfurt.

Zeigler, B.P. (1979). Structuring principles for multifaceted systems modelling. In Zeigler, B., Elzas, M. *et al.* (Eds) *Methodology in Systems Modeling and Simulation*. North-Holland, Amsterdam, 93–135.

14 Long-term Ecological Research: International Workshop II

E.H. TROTTER and JAMES R. GOSZ
Department of Biology, University of New Mexico, Albuquerque, New Mexico 87131, USA

14.1 BACKGROUND

The first Berchtesgaden workshop brought together scientists from various countries. Through the workshop, scientists were able to identify common ecological processes being studied at two or more sites, and to recognize that certain broad climatic phenomena were exerting similar influences on different ecological habitats. The scientists also realized the need for much broader geographical comparisons and analyses if ecological principles were to be identified and tested.

Workshop II built on the results of the Berchtesgaden workshop and on the personal and professional contacts that were fostered there. The goals of Workshop II were (1) to begin the formal development of international co-ordination and networking of collaborative long-term ecological research at specific sites, (2) to facilitate the introduction of the newest technologies available into these

Long-term Ecological Research. Edited by Paul G. Risser

collaborative research efforts, (3) to identify initial working groups in three selected biomes, and (4) to develop specific plans for the international collaboration necessary for long-term ecological research throughout each of these three biomes. Long-term research questions identified during the first workshop in the Federal Republic of Germany provided a starting point for Workshop II. The research plans developed from the biomes included the long-term ecological phenomena to be studied, the comparable data to be collected, and the analytical techniques to be used in designing the studies, collecting the data, and sharing the information.

A conclusion that was drawn from both workshops was that long-term ecological research on the international scale depends upon the dedication of individual scientists, but that enhanced communication of ideas by the scientists is vital to the research. This chapter presents a narrative transcription of the proceedings of Workshop II, preserving the identity of the participants and their ideas.

14.2 PURPOSE OF THE SECOND INTERNATIONAL WORKSHOP

The Albuquerque workshop (Workshop II) was considered a 'working' workshop; the number of participants was small, and each had very specific responsibilities. There was an implicit assumption that participation in the workshop mandated subsequent collaborative efforts to secure funding for and participation in long-term ecological research programs in three biomes identified at the Berchtesgaden workshop as being particularly amenable to international collaborative investigations. The three chosen biomes are (1) arid/semi-arid, (2) temperate forests, and (3) boreal forest/tundra. During the International Biological Program (IBP), some efforts were made toward international collaboration in these three biomes, but the efforts were largely *post-facto* comparisons of data which were not originally collected for international comparative purposes from projects which were not conceived in a collaborative fashion. In addition, at the time of the IBP, many of the current measurement, analytical, and communication technologies were not available; our concepts of global phenomena were more primitive, especially as they relate to ecological processes; and the use of models was not well developed. Subsequently, we have recognized the need to plan international research rather than just compare data from projects which were designed for another reason in different areas of the world. We have also learned that models are particularly effective in building collaborative research projects among scientists from various countries and ecological conditions. Thus, the development of specific, collaborative, international research by participants of the Albuquerque Workshop was designed to go beyond IBP, using research questions posed during the first workshop at Berchtesgaden, more sophisticated technologies, and greater understanding of the development and use of ecological models, in an additional decade or more of ecological research in the three biomes chosen for consideration.

14.3 WORKSHOP STRUCTURE

Before the Albuquerque workshop formally began, an all-day field trip was made to the Sevilleta Long-Term Ecological Research Site south of Albuquerque. The purpose of this field trip was to stimulate thoughts about collaborative long-term research among sites while actually being on a site, to demonstrate sophisticated methods of making large-scale field measurements, such as the Fourier Transform Infrared Spectrometer (FTIR), and to stimulate communication among the conference participants. The Sevilleta LTER site is one of the newest in the LTER network sponsored by the United States National Science Foundation. This site is unique in its location at the junction of at least four major biomes, a location which allows quantification of (1) gradient relationships with distance, (2) the scale-dependent or independent nature of spatial variability, (3) the influence of steep environmental gradients upon system properties, and (4) integrated responses across the region. Since research projects and the quantification of parameters at the Sevilleta occur at many different scales, what is learned about the studied phenomena, data management, and project synthesis should be helpful in the development of models for connecting local, regional, and global research efforts.

The first and second days of the Albuquerque workshop included presentations from discussions of the ecological questions that are amenable to long-term research on each of the three biomes. The presentations were not simply descriptions of current research, but rather made an attempt to identify ecological questions or hypotheses that must be addressed in an international and long-term context. These were followed by presentations of technologies that are integral to international long-term ecological research. The Great Plains modeling experience was presented by William Parton; the integration of small to large data bases to model regional and global CO_2 levels and predicted changes was presented by Michael Farrell of the Oak Ridge National Laboratory; geographical information systems (GIS) coupled with a dynamic modeling program (ARC/DYNAMO) were developed by Wolfgang Grossman and Jorg Schaller of Environmental Systems Research Institute (ESRI); remote-sensing, data management, and data networking, such as the NASA First Integrated Field Experiment (FIFE) project at the Konza prairie LTER in Kansas was presented by Donald Strebel of Versar, Inc.; and new communications technologies selected to address a series of spatial and temporal scales were presented by Steven Storch of BBN Systems and Technologies Corporation.

Three of the technologies are appropriate for multi-scale research: the GIS-model technology for studying site-specific processes and integration at the site level; remote sensing and management of data from several integrated technologies to expand from the site to the regional or the continental scale; and communications technologies for sharing the results from the first two technologies to ensure the active continuation of research projects. This plenary segment of the workshop acquainted participants with research efforts in each biome and with innovative technologies to facilitate data collection, management, sharing, and real-time and computer-to-computer communication among scientists throughout the global community.

Three working groups composed of scientists who were intimately involved in long-term ecological research and who have access to or control of a substantial dedicated research site were then established. These working groups, which met at the end of the first and second days of the workshop, were given five tasks. They were asked to:

(1) Develop one to a few research questions or hypotheses identified by the Berchtesgaden group that are international in scope, are also regionally important, and require long-term ecological research for a resolution;
(2) Plan an experimental approach which ensures that international comparisons and analyses are used to their greatest power;
(3) Describe technologies to be employed for both investigation and project communication purposes;
(4) Provide a plan for the analyses of the resulting data which co-ordinates structural components for the proposed study of the three sites, allowing for the later addition of new sites and investigators; and
(5) Develop strategies to obtain funding for the anticipated long-term ecological research project.

14.4 COMMUNICATIONS FROM THE WORKSHOP

The final segment of the workshop consisted of a televised presentation of research plans by each of the three working groups to an audience of scientists watching at most of the 17 LTER sites in the United States and at the Oak Ridge National Laboratory in Oak Ridge, Tennessee. Following the first hour's broadcast, scientists at these sites were encouraged to send comments and questions via electronic mail (E-Mail), FAX, or telephone. During the second televised hour, workshop participants addressed these comments and questions.

Workshops are a popular and common way for scientists to interact and advance the science of their discipline. Unfortunately, since workshops work best if discussions involve only small groups of individuals, a cost-effective and successful workshop can accommodate only a relatively small portion of the scientific community. Subsequent dissemination of information usually occurs as reports, books such as this one, or word of mouth. These forms of communication are often significantly delayed and can suffer from inaccurate interpretation. Small workshops also fail to involve the broad scientific community in the creative first steps of developing research plans. However, cost-effective technologies, such as the telecommunications already in use by large international companies, can allow near real-time interactions with a larger segment of the research community. This component was added to the Albuquerque workshop both to demonstrate its usefulness and to communicate and receive feedback from other scientists.

This input from many scientists at a wide range of sites by telecommunications was a secondary goal of the workshop. The television programs were duplicated for the additional audience of colleagues of the international ecological research community who could not be reached by the satellite down-link.

14.4.1 THE FIRST HOUR OF THE TELEVISION PRESENTATION

On the third day of the workshop, participants went on the air via satellite to the US LTER network and Oak Ridge National Laboratory to present the results of the discussions of three working groups concerning long-term research in the three biomes of arid/semi-arid, temperate forest, and boreal forest/tundra. Paul Risser began the program by stressing that scientists watching the televised program, and the studio audience, were all part of an experiment in communications among scientists who were at large distances on the planet from each other, but who were all concerned with and working on global-scale biological problems. Whereas 30 scientists attended the first Berchtesgaden workshop in 1988, and about 20 attended the Albuquerque workshop, participation by those in the scientific community was increased by the television link to over 100 individuals.

Jim Gosz described the presentations and discussion which had led up to the television presentation, and emphasized that since the presentation was a progress report only, and no final decisions had been made, input from the audience was extremely important. It was announced that there would be three reports, one from each of the working groups discussing research in the arid/semi-arid, temperate forest, and tundra/boreal forest biomes, followed by a discussion of research which could be performed across biomes.

14.4.1.1 The temperate forest biome

The temperate forests working group was chaired by Dr O.W. Heal (Scotland), and included Peter Beets (New Zealand), and Wolf Grossman (Austria). This group identified three interrelated issues (climate change, pollution, and management) for the temperate forest biome, and focused on climate change as a point of entry for research. The three sets of high priority questions were defined as follows.

First, what is the contribution of temperate forests to the global carbon budget? Can management of these forests help to control global carbon cycling? Research on these questions occurs at the site level where decisions are made about experiments on processes, parameters to measure, and how to validate results. Results from site-specific studies are aggregated to a regional information base and combined with other site-specific information in that region for global synthesis of information from other regions. Geographic information systems (GIS) and remote sensing techniques can be used to combine information from sites and regions. Long-term ecological research is appropriate because there is between-year variation, variation over longer natural or managed forest cycles (rotation), and variation between cycles in a biome consisting of long-lived dominant plant species.

Second, what is the contribution of forests to the global flux of the radiatively active gases such as methane (CH_4), nitrous oxide (N_2O), carbon dioxide (CO_2), the hydrocarbons, and ozone (O_3)? What are the feedback relations to the forest and to climate change, and can these relationships be quantified? Can management control or modify these interactions? This is a key topic because of the complex pollutant-changing climate interaction and because there is a range of management options for controlling the sources and sinks of the radiatively active gases (RAG). A simple physical study of sources and sinks is inadequate because biological processes are controlling these physical processes. A pollution–climate gradient can be used to examine environmental conditions that have existed or do exist as a result of pollution for exploring the variation in responses to climate change.

Third, what is the response of temperate forests to the change in CO_2 levels, temperature, and moisture resulting from climate change? How do the responses vary with different soil, nutrient, and moisture variability? Answering these questions will require manipulation of systems to examine the response of components of the systems, including the responses of individual species or groups of species. The temperate forest group concluded that a collaborative, cohesive world-wide program should focus on a few specific topics (for example, quantifying the sources and sinks of radiatively active gases and targeting a few selected responses of the forest). Since these fluxes will be different in each type of system, and will depend upon feedbacks to and from the changing climate, a network will be necessary to understand the global system.

14.4.1.2 The arid/semi-arid biome

The arid/semi-arid group was chaired by Uriel Safriel (Israel), and included Mary Seely (Namibia), Francisco Garcia-Novo (Spain), Carlos Montana (Mexico), Bill Lauenroth (Central Plains LTER, Colorado, USA), and Walt Whitford (Jornada LTER, New Mexico, USA). For biomes whose structure and function are driven by seasonal and multiple-year droughts, the central questions are (1) how does arid system stability interact with human impact to produce desertification? and (2) how will climate change influence the phenomenon of desertification? Finding answers to these questions depends upon understanding why human-impacted arid systems transform to non-exploitable, resilient states. What determines the resilience of impacted systems?

The general hypothesis is that the interaction between the stability properties of the system (resistance to change) and the type and strength of human impact (force applied to the components of the system) determine the transition of an arid system from one state to another. Theoretically, if global climate change is ignored, human impact on the world's arid systems (grasslands, shrublands, and scrublands) should produce divergence in the original system (the product of such influences as precipitation quantity and timing and temperature) to different secondary states (dependent upon the force of human impact on the same ecotype in different places on the globe). As human impact on the new system continues or changes, there should be a convergence of properties in severely affected arid lands. If human impact is like the force exerted on a ball to move it out of a pocket representing system stability, then the new impact of climate change will primarily affect system stability represented by the shape and depth of the pocket. Arid climates are believed to be much more variable than more mesic ones, but arid ecological systems are more stable because individual species have evolved adaptations to the climatic variability. The influence of climatic change upon arid system variability and the response of arid species to changed variability and a more rapid rate of change is entirely unknown. It is also important to identify those systems which are candidates for transformation and those which are already transforming. The model of transition from one state to another recognizes that system structure and function may transform independently.

To test this hypothesis, both world-wide experiments and long-term monitoring are necessary, and international co-operation is needed to identify as many states as possible, and to evaluate the relative contributions of climate change and human impact on the stability properties of states and the transformations from one state to another. The experimental manipulations of these arid systems are climate modifications of the amount and timing of water availability. These can be accomplished by using irrigation, rain-out shelters, redistribution of runoff, and the simulation of human impact by removal or addition of soil and/or vegetation. Three important physical variables to monitor are soil organic matter, nitrogen mineralization, and water infiltration. Measuring organismal responses might include changes in plant life forms; individual plant and community architecture (important to the animal community); plant primary production, growth, and phenology; physiological responses; and biodiversity represented by species composition and richness.

Long-term research is important to determine the effect of unusual, unpredictable natural history phenomena and environmental events which are of distinct and overwhelming importance to arid lands. An international effort is vital to allow major existing states of arid land systems to be monitored simultaneously in order to understand the impact of global change as it interacts with human impact and systems stabilities world wide. World-wide research will permit monitoring of predicted changes in albedo, dust generation, and the activity of decomposer communities (important contributors to radiatively active gas generation).

14.4.1.3 The boreal forest/tundra biome

Gaus Shaver (Tundra LTER, Alaska, USA) chaired this group, which included Kari Laine (Finland), Don Strebel (NASA, USA), Mike Farrell (Oak Ridge National Laboratory, USA), and John Yarie (Bonanza Creek, LTER, Alaska, USA). Four major long-term research issues were identified:

(1) The soil carbon budget and the quantity of carbon released as either CO_2 or CH_4 is virtually unknown in this biome, causing uncertainty about whether it is a source or a sink for carbon;

(2) Boreal forest/tundra biomes are strongly decomposition limited, and the constraints on decomposition through nitrogen and phosphorus cycling may be very important in regulating carbon cycling processes and net exchange of carbon with the atmosphere;
(3) Episodic, large-scale, and/or long-term phenomena such as animal population fluctuations and fire are important forcing functions for ecosystem dynamics; and
(4) Permafrost and soil hydrology are keys to system stability.

Research needs in this biome are:

(1) A network of monitoring stations based upon such guidelines as those just being instituted in the Nordic countries and presented in a publication called *Guidelines for Monitoring in Nordic Countries* by Kari Laine (Finland). This publication describes the standard protocols and close communications being set up in the Nordic countries to monitor biological change. The monitoring program was in part the result of the Nordic countries' lack of preparedness in the face of the Chernobyl catastrophe, which had disastrous effects on northern biological communities and organisms;
(2) Identification of common experiments in which the same manipulation is performed across the boreal/tundra region to determine whether limiting factors in this area are the same; and
(3) A well-developed boreal/tundra geographic information system which, combined with remote sensing, can monitor long-term change in response to fire or animal cycles. A network of monitoring stations, common experiments, and the GIS can all be used to model changes in response to global warming and new climate variations, and to validate and verify the model.

14.4.1.4 Cross-cutting research

Peter Beets (New Zealand) examined themes that were common to the recommendations of all three working groups. There were three of these. Biological responses to global climate change were recognized as species' responses, making it necessary, therefore, to understand how species' physiological responses to climate change are integrated from the community through the ecosystem level. The second theme was the role of biological systems in bringing about climate change, particularly by controlling the majority of atmospheric gases and by affecting water availability and distribution. Experiments in primary productivity and nutrient cycling were also themes common to all biomes. Last was the question of human impact on the biology of the globe, and the human responsibility to understand and to manage these impacts. Both scientists and policy makers must act together to become responsible for human impact and its implications to biological systems. Subsequent discussions noted that biodiversity and the frequency and intensity of disturbance were also common topics.

14.4.1.5 Reactions to the reports

Bill Lauenroth and Jerry Melillo expressed the belief that the importance of modeling in world-wide research efforts could not be emphasized enough. Lauenroth pointed out that global-scale research is now enhanced by improvements in technology such as remote sensing and computer power. Melillo emphasized that although links between GIS and dynamic models are difficult at present, they are possible. He noted that new techniques for research, monitoring, modeling, and validation must be put very clearly before professional scientists by those adept at these technologies, so that scientists can creatively conceive the uses for these technologies. Wolf Grossman encouraged the scientific community to be bolder in asking for immediate support for long-term global-scale research, pointing out that even if all polluting emissions could be stopped now, the situation world wide would improve only very slowly. Grossman expressed the belief that management at the ecosystem level was the only means of rapidly improving the situation once emissions were controlled, and this urgency should lead to more aggressive requests for research support. Walt Whitford made a plea for imaginative and large-scale experiments which, although expensive, can provide insight achievable only through whole-system manipulations. Don Strebel was asked whether NASA has produced enough information through remote sensing to use in making models. Strebel replied that there is much that NASA is doing in terms of modeling soils, vegetation, atmospheric behavior, and the links between these. Strebel emphasized that model verification and validation is needed. This requires intensive fieldwork to check links between components of systems over the short term, and to check the accuracy of predictions over the long term. Francisco Garcia-Novo also stressed validation, but suggested that historical and paleohistoric information be used to validate and verify model predictions. In such historical information, it is possible to trace past vegetation responses to climate change and human impact and management in order to predict landscape changes. Climate has changed in the past, and there is information in tree rings and in fossilized pollen to help us understand current changes.

Mary Seely pointed out that since there is a diversity of habitats and differing degrees of impact felt or predicted as a result of climate change, the few large and expensive experiments should be augmented by an extensive range of experiments that were site-specific. John Yarie focused on the problem of not having basic information in the boreal/tundra biome with which to build a predictive model.

The idea of large, whole-system manipulations or flagship experiments performed in all three biomes was important to Jerry Melillo because it allowed for comparison and contrast of a few common manipulations. Kari Laine said that it is sometimes difficult to predict which monitoring variables are going to turn out to be important in the future, and Bill Heal urged boldness in mounting large experiments in several biomes. In Europe, scientists should build on the Nordic Monitoring system and on successful German modelling of effects of air pollution on forests. In concluding the first hour, Paul Risser emphasized the significance of this time in the history of the world—a time when human impact has accelerated climate change through such things as land use—and pointed out that the new research is not just a question of understanding individual systems but is more a question of keeping the globe working. The central problem of arid systems is a lack of understanding of transitions from one state to another, in the temperate forests it is trace gases, and in the boreal/tundra regions it is soil processes.

14.4.2 THE SECOND HOUR OF THE TELEVISION PRESENTATION

14.4.2.1 Comments from viewers

Bruce Milne (Sevilleta LTER, University of New Mexico) reminded the panel that it is likely that plant communities may not respond as a unit to climate change, and that, according to paleoecological data, extant communities found in different biomes are recent assemblages of individual species. Lauenroth agreed that the response of individual species, rather than the entire plant community, should be a focal point of research, but thought that although extant plant communities may be quite transitory assemblages of species, responses to global climate change will first be seen in changed dynamics within an existing system structure. He also expects dynamics to change substantially, whereas community structure may not change for a very long time. The Central Plains Research LTER found the arid zone working group's conceptual model of arid ecosystems moving from one stability state to another to be unrealistic. They argued that ecosystems, especially arid ones, are not stable, and therefore the arid/semi-arid group's conception of an ecosystem transforming from one set of stability conditions to another was incorrect. Uriel Safriel stated that arid systems always have some properties of structure and function which are measurable and are known to change. Therefore, the question being asked in the conceptual model is, what is the kind of stability seen at present? It might be that a present system is at an extreme position of stability or lack of stability, but the important issue is how systems move under human impact from the present perceived state to another state. Milne suggested a refinement of the ball and pocket model presented by the arid working group. The ball is a golf ball which is moved in and out of a soft clay pocket. While it is in the pocket, however, the golf ball leaves small impressions in the soft clay which represent a different scale of stabilities within the pocket. Differences in variations in the environment which cause the ball to vibrate between different smaller states (the little depressions) within the pocket are not the variations in the environment responsible for moving the ball into or out of the pocket. Safriel agreed that the issue of scale brought up by Milne is predominant in all areas of ecology.

Jim Ellis (Central Plains LTER, Colorado State University) expressed the belief that the choice of desertification as the leading problem of arid/semi-arid systems is an overstated or highly questionable focus, whereas evaluating stability or shifts in system state resulting from human or climatic influence is quite reasonable. Francisco Garcia-Novo agreed that the overstatement of 'desertification' problems by some countries for political reasons was absolutely true. The word was chosen because it encompasses the acute problems that occur in arid and semi-arid areas. Usually 'desertification' applies only to the encroachment of arid areas by the desert, but this word was used by the working group to cover much broader problems in tropical, subtropical, Mediterranean, and arid areas that experience a dry period. These areas are suffering acutely from human impact. Where this impact has been very strong, it has produced very different results in different systems. A tremendous gradient of degrees of impact and effects exists from central Europe to central Africa, and obviously all impacts cannot be described as desertification.

Diane Marshall (Sevilleta LTER, University of New Mexico) suggested that the potential exists for the evolutionary response of individual species to alter and be altered by global climatic change, and this phenomenon deserves attention in the organization of long-term studies. Peter Beets answered that the temperate zone working group did touch on evolutionary studies when it discussed the physiological responses of individual organisms, but he agreed that evolutionary studies should be incorporated into LTER sites.

James Halfpenny (Niwot Ridge LTER, University of Colorado) pointed out that there is a difference in sensitivity to environmental change among various ecosystems, and suggested that alpine and arctic tundra systems are probably the most sensitive. Therefore, LTER sites within these systems could serve as early warning sites for the rest of the LTER network. Halfpenny also stressed the importance of modeling, and agreed with Garcia-Novo that historical and paleoecological data are useful in predicting responses to changing climate. Scientists watching the broadcast at the Bonanza Creek LTER also stressed modeling, and recommended the use of transects along climatic gradients and across ecotones.

Mark MacKensie (Temperate Lakes LTER, University of Wisconsin), Bob Gardner (Oak Ridge National Laboratory), and Judy Meyer (Coweeta LTER, University of Georgia) noted the total absence of aquatic systems from the three working group reports, and pointed out that aquatic systems are important parts of the global carbon budget because they store or release large amounts of carbon, especially CH_4, which is a potent greenhouse gas.

David Greenland (Niwot Ridge LTER, University of Colorado) suggested that more careful thinking needs to be done concerning the way in which climate and ecosystem change is studied. He was in favor of letting the ecosystem define the time scale of change rather than imposing conventional statistics of climate change on the ecosystem. Greenland called for new indices of climatic variability, and gave as an example changes in dominance of various air masses moving across continents.

More information on data management and on-line access to data was requested by a number of respondents. Kari Laine described the Inter-Nordic efforts in long-term monitoring, in which each country has a program director and a small group of scientists who act as the steering body for that country's monitoring. All sites send results to a data handling center in each country. On-line access between sites and from sites to the data handling centers is under development. Scientists at the Konza Prairie LTER (Kansas State University) were impressed that the reports all suggested a balance between modeling, experimentation, and monitoring, but believed that a concurrent need to pull these components together through innovative data management was underemphasized. Don Strebel (Versar, Inc., NASA) agreed that this was a good comment. He then described four different types of data management systems: (1) that which makes raw instrument data usable, (2) a data management system that supports an active field experiment, (3) operational support of a data base or archive beyond the field stage, and (4) deep archives or self-contained data bases. During the active experimental stage it is important to use on-line data bases which are connected through a network. Scientists can then proof data and work collaboratively with them. At this stage, the data system becomes part of the scienctific research, as well as determining quality assurance. Such a system involves many non-standard approaches, and requires interactive development technology which will include scientists, be iterative, and will allow the system to evolve as the science emphasis changes.

John Vande Castle (LTER Network Office, University of Washington) wanted discussion of the information content of remote sensing data, how it relates to the kinds of long-term ecological research proposed, and what kind of archiving and distribution methods are necessary. Strebel responded that information varies with the scale of the remote sensing data used. Advanced very high resolution radiometer (AVHRR) data has a resolution of 1 km and shows the broad character of a region at a scale larger than any single LTER site can show. Higher-resolution data can have a resolution of 10-20 meters, which allows examination of a given stand of trees, for example. This scale of information is very important in dealing with levels of change over a long period. Through the capability of remote sensing, a long-term historical archive is now being accumulated that will eventually allow scientists to know what change occurred in a given stand of trees over a period of 10 to 20 years. It is also important to build an integrated data base at each LTER site based on extant data bases of all kinds, including remotely sensed images. Data screening is necessary to produce a distributable and usable amount of data. Strebel gave as an example a five-fold reduction in data that occurred at the FIFE experiment at the Konza Prairie LTER; 100 gigabites of remotely sensed data were reduced to 20 gigabites of usable data.

Partially in response to the comments of Caroline Bledsoe of the National Science Foundation (NSF) that common experiments which could be performed across global LTER sites seemed to be missing, the focus shifted back to the three working groups. During the break between the first and second programs, each group had discussed just such cross-LTER experiments, which they decided to call flagship experiments. The temperate forest group wanted to develop a Forest Ecosystem Manipulation Experiment (FEMEX) or Whole Ecosystem Manipulation Experiment (WEMEX). This experiment would entail placing a controlled environmental chamber over a forest to examine whole-system responses to enhanced temperature, moisture, and CO_2 levels. The group acknowledged that while this type of experiment is expensive, it is needed to define the changes in response to global warming. The second type of experiment is to measure the fluxes of the radiatively active gases at a given site, not just instantaneously but consistently over a period of time at a number of sites. Heal called this sophisticated measuring apparatus a RAGRIG (rig for measuring the radiatively active gases). The RAG measurements should not be done in isolation, however; a suite of experiments examining the biological processes producing the gases should first be performed in order to predict what is to be measured by the RAGRIG. The combined data should then be integrated with remotely sensed data. This integration of different scales of collected data will provide the opportunity for validation of predictions and for expansion and extrapolation to other sites. The third experiment the temperate forest group recommended was a Biology of Forest Growth (BFG) experiment to produce maximum growth at a given site to understand the maximum carbon uptake of a given system under conditions of enhanced CO_2 levels world wide.

Small-scale experiments are important to enhance the information gained in the big flagship experiments. These small experiments may not be very exciting in isolation, but when performed in concert with the big experiments across sites, they become powerful means of comparing different sites. Suggested experiments included forest floor warming and wetting, soil organic matter—removal or addition of litter, and transplantation of organisms over a range of sites to allow better prediction of species' responses to new conditions they may encounter as global warming proceeds.

The arid/semi-arid zone group determined some rules of water manipulation, including real-time addition or removal of precipitation, i.e. to lengthen a natural event by continuing to water after the storm or to shorten the period of precipitation by introducing the rain-out shelter before the natural storm ended; water manipulation based on rainfall intensity, time interval, length, or time between intervals and seasonality of precipitation; and water manipulation based on simulating historical climate by augmentation or reduction of water. These rules could be based on calculated means and variances of historical precipitation records from such sources as tree ring analysis. An important constraint on the experimental design is the choice of a precipitation regime which will not generate runoff if the topic of interest is soil water storage and water infiltration. The quality of water used to augment a precipitation regime is also an important consideration. Two kinds of sites could be used for these experiments: sites with different known histories of human impact, or those where different human impacts can be simulated at a given site. The arid zone working group considered fertilization experiments, but decided that manipulation of water should be the first close focus of experiments in the world's arid lands.

The boreal/tundra zone group wanted to stress the research opportunities in each ecosystem type, believing that one of the prominent characteristics of this system was that of gradients from tundra underlain by permafrost across the tree line into boreal forest, and on into the temperate forest. The group suggested several sites or nodes where intensive experimental manipulation had been done, such as fertilization of plants and the effects on plants of air and soil temperature, moisture, herbivory, fire, and CO_2 addition. Gus Shaver thought that the controlled environment experiments might work well in tundra because of the low stature of the plants growing in the tundra system. Measurements here could include changes in such state variables as biomass, controls on mineralization, or changes in process after 10 years of fertilization. These changes could be followed by remote sensing. Between the experimental nodes lined up along a gradient, climate and microclimate variables should be recorded, and nutrient levels in soils sampled with resin bags or buried bags. Decomposition and mineralization measurements would use litter bag experiments. There should also be some land/water manipulation.

The group responded to comments from watching LTER scientists who wanted to know why the group believed a georeferenced data base was important in the boreal/tundra system in particular. Shaver answered that it was important to know where these systems are, and how much of the earth is in this condition, so that the contribution of the boreal/tundra system to global elemental cycling and the carbon budget can be understood. The group was also asked if there was a need for them to work with non-biologists—hydrologists, geologists, etc. There was agreement that soil chemists, paraglacial geologists, and hydrologists were vital collaborators in understanding this system. The phenomena of glaciation, permafrost, and the permafrost/soil moisture relationship are poorly understood, and exert strong controls over biological processes, particularly those of the tundra.

Steve Storch discussed the technology available for communication within the proposed international research network. Storch had previously given a presentation to the members of the Albuquerque workshop. He pointed out that telecommunications are essential for such distributed research efforts. Although this is a new way of working for ecologists, interactive communication of the kind which will be required to build an international research network has been used well in other scientific disciplines. He mentioned two kinds of tools: the interpersonal or group interactions which are one-on-one in real time, and the international computer-to-computer network infrastructure. These tools must be approachable, and must be documented and supported by a technical staff experienced in communications technology and in helping the user to conceive of creative uses of the technology.

Paul Risser concluded the second hour's presentation by reminding all participants that the time is unique in the history of the earth and offers unique opportunities for our understanding of the functioning of the globe. In the face of global warming and the effects of human impact, not only are we offered a unique responsibility for managing an individual forest, portion of tundra, or expanse of arid grassland, but we are confronted with the opportunity to manage the sustainability of the entire globe. In the second workshop, Paul Risser continued, important and specific research topics have been identified, and projects, small to large, have been suggested for research. A single experiment will not work to answer the questions posed at this Albuquerque workshop; these questions demand a plurality of approaches, for which three steps must be followed:

(1) As a minimum, the workshop should stimulate ideas for further proposals in the larger circle of scientists doing long-term research;
(2) Proposals from the workshop should define programs and convince funding agencies and governments that the research needs to be done; and
(3) Entire programs, identified in the workshop, must be funded so that the research can go forward.

The second television broadcast ended with an open discussion by the workshop participants about the needed research. Bill Heal urged everyone to think of how groups should be put together. In Europe the land area is small, but the differences between people are large. There is funding right now, however, for integrative projects in Europe. Peter Beets also thought the time is right to press for long-term research in New Zealand; the government is presently placing a strong emphasis on multidiciplinary research in highly relevant programs, especially those involving global change. Mary Seely noted that there is growing communication across political entities in southern Africa, and that biome projects are reorganized. Here too, the time is propitious for collaborative research. Uriel Safriel urged everyone to remember that sound research proposals are important if funding is to be forthcoming, especially in smaller countries that would be flattered to be included in collaborative projects which the global community believes essential. Francisco Garcia-Novo ended the discussion by suggesting that this effort will require large amounts of money and a lot more co-ordination than it has at present. He thought that great attention should be placed on land use and suitable management for conservation. Global co-operation in research also requires global co-operation to provide funding.

When the televised portion of the workshop ended, the formation of an international community of collaborators and research projects seemed to have been established; a valuable achievement, provided that international communication, momentum, and funding continue.

14.6 RETROSPECTIVE ON THE VALUE OF THE ALBUQUERQUE WORKSHOP

Three months after the Albuquerque Workshop, a questionnaire was sent to all workshop participants and to scientists at each LTER site receiving the televised portion of the workshop. The purpose of the questionnaire was two-fold. First, it was important to find out what was happening in the development of international, collaborative research on global-scale problems as a result of the workshop; and second, it was important to give both those in the audience and those who participated in the workshop a chance to reflect on what had or had not happened since the workshop. The three questions on the questionnaire were:

(1) Did the telecommunications workshop influence your thinking about global-

scale problems and international co-ordination of long-term ecological research?

(2) Did the workshop influence the research you are doing or will do? Did it suggest research questions you are exploring, techniques you are using, or contacts in other parts of the world?

(3) Did the workshop and/or television broadcast stimulate your collaboration with other researchers to study phenomena of global warming which may be expressed differently in various habitat types? Are you involved in any international collaborations that grew out of this workshop?

Responses ranged from descriptions of collaborations among a few individuals to international efforts to approach and study global-scale ecological problems. The workshop inspired some researchers to develop projects that would give students experience in the co-operation among different diciplines required to tackle global-scale ecological problems. The workshop also generated a thesis which used historical records to examine the record of climate anomalies in Namibia.

Within countries, collaborative research is being nurtured. For example, the issue of global climate change has been introduced into all regional and local planning in which the Nature Reserves Authority in Israel is involved. Areas where there is a steep climatic gradient seem ideal for exploring the effects of warming, and also the independent effect on productivity of increased carbon dioxide concentrations. Desert meteorologists, plant environmental physiologists, and desert ecologists are reported to be collaborating along such a gradient in Israel.

For researchers in Third-World countries, the workshop reinforced the perception that much of the large-scale research on global problems is financially out of reach for these countries. An important addition to a few large flagship experiments are small-scale, inexpensive projects to be conducted at many places on the globe. Smaller efforts stimulate thinking and training, and have a potential for local funding. Examples of some of these efforts are a collaborative project to study surface-active faunas of African, North American, and Central American arid areas using similar techniques of data collection and analysis; and testing ideas concerning multiple origins of weather affecting interannual rainfall variations seen in the arid South-western United States and in Africa. Collaborative work at the Sevilleta and Jornada LTER sites in New Mexico and at Mapimi, Mexico, were also reported. Research themes there are (1) variability of the influence of the ENSO (El Niño Southern Oscillation) phenomenon along the Chihuahuan Desert latitudinal gradient, (2) importance of abiotic decomposition of litterfall along the same gradient, and (3) experimental studies of the variability of plant genotypes from different geographic origin.

It is paradoxical that more collaborations were reported for the arid/semi-arid biome than for the boreal forest/tundra biome which was singled out by one respondent as more amenable to global-scale research than either the arid/semi-arid or temperate forest biomes. The physical continuity of the boreal forest/tundra biome should produce more uniform responses to global changes across the biome, whereas the varying surrounding environments of the arid/semi-arid and temperate forest biomes should make them sufficiently dissimilar that they would not show uniform responses to global changes.

Global-scale problems require global networking as an essential methodology. Several respondents believed that the creation of financial and other resources had to precede the formation of international links. Just as within each LTER site there is 'networking' between sampling, monitoring, and experiments, and between the specialists that operate them, so must there be similar networking between various sites on the globe.

Co-ordination of research and effective international communication are necessary in the areas of (1) site selection, output objectives of LTER, and parameters to be measured to ensure global representation of key processes; (2) co-ordination of research program development to ensure that core areas are covered to the extent and precision agreed upon; and (3) analysis and synthesis, so that the research and monitoring programs at LTER sites are based on up-to-date information. Innovative communication is essential to maintain the network of international research, and international co-operation will be especially valuable in the exchange of techniques and models. One participant said the workshop convinced him that solving ecological international problems requires international teams of scientists and others in addition to new ways of co-ordinating such an effort. Sophisticated telecommunications, such as that demonstrated by the workshop, are required, and the respondent is designing computer modeling programs to handle global-scale problems and to foster international co-ordination. It was suggested that long-term ecological research should be a partnership between researchers, industry, government policy developers, and the public. The involvement of all these groups will encourage co-operation arising from the improved understanding of the global nature of environmental issues. Information exchange should be done so that the real costs of resource management are clearly evident. The implications of global policy at local and regional scales are far reaching, and research results need to be convincing before governments are likely to agree to act. Presently, the governments of New Zealand, the United Kingdom, and Australia have agreed to joint climate change research.

The television broadcast and the responses from the audience indicated a willingness by the wider scientific community to get on with the job. New Zealand in particular shares this desire to do something. An international LTER effort could be splintered, however, by the individual initiatives emerging in different countries unless a truly co-ordinated effort is made to integrate them. An overall co-ordinating group must be formed now to manage the several research proposals under consideration. Other international initiatives, such as IGBP's GBO network headed by Rafael Herrera, are underway. It was suggested that the international network of long-term research groups should be represented on the IGBP international committee.

In summary, workshop participants reported the beginning of research collaborations between individuals and between sites in a given country, and plans for co-ordinated research within countries. It seems clear that the components for internationally co-ordinated long-term research at a global scale are present. The next task is the assembly of these components and the use of sophisticated remote sensing, geographic information systems, and both person-to-person and computer-to-computer communication, if there is to be true collaborative, co-ordinated, and international ecological research.

15 Current Status and Future of Long-term Ecological Research

PAUL G. RISSER*, JERRY M. MELILLO† and JAMES R. GOSZ*
**Department of Biology, University of New Mexico, Albuquerque, New Mexico 87131, USA*

†Ecosystem Center, Marine Biological Laboratory, Woods Hole, Massachusetts 02543, USA

15.1 INTRODUCTION

Ecological research is the investigation of processes and patterns that explain the relationships between living organisms and their environment. Such research involves a multitude of approaches, including laboratory experiments, field observations and tests, theoretical explorations, and the evaluation of conceptual and mathematical models. Explicit ecological research of various kinds has been conducted since before the beginning of this century, and a large and rich literature now exists.

Because of the complexity of the relationships between living organisms and their environment, definitive experiments are rarely possible. Consider, for example, the magnitude of the variation in weather on a daily basis, and then the seasonal, annual, and decadal variations, and finally the long-term changes in climate. Then consider

Long-term Ecological Research. Edited by Paul G. Risser

the enormous variation in the genotypes and phenotypes of organisms, and the broad array of their responses to weather and climate conditions. These combinations, superimposed upon the interactions among organisms, lead to exceedingly dynamic ecological systems that make definitive ecological experiments difficult to design and conduct. As a result, specific laws or principles are relatively infrequent in ecological research. Moreover, 'certainties' in ecology are usually defined in the specific conditions under which the expected observations occur and/or are defined in terms of the probability under which the observations will take place.

Many fundamental processes in ecological systems occur over relatively long periods of time. For example, the generation times of some insects are a matter of days, but some species of trees live for hundreds of years. Similarly, soil erosion may occur rapidly with one, possibly unpredictable, storm, while soil-building processes may require centuries. Thus, ecological processes, consisting of both organisms and their environment, are driven by dynamics with periodicities and durations spanning many time and space scales.

The complexity of ecological systems and the various temporal and spatial dimensions have led to the recognition that ecological research must be conducted on long time scales (here long-term ecological research is defined as decades to longer). The argument is that long-term study is required not only to understand ecological complexity, but also to ask important or meaningful questions (Likens, 1983, 1989). Moreover, it is argued that long-term studies are valuable because they encompass unusual or interesting events that allow understanding that would not otherwise be possible. Careful analysis of the published literature, however, suggests that these unusual events have not been a commonly used tool in ecological research (Weatherhead, 1986). Previous evaluations of ecological research have noted that classes of ecological phenomena requiring long-term studies include slow processes, rare events or episodic phenomena, processes with high variability, subtle processes, and complex phenomena (Likens, 1983, 1989; Strayer *et al*., 1986), and there are examples of each class (Franklin, 1989; Pickett, 1991).

Despite the recognized values that have been derived from previous long-term ecological studies, it is important to consider whether there are alternative approaches for investigating ecological phenomena. That is, a significant portion of our current ecological understanding has arisen because of planned or serendipitous measurements taken over long periods of time. The question arises as to whether there are other methods for arriving at the same understanding.

The chapters in this volume present numerous examples of long-term ecological research that have been conducted throughout the world. In these instances, the authors have described the resulting information, and have usually attributed the understanding to the long-term measurements. For the purpose of this summary discussion, we have selected examples from several general topics. These examples are then examined to see whether there were alternative approaches that would have produced the same information.

15.2 EXAMPLES OF LONG-TERM ECOLOGICAL RESEARCH

15.2.1 ENVIRONMENTAL QUALITY

Experiments begun in the 1840s to 1860s at Rothamsted have permitted the detection of several long-term changes in agricultural systems (Johnston, 1991). In the last 40 years, there have been large increases in the amounts of cadmium and polynuclear aromatic hydrocarbons in the soil. The experimental treatments allowed differentiation of the sources of these substances (for example, of cadmium from aerial sources, from farmyard manure, and from phosphate fertilizer). These studies indicated that increasing soil organic matter that accompanied manure additions may have contributed to the retention of cadmium in the soil. Also, cadmium did not accumulate in the seeds, but did accumulate on or within the herbage where it could be ingested by herbivores.

The polynuclear aromatic hydrocarbon (PAH) burden in the soil increased five-fold in the last hundred years. Although there are several proposed mechanisms for the loss of PAHs (microbial breakdown, photo-oxidation, vaporization, crop removal and leaching), only very small proportions of the annual input were lost, and the PAHs were shown to have long residence times in the soil (Johnston, 1991). Moreover, compounds with complex structures increased in the soil more than those with simpler structures, suggesting that microbial decomposition and soil retention of PAHs may depend upon molecular structure.

In another set of experiments at Rothamsted, nitrogen concentrations were monitored for the past 100 years in the soils of manured and unmanured fields growing wheat, and in fields that were allowed to revert to woodlands. The manured fields demonstrated an increased soil nitrogen content which equilibrated only after approximately 80 years. The soils of the woodlands now have about the same nitrogen content as the manured wheat fields, although the equilibrium point has not yet been reached. It is possible that once the equilibrium point is reached, these forest soils will also lose nitrogen due to leaching, as do the manured wheat fields.

In all three examples, long-term measurements were necessary for detecting trends in the materials of interest. It is difficult to design experiments that would have predicted the specific accumulation rates of cadmium, polynuclear aromatic hydrocarbons, and nitrogen. Cadmium accumulation rates also apparently depend upon organic matter increases, PAH concentrations depend upon several climate-controlled processes and also the chemical structure of the compounds themselves, and even after measuring nitrogen for decades it is impossible to predict the equilibrium concentration in the woodlands. Of the three examples, cadmium may now be the most amenable to prediction since it appears to depend upon input rates and organic matter content. In addition, it should be possible to calculate ingestion rates by herbivores.

15.2.2 SOIL PROCESSES

Threshold-controlled responses have not been considered common in soil dynamics, although Anderson (1991) describes several examples of this phenomenon. In a rare but large rainstorm in Alaska, formerly stable horizons in a Spodosol soil were disrupted, and normally insoluble organo-metallic bodies were dissolved. These materials moved down the profile in association with suspended particulates to form unusually deep B horizons enriched in sesquioxides and humus (Stoner and Ugolini, 1988).

There is a common perception that soil processes occur at relatively slow rates when compared to other parts of ecological systems. This perception may arise because soils are mapped according to standard descriptive characteristics and the subsequent soil classifications are regarded as the result of gradual development that appears stable for long periods of time. This classification scheme has led to the acceptance of soils as a rather stable component of the ecosystem. However, Anderson (1991) has shown in Canada that there are dynamic components to the soil and that these dynamic components respond rapidly and may show threshold responses. Specifically, soil salinity at the soil surface changes rapidly depending not only upon agricultural practices, but upon the movement of soil water, i.e. downward flux caused by leaching and upward flux by capillary rise or artesian pressure. The surface soil water dynamics are driven by precipitation and evaporation—processes that themselves change rapidly through the season and are subject to changes in global climate patterns. On the other hand, the total salt content of the entire soil profile changes much more slowly and is controlled by larger-scale changes in regional groundwater flows, often through deep aquifers. Organic matter dynamics in soils include processes of different time scales ranging from rapid changes associated with microbial processes to slow ones controlled by physical sorption to clay or chemically-stabilized humus (Anderson, 1979). Among the most instructive studies of soil organic matter are the long-term crop rotation studies at Rothamsted and elsewhere. As an example of the multiple time scales, the short-term dynamics of microbial biomass over the course of the growing season and the longer-term processes affecting total soil organic matter content were studied in Alberta, Canada (McGill *et al.*, 1986). Soils under a five-year cereal-forage rotation contained 38% more total nitrogen, but 117% more microbial nitrogen than the comparative wheat–fallow soils. The latter soils had average organic matter turnover rates 1.5 to 2.0 faster than the cereal–forage rotation. Furthermore, the faster turnover rates in the wheat–fallow soils could not be accounted for by the average carbon additions and may indicate that native soil organic matter is being mined (Anderson, 1991). Since the total organic matter contents changed very slowly, these conclusions depended upon studies established more than five decades earlier.

Again it is clear that all three questions about soil properties depend upon long-term processes. In the first instance, the unusual event that permitted the leaching of organo-metals would not be understood as unusual if the usual conditions were not known. The surface and subsurface salinity dynamics perhaps could be predicted, but the variability in the climate-driving variables makes prediction tenuous, and validation data would be necessary. In the last case, the small changes in organic matter would only be detectable over a long period of time.

15.2.3 FLUVIAL SYSTEMS

Rivers in human-inhabited regions have undergone two types of major modifications. First, the flow dynamics of these rivers have been modified by control structures on the rivers themselves and on the secondary channels and backwaters. Second, urbanization and intensified agriculture in the watershed have changed water quality in the entire system, fragmented the riparian forests, and affected the flow of organic matter to and from the river. The River Garonne in southern France has been influenced by flood and erosion control devices since the seventeenth century and in the river as it flowed through Toulouse since the twelfth century (Décamps and Fortuné, 1991). The results of these human activities include a reduction in the secondary arms of the river, a straightening of the river channel, and stabilization of the river bed. Ecological consequences of these changes include alterations in the successional sequences of the riparian vegetation and of the associated biota, a reduction in the retention of nutrients along the edges of the river, changes in richness of the fish communities, and depression of the water table near the alluvial rivers.

River ecosystems and their responses to human activities represent a special case in the analysis of long-term ecological research. These circumstances arise because of the great inertia to reversal of geomorphic changes (Munn, 1987). These changes are caused both because of the inertial strength of the river ecosystems and because the management of the rivers is so closely tied to socio-economic development. Thus, these fluvial systems respond not only to ecological processes in the usual sense, but also in a cumulative way to local, regional, and global economic conditions. However, unlike a farmer who can simply change the choice of a crop depending upon last year's commodity market, changes in fluvial systems persist for decades and even centuries.

15.2.4 SPECIES RESPONSES

Coastal dunes in Australia, near Myall Lakes National Park, are mined for heavy metals (Westoby, 1991). Small mammal patterns have been investigated on the recovering mined sites (Fox and Fox, 1978, 1984). The abundance of *Mus musculus* reaches a peak three to five years after the mining, but *Pseudomys novaehollandiae* becomes more abundant after five to seven years. On sites regenerating after fire, the sequence is compressed and *P. novaehollandiae* becomes more abundant after only the second year. This is probably because many plant species regenerate vegetatively after fire, but the regeneration, especially of rhizomes and tubers, is much slower after mining. *P. novaehollandiae* abundance is positively correlated with vegetation cover.

Experimental manipulations done in concert with these longer-term field measurements confirm that *M. musculus* declines because it behaviorally avoids *P. novaehollandiae*. Measuring and predicting the response of organisms to climate change will depend upon the location of the sample sites relative to the ecology of the organisms in question (Heal, 1991). For example, repeated annual sampling of the homopteran *Neophilaenus lineatus* showed that the population regulation was density dependent at a low-altitude site, but density independent and climatically controlled at a more northern high-elevation site (Whittaker, 1971). Thus, the behavior of these two populations was only known from repeated sampling, but just as important was the realization that population control could be density or climate controlled.

These last two examples again depend upon long-term measurements. In the first case, the species response to disturbance depended upon the recovery rates of vegetation and the species–species interactions. These first- and second-phase responses were found because field sampling occurred over a long enough period to detect the change in small mammal dominance; and the associated experimental studies identified the actual mechanisms. The homopteran example might have been predictable from carefully controlled experiments, but again, testing the results of experiments requires a sequence of field observations.

The chapters in this book arise from various ecosystems around the world and they focus on different parts of the ecosystems. Throughout, there is a consistent recognition that long-term studies are necessary to understand ecological processes. The examples given are not idle curiosities, but rather they exemplify many of the most vexing ecological issues facing us today.

15.3 INTERNATIONAL NETWORKS FOR STUDYING GLOBAL CHANGE

During the Berchtesgaden meeting there was a special effort to identify important long-term ecological questions which might form the basis for studying ecological processes on a global scale. The specific objective was to 'further define the rationale for long-term ecological research and identify important existing and emerging scientific questions which could most appropriately be addressed at long-term ecological research sites, particularly those related to environmental changes at the global scale'. The working group addressed two related aspects: the scientific questions to be addressed at long-term research sites and the networking of those sites.

Understanding the issues of global and regional change depends upon ecological processes at all smaller spatial scales. Achieving the detection, understanding, and prediction of global change and the prescription of human response to global change requires the formation of interactive, real-time networks of ecological studiesaround the globe. Explicit goals of international and regional networks are (1) to identify common patterns of change, (2) to develop regional and global syntheses of change and a mechanistic understanding of these changes, (3) to partition the variances between global and site-specific forcing functions, and (4) to develop approaches to scaling up (and down) from local to regional to global spatial scales. Networks will accelerate the rate of achieving these goals by reducing the time lags both for information transfer and the planning and conduct of appropriate long-term ecological research.

Networks should be flexible and form around scientists studying similar systems, on the one hand (e.g. those who study grasslands), to those studying markedly different systems (e.g. grasslands, lakes, urban ecosystems, coastal zones). These networks are essential when recognizing the short times over which global change may occur, the need for implementation of new techniques, and the spatial scale and diversity of ecosystems involved in these global problems.

The working group represented many disciplines, study sites, and countries. Discussions soon revealed that research priorities and objectives differed markedly, and that a generic research program could not be forced on sites/countries with different priorities. However, there were concepts that were common to all. For example, all individuals recognized the benefits of long-term research since the fundamental approach was to explain or account for variation in the parameters measured. The resulting measurements form the basis for justifying a global network and extrapolating results to regional and global scales. Only a network of sites can represent the large spatial scales needed to document changes that occur at those scales. The working group envisioned a global scale Geographic Information System (GIS) that could map and project results at appropriate broad scales. Networking allows sites to work on a common time scale and to synchronize efforts at quantifying responses to broad-scale environmental phenomena.

For example, an El Niño event may cause very different responses among different ecological parameters at one site and among different sites. This results from the many interacting factors at a site and the site-specific nature of those interactions. However, the initiation of the different responses may well be caused by the El Niño phenomena, a triggering effect, which could be common at many sites. The network could identify the scale of the triggering phenomena, the times of initiation and conclusion, and the types of ecological responses that were similar and dissimilar. For some sites, the network could separate the triggering response from the subsequent chain of events (succession of events) until a new triggering event occurred. Some regions may respond to certain types of triggering events while others may not. This information may help to identify the significance of various types of constraints on systems, the regional extent of those constraints, and the sequence of ecological phenomena which are typical following changes in those constraints. Networking is essential for a rapid information transfer, and facilitates early recognition of events that are common to large areas.

15.4 PROPOSED EXPERIMENTS

During the second workshop, investigators attempted to build on the results of the first workshop by identifying important issues that required long-term ecological studies. The discussion involved three biome types: arid/semi-arid, temperate forests, and boreal forest/tundra. Important research questions for the arid/semi-arid biome focused on the stability of these systems and their responses to changing climate and human use. Climate change, pollution, and management issues were deemed most important in boreal forests, particularly the relationships between vegetation and the fluxes of radiatively active gases. In the boreal forest/tundra biome, research topics of highest priority dealt with the soil carbon budget, decomposition-limited nitrogen and phosphorous processes, episodic phenomena such as animal population fluctuations and fires, and permafrost and soil hydrology. Recommended experimental approaches included large 'flagship' studies, networks of smaller experiments, and various manipulation tests depending upon the research question itself.

A significant component of the second workshop involved the use of data management, and observational and communication technologies. These topics are important for organizing and conducting international-level long-term ecological studies, especially those addressing scientific questions at regional and global scales. As described in Chapter 14, the workshop included televised sessions where scientists from throughout North America participated in the workshop.

15.5 CONCLUSIONS

Over the past century, ecological research has made significant, even remarkable, progress toward providing an understanding of many fundamental ecological processes. A portion of this success derives from the results of 'experiments' in which important variables have been manipulated and compared with controls in the classical sense. Because of the success of these experiments, ecological research has eagerly adopted this approach as the *sine qua non* for masterful and clever research. However, as described in this volume, many important discoveries have arisen from long-term measurements—both with and without experimental manipulation. Thus, there is great jeopardy in focusing exclusively on just one approach (Taylor, 1989) and losing the value of multiple research strategies. In addition, the natural history observations possible through long-term research projects provide valuable explanations of both specific processes and the evolutionary context of the current observations (Gould, 1989).

The greatest value of this book is the rich array of described long-term ecological studies and the insights provided by the authors. Selecting certain summary points detracts from this richness. However, one of the major purposes of the SCOPE project was to provide recommendations concerning the establishment of international networks for long-term ecological research. It is in this spirit that the following ideas are presented. In many cases, several authors made the same or related points, but at least one citation to a chapter in the book is made for each point.

Long-term ecological studies judged to be successful have been (1) associated with one dedicated leader who was responsible for the project and (2) designed simply, so that the experiments were easy to operate, the data could be used in various ways, and ancillary studies could be developed (Strayer *et al.*, 1986). In addition, many of these successful long-term studies have included one or more of the following characteristics: they involved experimentation as part of the design, had clearly defined objectives, were conducted on protected sites, included provisions for archiving samples for future analysis and for comparing new methodologies, provided short-term justification by a continuing set of publications, responded to recognized policy or society needs, and included modeling and synthesis during the course of the study (Pickett, 1991).

In many cases, data to address long-term ecological phenomena will be acquired from many sources and will cover different time scales. An aspiration of an integrated long-term ecological research approach will be to minimize this *ad hoc* approach. However, past research should not be ignored since it possesses enormous amounts of information, and significant efforts should be devoted to the incorporation of meaningful and useful information from past studies for addressing specific ecological questions. When the data sets are incomplete or it is difficult to link data with specific management strategies, data from one or more levels may be 'soft coupled' by employing data management techniques combined with expert judgement systems (Grossmann, 1991).

Long-term ecological studies can be most useful for management of natural resources in three ways: (1) by providing records of subtle, chronic changes in ecosystems, (2) providing early warning of the onset of acute changes, and (3) suggesting management strategies for dealing with both of these types of changes (McNaughton and Campbell, 1991).

The substitution of space for time, or the chronosequence, has been used to explore time-dependent processes such as succession. In many cases the results from chronosequences have been misleading, especially because of different initial conditions, and thus long-term studies may be necessary (Pickett, 1991).

Designs for long-term experiments should include estimations of the proportions of variation in an ecological measurement which are attributable to differences between years, to differences between places, or to place–year variations, such as patterns which occur every year but are shifted in space (Westoby, 1991).

Although studying rare events is a valid reason for instituting long-term studies, some rare events are predictable and can be studied with short-term observations. Unique events, such as invasion by exotic species or cases where the system moves to a new set of conditions, are more difficult to predict, and understanding them may require long-term studies (Franklin, 1989; Pickett, 1991).

Long-term studies provide a perspective on rare events, and experimental designs must include sufficient duration to estimate return times, recognizing that return times may change as a result of the phenomenon itself (Westoby, 1991).

Long-term ecological studies must provide sufficient time to study second-phase events, such as a species arriving subsequent to the first conditions after disturbance or processes that are quite slow (Westoby, 1991). Not all studies can be extended for a longer duration because of the expense. Thus, the investigator should anticipate some of the likely changes and include these as experimental treatments. Also, modest funding for periodic returns to field sites would be an inexpensive approach for identifying and measuring second-phase and slow processes.

Establishment of a network of long-term ecological sites in any one country must recognize the specific characteristics of the research support structure of that country and, if the network becomes international, of the collection of funding agencies (Callahan, 1991).

The existence and expansion of long-term ecological research studies is constrained by the ability of the scientific community to perform the necessary observations and experiments, the availability of appropriate research sites, and the capacity of the sponsoring agencies to support and administer long-term research projects (Callahan, 1991).

Networks of sites allow more intensive narrow-scale studies to be put in context, and also permit estimates of space and time scales over which aggregated variables become relatively predictable (Westoby, 1991).

If unusual events are considered biologically important, merely monitoring them will not advance knowledge fast enough and, therefore, science funding must be organized so that hypotheses can be tested experimentally during unusual events (Westoby, 1991).

For many problems, the influence of humans is most likely to be important and detectable under conditions which are marginal to the species or system, or which are transitional between states. Thus, selection of sites for long-term measurements on ecotones or threshold sites can be as important as selection of sites which represent more average or typical conditions. Also, selection of sites should include careful analysis of the range of variation to be covered and should build on a planned network of intensive and extensive sites designed to accumulate the expected range of variation (Heal, 1991). A great challenge for long-term ecological research is to begin to measure variables over time, on the premise that we must begin collecting a coherent record now in order to have data against which to assess understanding and models of global change that may be developed in 10, 30, or 50 years time (Pickett, 1991; Westoby, 1991). Measuring decade-to-century level ecological phenomena requires serious time and financial commitments to data documentation (Seastedt and Briggs, 1991).

15.6 REFERENCES

Anderson, D.W. (1979). Processes of humus formation and transformation in soils of the Canadian Great Plains. *Journal of Soil Science*, **30**, 77–84.

Anderson, D.W. (1991). Long-term ecological research: a pedological perspective. Chapter 7, this volume.

Callahan, J.T. (1991). Long-term ecological research in the United States: a federal perspective. Chapter 2, this volume.

Décamps, H. and Fortuné, M. (1991). Long-term ecological research and fluvial landscapes. Chapter 8, this volume..

Fox, B.J. and Fox, M.D. (1978) Recolonization of coastal heath by *Pseudomys novaehollandiae* (Muridae) following sand-mining. *Australian Journal of Ecology*, **3**, 447–465.

Fox, B.J. and Fox, M.D. (1984). Small-mammal recolonization of open-forest following sandmining. *Australian Journal of Ecology*, **9**, 241–252.

Franklin, J.F. (1989). Importance and justification of long-term studies in ecology. In Likens, G.E. (Ed.) *Long-Term Studies in Ecology. Approaches and Alternatives*. Springer-Verlag, New York, 3–19.

Gould, S.J. (1989). *Wonderful Life. The Burgess Shale and the Nature of History*. W.W. Norton, New York.

Grossmann, W.-D. (1991). Model- and strategy-driven geographical maps for ecological research and management. Chapter 13, this volume.

Heal, O.W. (1991). The role of study sites in long-term ecological research: a UK experience. Chapter 3, this volume.

Johnston, A.E. (1991). Benefits from long-term ecosystem research: some examples from Rothamsted. Chapter 6, this volume.

Likens, G.E. (1983). A priority for ecological research. *Bulletin of the Ecological Society of America*, **64**, 234–243.

Likens, G.E. (Ed.) (1989). *Long-Term Studies in Ecology. Approaches and Alternatives*. Springer-Verlag, New York.

McGill, W.B., Cameron, K.R., Robertson, J.A. and Cook, F.D. (1986). Dynamics of soil microbial biomass and water soluble organic C in Breton L after 50 years of cropping to two rotations. *Canadian Journal of Soil Science*, **66**, 1–20.

McNaughton, S.J. and Campbell, K.L.I. (1991). Long-term ecological research in African ecosystems. Chapter 10, this volume.

Munn, R.E. (1987). *Environmental Prospects for the Next Century: Implications for Long-Term Policy and Research Strategies*. Research Report 15. International Institute for Applied Systems Analysis. Laxenburg, Austria.

Pickett, S.T.A. (1991). Long-term studies: past experience and recommendations for the future. Chapter 5, this volume.

Seastedt, T.R. and Briggs, J. M. (1991). Long-term ecological questions and considerations for taking long-term measurements: lessons from the LTER and FIFE programs on tallgrass prairie. Chapter 9, this volume.

Stoner, M.G. and Ugolini, F.C. (1988). Arctic pedogenesis. 2. Threshold-controlled subsurface leaching episodes. *Soil Science*, **145**, 46–51.

Strayer, D.J., Glitzenstein, S., Jones, C.G., Lolasa, J., Likens, G.E., McDonnell, M.J., Parker G.G. and Pickett, S.T.A. (1986). *Long-Term Ecological Studies: An Illustrated Account of Their Design, Operation, and Importance to Ecology*. Institute of Ecosystem Studies Occasional Publications 1.

Taylor, L.R. (1989). Objective and experiment in long-term research. In Likens, G.E. (Ed.) *Long-Term Studies in Ecology. Approaches and Alternatives*. Springer-Verlag, New York, 20–70.

Weatherhead, P.J. (1986). How unusual are unusual events? *American Naturalist*, **128**, 150–154.

Westoby, M. (1991). On long-term ecological research in Australia. Chapter 11, this volume.

Whittaker, J.B. (1971). Population changes in *Neophilaenus lineatus* (l.) (*Homoptera:Cercopidae*) in different parts of its range. *Journal of Animal Ecology*, **40**, 425–443.

16 Summary

PAUL G. RISSER
University of New Mexico, Albuquerque, New Mexico 87131, USA

The importance of long-term ecological studies has been recognized throughout the world. Several examples include the famous agricultural plots at Rothamsted in Great Britain and the network of forest research stations in Sweden. The more recent Chinese Co-operative Ecological Research Programme (CERP) with the Federal Republic of Germany focuses on research in forest ecosystems, biosphere reserve management, urban ecosystems and water pollution. In Poland, long-term regional research projects, such as at the Bialowieza Geobotanical Station of Warsaw University in the Bialowieza Primeval Forest, are testing the assumptions of equilibria in disturbance regimes. In the United States, much of the current effort in long-term ecological research emanates from the long-term studies at Coweeta and Hubbard Brook Experimental Forests.

Despite these world-wide examples, there has been little attempt to critically analyze long-term ecological studies to determine when such studies are really needed, what are the characteristics of successful studies, and how they might be co-ordinated on a global scale to provide greater value to individual projects and to the total array of long-term ecological studies. This was the task of the two SCOPE workshops described in this book. The first workshop, held in Berchtesgaden, Germany, emphasized the issues of analyzing long-term ecological studies and their successes. The second workshop, held in Albuquerque, New Mexico, USA, focused on the important questions that require long-term studies for answers and on the necessary organizational structure for international co-ordination.

From the chapters in this volume, it is clear that long-term ecological studies are useful when the ecological phenomena are themselves long term in their dynamics, when the phenomena are episodic, rare, complex or subtle and long-term measurements are needed to isolate their dynamics and control processes, when the phenomena are poorly understood and cannot be predicted from short time scales, and when long-term records are needed to make and justify making policy decisions. Each chapter addresses specific issues about long-term ecological studies and together they bring together a world-wide set of experiences and points of view.

Long-term Ecological Research. Edited by Paul G. Risser

As discussed by Callahan, the United States Long-Term Ecological Research (LTER) program consists of 17 sites (the 18th site was added in 1990 in the Antarctic). Though the program began in 1980, most of the sites had substantial long-term ecological measurement already underway. The focus of these studies is on understanding long-term ecological phenomena, not simply on monitoring or inventory activities. Each site is funded separately by the National Science Foundation and initial selection of sites and continued funding decisions are made on a competitive basis. There are, however, funds committed to encourage inter-site co-ordination and research co-operation.

Heal considers the establishment of a network of long-term ecological reference sites in Great Britain, noting the importance of using well-established sites and considering whether there are existing data bases. From the experience at Rothamsted and other sites, he identifies the type of topics that require long-term studies: climate change, chemical pollutants, management effects and invasions and extinctions of species. From examination of the research needed on these topics, it is possible to develop necessary criteria for selecting sites as well as the types of measurements and experiments that can be accomplished at individual sites and throughout the network.

Within the past decade it has become quite obvious that both the spatial and temporal scales of ecological measurements are extremely important. Magnuson and his co-authors discuss the consequences of choices among spatial and temporal scales for measurements and observations. In particular, choices of time scales are affected by life-cycle characteristics of the organisms, dispersal patterns and by ecological processes that operate with time lags. Similarly, heterogeneous spatial patterns in habitats and resource bases have enormous impacts on ecological systems. Lastly, these authors draw interesting comparisons of long-term ecological processes across quite disparate ecosystems (for example, treating lakes as islands and examining the consequences of ecological processes).

Much of the theoretical basis for considering long-term ecological phenomena has come from Pickett and his colleagues at the Institute of Ecosystem Studies at Millbrook, New York, USA. Pickett points out that the motivation for long-term studies comes from theoretical questions, e.g. the study of succession, empirical investigations on selected ecological patterns, and from the need for answering policy and political issues. Evaluations of successful long-term ecological studies point to the importance of conceptual models to organize the research questions, of adequate methods for handling data, of ensuring that the experiments that require long-term support have some immediate applications, and of the need to give attention to standardization and documentation of the methodology.

The Rothamsted studies in Great Britain are recognized as the model for many subsequent long-term study sites. These studies emphasized soil fertility over more than 100 years, particularly processes such as soil acidification, effects of soil pH on soil properties and on the soil and above-ground flora, and on the consequences of various management techniques. Eight of the original experiments begun by Lawes and Gilbert in the 1840s and 1860s continue with various amounts of modification. These studies have proven invaluable in understanding the dynamics of soil microbial biomass and soil organic matter, and in defining when predictions can be made about their future status. Successes at Rothamsted can be attributed to the dedication of the site and that there has been continuous financial support, that each research group consists of several disciplines, that data and samples are archived, and that experiments are planned and changed only after consideration by these multidisciplinary groups of scientists.

Considerable portions of the central Great Plains of Canada are former grasslands that have been converted to farmland. Because of these conversions from natural grasslands, there is great concern for the sustainability of the cropland systems. Of greatest concern is the question of soil deterioration due to erosion, loss of organic matter and fertility and increases in salinization. Anderson uses specific field studies to analysis the spatial and temporal properties of these processes, particularly as they are influenced by changing climatic conditions. Finally, Anderson synthesizes this information, arguing for the need for not only understanding the processes themselves, but also for monitoring and modeling of the soil systems.

Although the previous chapters have emphasized terrestrial systems, aquatic systems are equally amenable to long-term ecological studies. Décamps and Fortuné use the River Garonnne in southern France to evaluate the changes in the river system caused by urbanization and changes in the fluvial landscape. From these analyses, the authors discuss the global and comparative necessity of considering socio-economic development, land-use change, fluvial dynamics and the associated ecological properties and consequences in long-term ecological studies.

Seastedt and Briggs consider long-term ecological processes in the tallgrass prairie, primarily from the Flint Hills in east central Kansas, USA. Their observations describe the long-term dynamics of primary production, the interaction between productivity and surface climate, and the effects of soil nutrients and soil organic matter on system behavior. A significant portion of this discussion focuses on the importance of spatial and temporal scaling of experiments designed to evaluate these processes and for making appropriate field measurements. In the subsequent chapter, McNaughton and Campbell discuss long-term ecological research in the African savanna, especially related to the Serengeti National Park. Although parts of the discussion involve the ecological processes, much attention is paid to the value of monitoring and the need for the organization of the research effort to meet the needs of resource managers. These two chapters together demonstrate the value of multidisciplinary research on long-term ecological processes, and just how these research efforts must be organized and maintained if they are to ultimately influence the management of grassland and savanna ecosystems.

Australian ecological research has considerable potential applicability to tropical and subtropical areas of the world. Westoby discusses the nature of science in Australia and then describes several examples of long-term ecological studies involving selected species where the results are essential for strategic decisions regarding the management of specific habitats. From these discussions come several recommendations for optimizing measurements directed toward long-term studies, such as durations of measurements, focusing on unusual but biologically important events, examining secondary consequences of experimental manipulations and, like Magnuson in Chapter 4, ways of relating time and space scales.

Chapters 12 and 13 address the general topic of modeling, but from different points of view. Shugart and co-authors discuss several simulation models and their particular use in long-term ecological research. The need to scale known ecological systems to larger spatial and temporal scales, especially to continental scales, means that models are necessary. In Chapter 13, Grossmann uses the application of geographical information systems to the project at Berchtesgaden to demonstrate a hierarchical three-layer system of models. The lowest layer consists of specific ecological processes which are well-studied and for which there are empirical data. The second layer is the dynamic dimension which concerns the changing system, as, for example, effects of climate change. Finally, the highest level incorporates the strategic decisions of managing natural resources over the long term.

The second workshop described in this book was devoted to identifying the types of questions that require long-term ecological measurements and which require networks for collaborative studies comparing ecosystems throughout the world. For the purpose of focusing the efforts, three types of ecosystems were chosen: temperate forest biome, arid/semi-arid biome and the boreal forest/tundra biome. For each biome key long-term ecological questions were defined as were the most appropriate collaborative experiments and measurements. These ideas were then evaluated using a television network with other scientists throughout the United States. Chapter 15 summarizes the primary points from all the chapters in the book and presents the most important recommendations for long-term ecological research.

Index